IFMBE Proceedings

Volume 82

Series Editor

Ratko Magjarevic, Faculty of Electrical Engineering and Computing, ZESOI, University of Zagreb, Zagreb, Croatia

Associate Editors

Piotr Ładyżyński, Warsaw, Poland

Fatimah Ibrahim, Department of Biomedical Engineering, Faculty of Engineering, University of Malaya, Kuala Lumpur, Malaysia

Igor Lackovic, Faculty of Electrical Engineering and Computing, University of Zagreb, Zagreb, Croatia

Emilio Sacristan Rock, Mexico DF, Mexico

The IFMBE Proceedings Book Series is an official publication of *the International Federation for Medical and Biological Engineering* (IFMBE). The series gathers the proceedings of various international conferences, which are either organized or endorsed by the Federation. Books published in this series report on cutting-edge findings and provide an informative survey on the most challenging topics and advances in the fields of medicine, biology, clinical engineering, and biophysics.

The series aims at disseminating high quality scientific information, encouraging both basic and applied research, and promoting world-wide collaboration between researchers and practitioners in the field of Medical and Biological Engineering.

Topics include, but are not limited to:

- Diagnostic Imaging, Image Processing, Biomedical Signal Processing
- Modeling and Simulation, Biomechanics
- Biomaterials, Cellular and Tissue Engineering
- Information and Communication in Medicine, Telemedicine and e-Health
- Instrumentation and Clinical Engineering
- Surgery, Minimal Invasive Interventions, Endoscopy and Image Guided Therapy
- Audiology, Ophthalmology, Emergency and Dental Medicine Applications
- Radiology, Radiation Oncology and Biological Effects of Radiation

IFMBE proceedings are indexed by by SCOPUS, EI Compendex, Japanese Science and Technology Agency (JST), SCImago.

Proposals can be submitted by contacting the Springer responsible editor shown on the series webpage (see “Contacts”), or by getting in touch with the series editor Ratko Magjarevic.

More information about this series at http://www.springer.com/series/7403

Yasuyuki Shiraishi · Ichiro Sakuma ·
Keiji Naruse · Akinori Ueno
Editors

11th Asian-Pacific Conference on Medical and Biological Engineering

Proceedings of the Online Conference APCMBE 2020, May 25–27, 2020

Springer

Editors
Yasuyuki Shiraishi
Institute of Development, Aging and Cancer
Tohoku University
Sendai, Japan

Ichiro Sakuma
School of Engineering
The University of Tokyo
Tokyo, Japan

Keiji Naruse
Graduate School of Medicine, Dentistry
and Pharmaceutical Sciences
Okayama University
Okayama, Japan

Akinori Ueno
School of Engineering
Tokyo Denki University
Tokyo, Japan

ISSN 1680-0737 ISSN 1433-9277 (electronic)
IFMBE Proceedings
ISBN 978-3-030-66168-7 ISBN 978-3-030-66169-4 (eBook)
https://doi.org/10.1007/978-3-030-66169-4

This Springer imprint is published by the registered company Springer Nature Switzerland AG
The registered company address is: Gewerbestrasse 11, 6330 Cham, Switzerland

Preface

APCMBE 2020 has been the 11th edition of the Asian Pacific Conference on Medical and Biological Engineering. The conference was organized by IFMBE Asian Pacific Working Group and the Japanese Society for Medical and Biological Engineering. The conference was supposed to be held at Okayama Convention Center, Okayama, Japan, from May 25 to 27, 2020.

However, due to global pandemic of COVID-19, the safety and well-being of all conference participants has become our priority. After evaluating the current COVID-19 situation, the decision has been made to transform APCMBE 2020 into an all-digital conference experience. The conference was held together with the 59th Annual Conference of the Japanese Society for Medical and Biological Engineering. This book gathers the proceedings of APCMBE 2020. Contributions have been carefully reviewed according to the criteria of originality, innovativeness, clinical significance, and level of completion of the study. All traditional medical and biological engineering (MBE) areas as well as new and emerging fields, such as artificial intelligence in health care, medical IoT, medical/healthcare robotics, and regenerative medicine, are covered.

The role of medical and biological engineering has become more and more important to provide basis for innovation for new industries and solution to problems in health care. We would like to exchange our knowledge on MBE and to discuss what MBE can contribute to improve quality of life of people. We hope that this book, which is based on the works selected for the conference, will promote future international collaboration in MBE research and development.

The research topics above may be relevant to counter against COVID-19. For example, implementation of continuous monitoring of patients does require wearable sensing; telemedicine does require modern technologies combined with medical and biological knowledge, etc. Moreover, we hope that discussions generated during APCMBE 2020 will foster more collaborations and exchanges among BME researchers, helping to find solutions to various healthcare problems, including but not limited to issues emerged during the current coronavirus pandemic.

We sincerely appreciate the active participation of all people at APCMBE 2020 and their commitment in writing the outstanding contributions presented in this book.

May 2020

Ichiro Sakuma
Akinori Ueno
Yasuyuki Shiraishi
Keiji Naruse

Contents

Imaging

Background Noise Reduction of OCT Images Based on Region Filling 3
Yingwei Fan, Chengquan Hu, Hongxiang Kang, and Hongen Liao

Monte-Carlo Simulation and Clinical Image Validation for Vessel Size Imaging with Twice and Single Refocus Spin-Echo Diffusion Sequence 10
Kazuhiro Nakamura, Shin Minakata, Hideto Toyoshima, and Toshibumi Kinoshita

Medical Devices

Surgical-Tools Tracking Based on Convolutional Neural Network and Long Short-Term Memory 19
Yu Yang and Zijian Zhao

Development of a Novel Laparoscopic Palpation System Using a Wearable Motion-Sensing Armband 25
Kerwin Caballas, Harold Jay Bolingot, Ingrid Salud, Lawrence Ibarrientos, Lea Cristina Macaraig, Nathaniel Joseph Libatique, and Gregory Tangonan

Surface-Based Electromyography Gesture Profiling of Laparoscopic Tools Using a Wearable Sensor 31
Ingrid Salud, Harold Jay Bolingot, Lea Cristina Macaraig, Nathaniel Libatique, and Gregory Tangonan

Development of Rehabilitation System with Bilateral Upper Extremity Training Robot and Virtual Reality for Stroke Patient 38
Thitikorn Kaewlee and Panrasee Ritthipravat

Therapeutic Medical Engineering

Numerical Analysis of Aortic Valve Stress for Phantom Experiments 47
Takuya Watanabe, Asuka Hatano, Satoshi Izumi, Naoki Tomii, Eriko Maeda, Haruo Yamauchi, Kan Nawata, Shu Takagi, Ichiro Sakuma, Minoru Ono, and Hiroshi Tsukihara

Development of an Ultrastructural Model of Cardiomyocytes Based on Electron Microscopy 52
Makoto Someya, Hiroaki Tanaka, Asuka Hatano, Satoshi Izumi, Masahiko Hoshijima, and Andrew D. McCulloch

Fluid-Structure Interaction Analysis of 3D Human Aortic Valve Model Constructed from CT Images 60
Sakurako Miyazaki, Asuka Hatano, Satoshi Izumi, Yusuke Morishita, Bowen Fan, Shu Takagi, Naoki Tomii, Hiroyuki Tsukihara, Eriko Maeda, Haruo Yamaguchi, Kan Nawata, Ichiro Sakuma, and Minoru Ono

Experiment on Flow Estimation of Simulated Urination in a Western Style Toilet with Built-in Image Sensor 67
Yusuke Niwa, Haruki Kawanaka, Eiichi Watanabe, and Koji Oguri

Diagonstic Evaluation

Effect of a Television Commercial on the Autonomic Nervous System 77
Tomoyuki Yambe, Yasuyuki Shiraishi, Yusuke Inoue, Akihiro Yamada, and Makoto Yoshizawa

Modeling of Pupillometric Signals for Studying Children's Rare Diseases 83
Rachele Fabbri, Ernesto Iadanza, Monica Gherardelli, Paolo Melillo, Francesca Simonelli, and Leonardo Bocchi

Current Technology in Prosthesis, Orthosis and Rehabilitation

Introduction of Prosthesis and Orthosis Evaluation System that Utilizes Human Models 93
Akihiko Hanafusa, Mohd Syahmi Jamaludin, Le Van Tuan, Hayato Ikebata, Ryogo Suzuki, Taiki Kawamura, Shin-ichiro Yamamoto, Yukio Agarie, Hiroshi Otsuka, and Kengo Ohnishi

Analyzing the Shear Stress on the Contacting Interface During Ambulation with Transfemoral Prosthesis Socket 102
Tran Van Thuc, Shin-ichiroh Yamamoto, Akihiko Hanafusa, Kengo Ohnishi, Hiroshi Otsuka, and Yukio Agarie

Lower Limb Muscles Activation of Endurance and Power Athletes 111
Maisarah Sulaiman, Aizreena Azaman, Noor Aimie Salleh, and Azli Yahya

Experimental Analysis of Ankle Foot Orthosis Using Pneumatic Artificial Muscle 118
Nurhanna Zulaikha Ishak, Shahrol Mohamaddan, Annisa Jamali, Shinichiro Yamamoto, Helmy Hazmi, Low Cheng Yee, and Akihiko Hanafusa

Conceptual Design and FEM Analysis of an Exoskeleton Suit for Post-stroke Patient: A Lower Limbs Exo Suit 126
Muhamad Fikhri bin Abdul Hamid, Mohd Hanif bin Mohd Ramli, Noor Ayuni Che Zakaria, and Zulkifli Mohamed

Preliminary Study on the Use of Kinect Camera for Observational Gait Analysis System 135
Aizreena Azaman, Husnir Nasyuha Abdul Halim, Muhammad Fariz Shafiq Abd. Aziz, and Sagida M. A. Bilal

Finger Exoskeleton for Early Acute Post Stroke Rehabilitation: Control Design and Performance Analysis 141
Mohd Nor Azmi Ab Patar, Muhammad Naim Abdu Salam, Muhammad Amir Hazim Jamaluddin, Hokyoo Lee, and Akihiko Hanafusa

Biomedical Database

Analysis of Occupational Injury Reports Related to Patient Care Activities Using Text Mining Technique 153
Kodai Kitagawa, Takayuki Nagasaki, Sota Nakano, Mitsumasa Hida, Shogo Okamatsu, and Chikamune Wada

Proposed Module for Establishment of National Cardiovascular Database in Low Middle Income Countries 159
Md. Ashrafuzzaman, Md. Sakib Abrar Hossain, Monjurul Ahsan, and Nabil Islam

Diagnostic Evaluation

Time-Frequency Analysis of the First-Order Coefficient of Auto-Regressive Approach to Recognize the Normal Breath or Wheeze 175
Bing-Yuh Lu and Chin-Yuan Lin

Analysis of Representations of 3-Dimensional Objects in the Cell Populations in the Temporal Association Area Using Machine Learning 181
Jun-ya Okamura, Yusuke Yamamoto, Lulin Dai, Yoshihiro Uto, Yousuke Yamada, and Gang Wang

X-ray Projection and Fluid Analysis of Contrast Agent Dynamics Through Stenosis 188
Ryotaro Kaneko, Asuka Hatano, and Satoshi Izumi

Medical Devices

Development of a Durable Extracorporeal Ventricular Assist Device Consisting of a Centrifugal Pump with Hydrodynamic Bearings 197
Tomonori Tsukiya, Toshihide Mizuno, Tomohiro Nishinaka, and Eisuke Tatsumi

Sliding Mode Control Based on Nonlinear Disturbance Observer for Pneumatic Artificial Muscle 202
Thuy-Nguyen Vu, Quy-Thinh Dao, Van-Thuc Tran, and Shin-ichiroh Yamamoto

A Novel OCT Image-Guided Laser Automatic Ablation Method Based on Non-common Optical Path Structure 215
Chengquan Hu, Wei Chang, Yangxi Li, Yingwei Fan, Xinran Zhang, Hongxiang Kang, and Hongen Liao

Power Transmission Efficiency Evaluation of Spiral Coil with Center Tap for Transcutaneous Energy Transmission System 223
Tomoki Okinaga, Takahiko Yamamoto, and Kohji Koshiji

Machine Learning

Initial Study of Emotion Recognition Using Brain Rhythm Sequencing Method 231
Jia Wen Li, Shovan Barma, Sio Hang Pun, Mang I. Vai, Pan Ke Wang, Xin Liu, Yuan Yu Yu, Jiu Jiang Wang, Shuang Zhang, and Peng Un Mak

Fast Learning from Imperfect Labels to Segment Brain Based on Active Contour Model and 3D U-Net 240
Honglei Ji, Yi Lv, and Junchen Wang

Comparison of Reconstruction Performances Between Deep Learning and Iterative Methods in MR Compressed Sensing 249
Shohei Ouchi and Satoshi Ito

Tissue Engineering

Ceramide: An Extravagant Prospective for Treatment of Burn Patients 257
Md. Ashrafuzzaman, Faiza Ahmed, Afsana Mimi, Farahdiba Zarin, and Sabrina Shetu

Development of Elemental Technologies for Magnetic Hyperthermia in Cancer Treatment 272
Loi Tonthat, Fumitaka Aki, Kazutaka Mitobe, Shin Yabukami, and Yoshiyuki Yamamoto

Immunogenic Cell Death Induced by Selective Intracellular Vaporization *in vitro* 278
Ayumu Ishijima, Satoshi Yamaguchi, Takashi Azuma, Etsuko Kobayashi, Kazuhiro Kakimi, Yoshikazu Shibasaki, Teruyuki Nagamune, and Ichiro Sakuma

Reconstruction of a Vascular Bed with Perfusable Blood Vessels Using a Decellularized Porcine Small Intestine for Clinical Application 284
Yusuke Tobe, Katsuhisa Sakaguchi, Jun Homma, Kazunori Sano, Eiji Kobayashi, Hidekazu Sekine, Kiyotaka Iwasaki, Tatsuya Shimizu, and Mitsuo Umezu

Therapeutic Medical Engineering

Tactile Cue by Touching the Agonist Muscle Increases the Muscle Activity During Arm Curl Exercise 295
Kazushige Oshita

Gait Characteristics and Pain Assessment of Symptomatic Knee Osteoarthritis 301
Husnir Nasyuha Abdul Halim, Aizreena Azaman, and Azli Yahya

Diagnostic Devices

PET-Laparoscope System with a Multi-layer Movable Detector: Feasibility Study with an Abdominal Phantom 311
Madhushanka R. Liyanaarachchi, Kenji Shimazoe, Hiroyuki Takahashi, Keiichi Nakagawa, Etsuko Kobayashi, and Ichiro Sakuma

Improvement of Force Sensor Accuracy Using Redundant Outputs for Biomedical Applications 319
Masaru Higa, Ryutaro Hayami, Masahiro Ono, Shizuki Nakamura, and Shiori Matsuda

Combining Surface-Based Electromyography and Motion Sensing for Detection of Human Falling 325
Lawrence Ibarrientos, Harold Jay Bolingot, Lea Cristina Macaraig, Nathaniel Libatique, and Gregory Tangonan

Wireless ECG Electrode to Improve Tele-cardiac Monitoring Usability 332
Norma Hermawan, Achmad Arifin, Fauzan Arrofiqi, Mohammad Nuh, and Yoshifumi Saijo

Author Index 341

Imaging

Background Noise Reduction of OCT Images Based on Region Filling

Yingwei Fan[1](✉), Chengquan Hu[2], Hongxiang Kang[1], and Hongen Liao[2]

[1] Beijing Institute of Radiation Medicine, Beijing 100850, China
fanyingwei@163.com

[2] Department of Biomedical Engineering, School of Medicine, Tsinghua University, Beijing 100084, China

Abstract. Optical coherence tomography (OCT) is widely used in biomedical imaging. However, background noise severely affects the diagnosis and identification of diseased tissues. Here, we propose a background-noise reduction method based on region filling. OCT images of *in vivo* skin tissues, *ex vivo* brain tumors and *ex vivo* brain normal tissues had been analyzed to remove background noise using the method. Results demonstrated that the mean of the noise reduction ratio (NRR) reached more than 95% and the image-to-noise ratio (INR) ranges from −0.5 dB to 2.5 dB. The denoised OCT images demonstrated a better three-dimensional (3-D) visualization effect. The proposed method is useful for removing the background noise of OCT images in medical imaging.

Keywords: Optical coherence tomography · Background noise reduction · Region filling · Biomedical imaging

1 Introduction

Optical coherence tomography (OCT) [1] is widely used in medicine and biomedicine. Noise seriously affects the quality of the OCT images. The noise removal of OCT images is still a hot topic, especially in the intraoperative OCT imaging [2].

Speckle noise [3] reduction is along the main focus. Currently, many methods are developed to remove speckle noises, such as the wavelet domain compounding [4], the A-scan reconstruction [5], the edge-sensitive cGAN [6] and the shearlet-based total variation algorithm [7]. However, some fixed noises still occur. Fixed mode noise and background noise usually remain in OCT images. Thus, these noises let original OCT images unclear and unsharp. Recently, we developed the fixed-pattern noise reduction method base on the Hough transform [8]. The proposed method reaches the high-efficiency noise reduction ratio for the fixed-pattern noise. Similarly, the background noise affects the post-processing of OCT images. The removal of the background noise is important for improving the quality of images and three dimensional reconstruction.

In this study, the background noise reduction method is proposed based on the region filling. The preprocessing of OCT provided the basis of the noise removal method. The region filling was conducted after the binarization of OCT images. The upper surface

Y. Shiraishi et al. (Eds.): APCMBE 2020, IFMBE Proceedings 82, pp. 3–9, 2021.
https://doi.org/10.1007/978-3-030-66169-4_1

of biological tissues was searched through threshold classification. The noise reduction ratio (NRR) and the image-to-noise ratio (INR) were computed to evaluate the proposed method. The validation experiments were conducted via the 3-D visualization of the denoised OCT images.

2 Methods and Materials

Noise reduction of OCT images is important, especially the background noise reduction. The region filling-based noise removal method was proposed, some *in vivo* and *ex vivo* experiments were conducted. The quantitative and qualitative parameters of noise removal were computed and visualized through the computation of NRRs and INRs and the visualization of 3-D volumetric datasets.

2.1 Biological Tissue Preparation and OCT Imaging

Three types of biological tissues were acquired, which included the *in vivo* mouse skin, *ex vivo* mouse brain tumor and healthy brain tissues. Conventional mice were anesthetized by an intraperitoneal injection of 1% sodium pentobarbital solution. The solution was injected into the mouse at a dose of 0.1 mm/10 g. Brain tumor cell was injected into the brain, and about ten days after injection, the tumor model was formed. This study was approved by the Beijing Institute of Radiation Medicine Experiment Animal Center-Approved Animal Protocols. All animal experiments were performed in accordance with the guidelines in IACUC-DWZX-2019-502.

One mouse was conducted to use OCT scan the *in vivo* skin tissue. One mouse was sacrificed to acquire the *ex vivo* healthy brain tissues. And one mouse was sacrificed to acquire the *ex vivo* brain tumor tissue. OCT was used to scan these kinds of the biological tissues. Each dataset of biological tissues included 200 B-scan images. One of OCT images was chosen to preprocess and reduce the background noise. The other OCT images were used to evaluate the performance of the proposed method via computing the NRRs and the INRs.

2.2 Preprocessing and Background Noise Reduction Based on Region Filling

OCT images included some noises and useful information of biological tissue. Preprocessing and background noise reduction were conducted. To reduce the fixed mode noise, our previous HTFPNR method [8] was used to process the OCT images.

The original OCT was histogram-equalized. Otsu algorithm was used to automatically choose the gray threshold for the binarization of a whole OCT image. Then, a 5 $\times$ 5 mask was applied to implement the median filtering for an OCT image. The region filling operation was conducted. The region was the hole in binary OCT image. This procedure is shown in Fig. 1(a).

After preprocessing, the upper surface was searched through comparing the gray threshold. The gray value of the binarized OCT image ranges from 0 to 255. Here, we chose the 245 as the gray threshold. Then, the upper surface was smoothed through the median filtering method. The gray value where located above the upper surface was set as zero. The useful-information depth was approximately 1.5 mm, and the 150-pixel-depth OCT image below the upper surface was used to post-process (Fig. 1(b)).

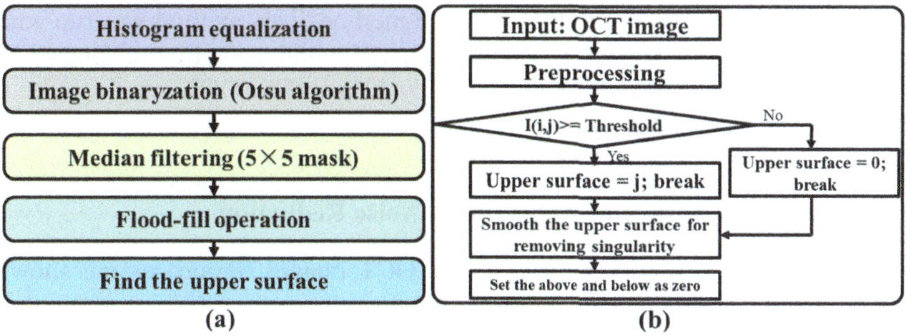

Fig. 1. (a) The preprocess procedure of background noise reduction. (b) The procedure of the background noise reduction method.

2.3 Validation of Background Noise Reduction

To evaluate the performance of the background noise reduction, we conducted the noise reduction ratio to compute the noise reduction index, which included the background noise that is reduced by this proposed method. The NRR is defined as in Eq. (1):

$$\mathrm{NRR} = \frac{10\log_{10}(\sum I^2_{noise\_filtered})}{10\log_{10}(\sum I^2_{original\_image})} \times 100\% \tag{1}$$

where $I_{noise\_filtered}$ and $I_{original\_image}$ are the intensities of the noise removed using the algorithm and the original image, respectively.

Moreover, the INR was used to validate the background noise reduction for noisy OCT images. The INR provided the power ratio between OCT image without background noise and its background noise. The INR is defined as Eq. (2):

$$\mathrm{INR}[dB] = 20 \times \log_{10}\left(\frac{\sqrt{\frac{1}{N}\sum I^2_{filtered}}}{\sqrt{\frac{1}{N}\sum I^2_{noise}}}\right) \tag{2}$$

where N is the total number of OCT images' pixels, and $I_{filtered}$ is the image intensity without the background noise, I_{noise} is the intensity of the noisy image.

Three groups of OCT images were used to validate the proposed method. The *ex vivo* brain tumor, normal brain tissue and *in vivo* skin tissue had been imaged with OCT. These OCT images were denoised and analyzed with the proposed method. The INR and NRR of each OCT image were computed and contrasted. The trends of NRR and INR were analyzed.

3 Results

OCT images of the three types of biological tissue were acquired. Each dataset included 200 OCT images. Then, the preprocessing, noise removal and validation of the proposed method were conducted and implemented. 3-D visualization of OCT images presented the effect of noise reduction with the proposed method. This method was run with the software of Matlab 2017. The OCT system had a horizontal resolution of approximately 22 μm and an axial resolution of approximately 12 μm. The OCT imaging speed was approximately 60 frames per second.

3.1 Image Preprocessing and Background Noise Reduction

Figures 2(a)–(d) shows the preprocessing of OCT images. Preprocessing shows the enhancement of OCT image (Fig. 2(b)). Figure 2(c) shows the OCT images with binarization using the Otsu algorithm. Figure 2(d) shows the hole-filling binarized OCT image.

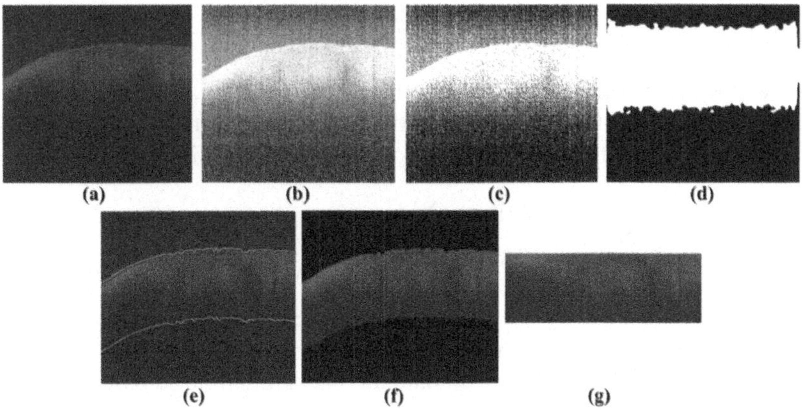

Fig. 2. The background noise reduction. (a) The original OCT image of the brain tumorous tissue. (b) The histogram equalization of OCT images. (c) The OCT images with binarization using Otsu algorithm. (d) The hole-filling binarized OCT image. (e) The upper surface and lower boundary. (f) The OCT images with background noise reduction. (g) The flatten OCT images without background noise.

After preprocessing, the background noise reduction was useful to enhance and filter OCT images. The useful information of the original OCT image presented the feature of biological tissue. Figure 2(e) shows the upper surface and the lower surface of OCT images. Then, the intensities of the OCT image above the upper surface and below the lower surface were set as zero. Figure 2(f) shows the OCT images without background noise. Figure 2(f) shows the flatten OCT after background noise reduction.

3.2 Validation of Background Noise Reduction Method

The validation of the proposed method included the 3-D visualization of OCT images for qualitative evaluation and the computation of INR and NRRs for quantitative evaluation. Figure 3 shows that the NRRs and INRs of OCT images of the three types of biological tissues.

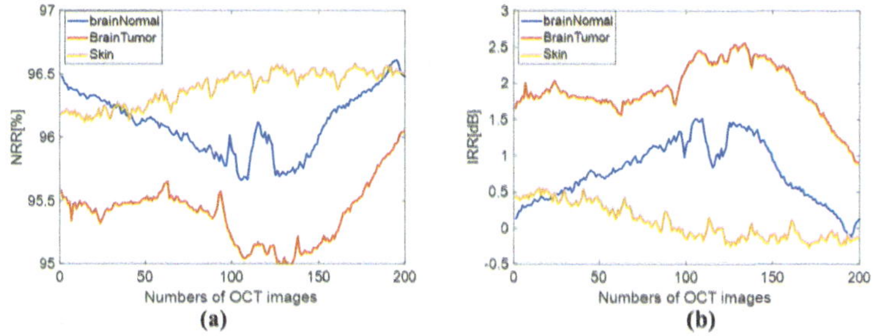

Fig. 3. The NRRs and INRs corresponding to different kinds of tissues. (a) The listed NRR of the three kinds of biological tissues. (b) The listed INR of the three kinds of biological tissues.

Table 1. The mean NRRs and mean INRs of the proposed method.

Tissue kinds	Mean NRR [%]	Mean INR [dB]
Skin tissue	96.40 ± 0.14	0.10 ± 0.24
Brain tumor tissue	95.41 ± 0.24	1.93 ± 0.36
Brain normal tissue	96.11 ± 0.25	0.78 ± 0.41

Figure 3(a) and Table 1 lists the NRRs and INRs of the proposed method. Three types of biological tissues demonstrated that the NRR and the mean NRRs both are more than 0.95%. The background noise was effectively reduced. Figure 3(b) shows that the range of INRs from −0.5–2.5 dB. The mean INR had changed significantly. The change of the standard INR also was the same with mean INR. The intensity of the background noise would be larger than the useful OCT images.

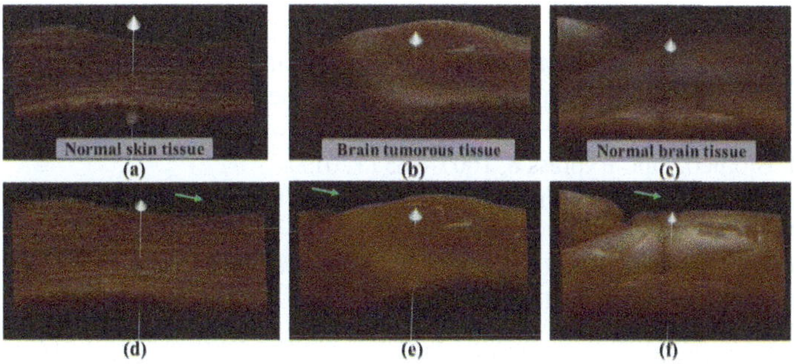

Fig. 4. The background noise reduction procedure. (a)–(c) The OCT images with background noise reduction. (d)–(f) The 3-D visualization of OCT images without background noise corresponding to the *in vivo* skin tissue, the *ex vivo* brain tumorous tissue and brain normal tissue.

The 3-D visualization of OCT images showed the effect of noise reduction and non-reduction. Figure 4 shows the 3-D visualization of the original and denoised OCT images. We can found that the denoised OCT images (Fig. 4(a)–(c)) have a better visualization effect than the original OCT image with background noise (Fig. 4(d)–(f)). Furthermore, the image quality of 3-D visualization of the denoised OCT image was obviously better than the original OCT images (green arrow labels to the removal background noise). Hence, the OCT images without background noise could be presented better 3-D visualization and image quality.

4 Discussions and Conclusion

The background noise reduction enhanced the image quality and 3-D visualization effect. The noise removal ratio reached more than 95%. The quantitative and qualitative evaluation demonstrated that our proposed method provides good performance of the removal of the background noise. This method could be used in endoscopic imaging [9]. Furthermore, the background noise reduction was not widely researched comparing with speckle noise [10], we just provided the pilot method to reduce the noise. Future research will try to novel intelligent algorithms [11], such as deep learning and some machine learning algorithm.

In conclusion, we propose a region filling-based background noise reduction method. The *in vivo* and *ex vivo* experiments are conducted to acquire OCT images and the proposed method is used to remove the OCT images' background noises. The quantitative and qualitative evaluations demonstrate that this method can effectively remove background noise.

Acknowledgement. This research was partially funded by National Science Foundation of China (NSFC) (No. 81901907) and China Postdoctoral Science Foundation (No. 2018M643846 and No. 2019T120982).

References

1. Huang, D., et al.: Optical coherence tomography. Science **254**, 1178–1181 (1991)
2. Carrasco-Zevallos, O.M., Viehland, C., Keller, B., Draelos, M., Kuo, A.N., Toth, C.A., Izatt, J.A.: Review of intraoperative optical coherence tomography: technology and applications [Invited]. Biomed. Opt. Express **8**, 1607–1637 (2017)
3. Schmitt, J.M., Xiang, S.H., Yung, K.: M: Speckle in optical coherence tomography. J. Biomed. Opt. **4**(1), 95–105 (1999)
4. Xu, J., et al.: Wavelet domain compounding for speckle reduction in optical coherence tomography. J. Biomed. Opt. **18**(9), 096002 (2013)
5. Cheng, J., et al.: Speckle reduction in 3D optical coherence tomography of retina by A-scan reconstruction. IEEE Trans. Med. Imaging **35**(10), 2270–2279 (2016)
6. Ma, Y., Chen, X., Zhu, W., Cheng, X., Xiang, D., Shi, F.: Speckle noise reduction in optical coherence tomography images based on edge-sensitive cGAN. Biomed. Opt. Express **9**, 5129–5146 (2018)
7. Xu, M., Tang, C., Chen, M., Qiu, Y., Lei, Z.: Texture preservation and speckle reduction in optical coherence tomography using the shearlet-based total variation algorithm. Opt. Lasers Eng. **122**, 265–283 (2019)
8. Fan, Y.W., Ma, L.F., Jiang, W.P., Luo, S.T., Zhang, X.R., Liao, H.E.: Optimized optical coherence tomography imaging with Hough transform-based fixed-pattern noise reduction. IEEE Access **6**, 32087–32096 (2018)

9. Fan, Y.W., Luo, S.T., Kang, H.X.: A comparison of the long working distance of the GRIN lens- and C-lens- based OCT probe. IEEE Access **7**(1), 93212–93218 (2019)
10. Chen, Z., Zeng, Z., Shen, H., Zheng, X., Dai, P., Ouyang, P.: DN-GAN: Denoising generative adversarial networks for speckle noise reduction in optical coherence tomography images. Biomed. Signal Process. Control **55**, 101632 (2020)
11. Devalla, S.K., et al.: A deep learning approach to denoise optical coherence tomography images of the optic nerve head. Sci. Rep. **9**, 14454 (2019)

Monte-Carlo Simulation and Clinical Image Validation for Vessel Size Imaging with Twice and Single Refocus Spin-Echo Diffusion Sequence

Kazuhiro Nakamura(✉), Shin Minakata, Hideto Toyoshima, and Toshibumi Kinoshita

Research Institute for Brain and Blood Vessels Akita, Akita 010-0874, Japan
knam@akita-noken.jp

Abstract. Vessel size imaging (VSI) was required an intravascular superparamagnetic contrast agent for vessel diameter estimation. Apparent diffusion coefficient (ADC) calculated from single refocused pulsed-gradient spin-echo (SRSE-DWI) was influenced by the magnetic susceptibility of vasculature. The influence of the ADC from the vessels can be reduced by using twice refocused spin-echo (TRSE-DWI) sequence. We evaluate the VSI with SRSE-DWI and TRSE-DWI using Monte-Carlo simulations in a vascular model. It was also evaluated in nine patients with cerebral infarction in the penetrating branch area within two weeks after stroke onset. The simulation results show that vessel radius was determined uniquely from the pair of SRSE-DWI and TRSE-DWI signal. Clinical images also suggested the vasodilation in the periphery of cerebral infarction might be detected in the difference image of SRSE-DWI and TRSE-DWI. It suggests that vessel radius might be estimated by acquiring SRSE-DWI and TRSE-DWI. It could be useful for the clinical diagnosis.

Keywords: MRI · Vessel size imaging · Diffusion weighted image · Stroke

1 Introduction

Vessel size imaging (VSI) was already reported based on transverse relaxation time images obtained from spin echo and gradient echo sequence [1]. VSI is required an intravascular superparamagnetic contrast agent for vessel diameter estimation. On the other hand, susceptibility-induced background gradient field interact to diffusion weighted signal intensity, especially with typical single refocused pulsed-gradient spin-echo (SRSE-DWI) sequence. Apparent diffusion coefficient (ADC) calculated from SRSE-DWI indicates an underestimation of the ADC proportional to the magnetic susceptibility of vascular radius [2]. The influence of the ADC from the vessels can be reduced by using twice refocused spin-echo (TRSE-DWI) sequence [3, 4]. The comparison between SRSE-DWI and TRSE-DWI will indicate vessel radius weighted image and be useful image for evaluating the several kinds of diseases. For example, cerebrovascular reserve of stroke patient could be estimated with VSI. It might be also explicit about

Y. Shiraishi et al. (Eds.): APCMBE 2020, IFMBE Proceedings 82, pp. 10–15, 2021.
https://doi.org/10.1007/978-3-030-66169-4_2

the difference of core and penumbra lesions in ischemic stroke. At first, we evaluate the VSI with SRSE-DWI and TRSE-DWI sequences using Monte-Carlo simulations in a vascular model [2, 5]. And then, we also examine the property of the difference image of two DWIs in early phase of stroke patients.

2 Methods

2.1 Monte-Carlo Simulation

In order to assess the TRSE-DWI and SRSE-DWI, Monte-Carlo simulations were performed to calculate the signal of molecules diffusing in a vascular network [2, 5] The brain vasculature radius was assumed 1, 2, 5, 10 and 20 μm as same as previous literature. The MC algorithm was implemented using homemade C language program performed with ubuntu linux distribution (Version 18.04) in personal computer (CPU: Intel® Core™2 Extreme QX9650). The intra-/ extra-vascular susceptibility difference was set to 0.8×10^{-7} assumed the normal brain tissue oxygenation level. Other susceptibility difference of 2.0, 5.0 and 10.0×10^{-7} were simulated. Diffusion constant of the proton was set at 1.0×10^{-5} cm^2/s assumed the normal brain tissue. ADC was calculated by simulated signal intensity with b-value of 0 and 1000. Signal reduced ratio was defined with the ratio between simulated signal value and initial value. ADC reduced ratio was defined with the ratio between simulated ADC value and assumed diffusion constant.

2.2 Clinical MRI Image Evaluation

Nine patients with cerebral infarction in the penetrating branch area were included. MRI was examined within two weeks after stroke onset (9 ± 3 days) with a 3T MRI(MAGNETOM Verio, Siemens). SRSE-DWI and TRSE-DWI were acquired with TR/TE = 6000/72 ms and 5 mm slice thickness. ADC were calculated from DWI of b values 0 and 1000. Difference image (ADC-Diff) between ADC of TRSE (TRSE-ADC) and SRSE (SRSE-ADC) was evaluated. For the evaluation of the healthy contralateral hemisphere, brain segmentation was processed by the Freesurfer v6.04 using anatomical T1-weighted image acquired by 3D MPRAGE sequence [6]. Coregistration to DWI from structural image was also processed by the Freesurfer software. For the evaluation of ischemic brain region, region of interest (ROI) was manually selected surrounding the ischemic region and almost the same size in the healthy contralateral hemisphere. Typical ROI selection was shown in Fig. 5. ADC-Diff value in ROI was compared as a function of SRSE-ADC threshold. This study was approved by the Ethics Committee of the Research Institute for Brain and Blood Vessels-Akita (reference number: 16-06).

3 Results

3.1 Monte-Carlo Simulation

Figure 1 shows the relation between given radius and ADC reduced ratio using SRSE-DWI and TRSE-DWI sequence. Simulation results with susceptibility difference of

2.0×10^{-7} was shown. As same as previous literature, the ADC underestimation is bigger in SRSE-DWI than TRSE-DWI and the ADC underestimation depended on the vessel radius [2]. The signal intensity relation between SRSE-DWI and TRSE-DWI was shown in Fig. 2. The pair of susceptibility and vessel radius were determined uniquely from the pair of SRSE-DWI and TRSE-DWI signal. This indicates that the vessel radius can be estimated by acquiring SRSE-DWI and TRSE-DWI.

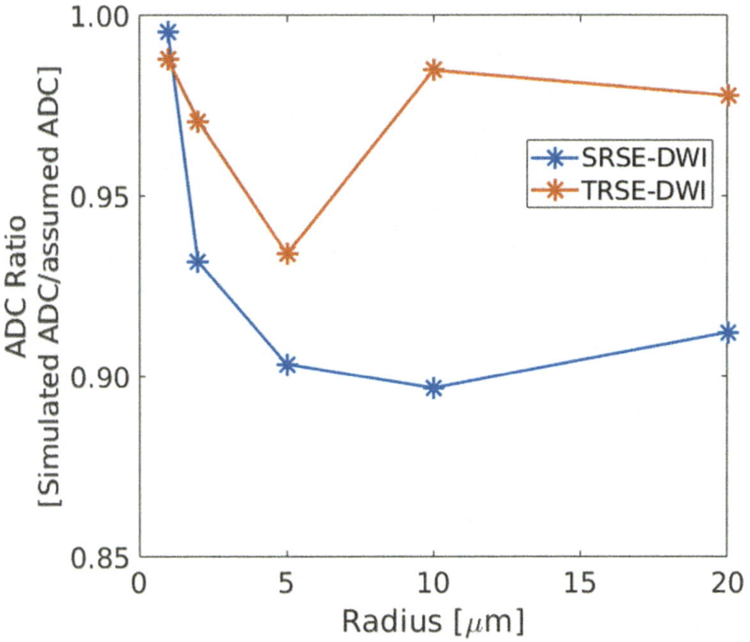

Fig. 1. Relation between given vessel radius and ADC reduced ratio using SRSE-DWI (blue) and TRSE-DWI (red) sequence. Simulation results with susceptibility difference of 2.0×10^{-7} was shown.

3.2 Clinical Image

Typical DWI and ADC images in a stroke patient were shown in Fig. 3. Even almost same contrast in SRSE-DWI and TRSE-DWI sequence, ADC-Diff shows a high value in the periphery of cerebral infarction and the basal ganglia region. Mean and standard deviation of ADC-Diff in several brain regions was shown in Fig. 4. ADC-Diff was large in the place where the larger magnetic susceptibility effect exists such as the pallidum and striatum. ADC-Diff in the ischemic ROI and healthy ROI were shown as a function of SRSE-ADC threshold in Fig. 5. Mean and standard error of ADC-Diff in each SRSE-ADC threshold was evaluated.

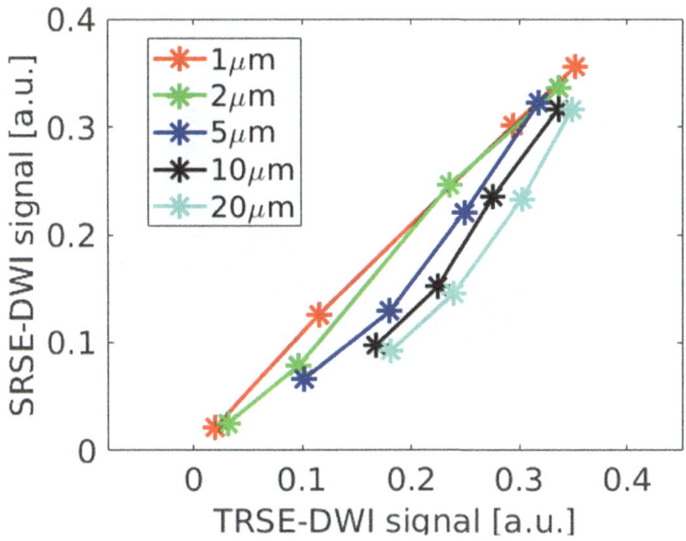

Fig. 2. The signal intensity relation between SRSE-DWI and TRSE-DWI. For each of the five vessel diameters of 1, 2, 5, 10, and 20 μm, the signal intensity relationship was shown. The change in magnetic susceptibility of 0.8, 2.0, 5.0 and 10.0 × 10^{-7} was simulated.

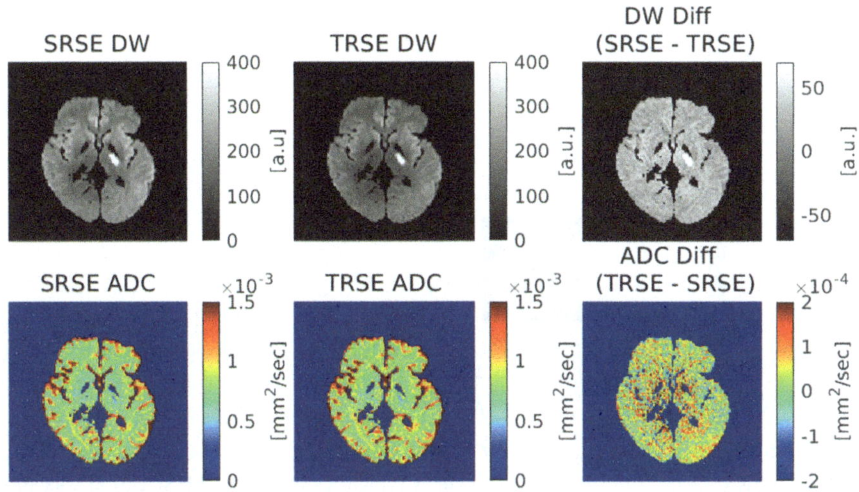

Fig. 3. Typical diffusion weighted, ADC and ADC difference images from a patient. The voxel value of TRSE-DWI was smaller than that of SRSE-DWI and the value of TRSE-ADC is bigger than that of SRSE-ADC. Therefore, difference image of DWI and ADC was subtracted by TRSE and SRSE, respectively.

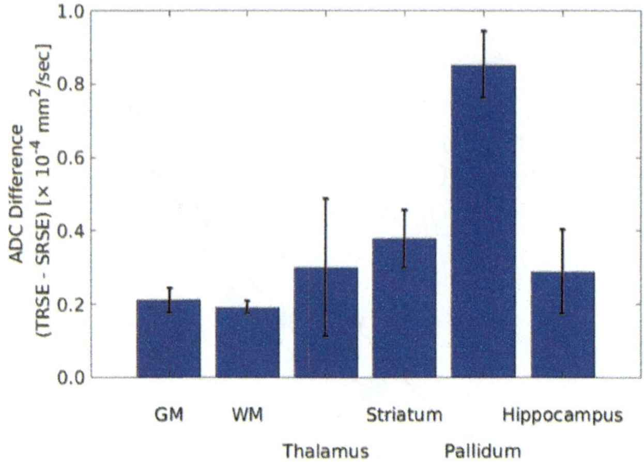

Fig. 4. The ADC difference in gray matter (GM), white matter (WM), thalamus, striatum, pallidum and hippocampus in the healthy contralateral hemisphere was shown. Bar height indicates the mean value of each segmented region and error bars indicate the standard deviation.

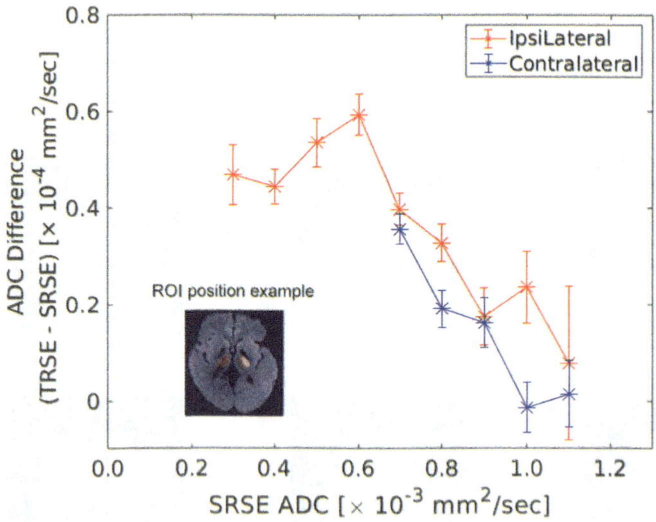

Fig. 5. ADC difference in the ischemic ROI (red) and healthy ROI (blue). Mean and standard error of each SRSE-ADC threshold was shown. A typical example of ROI selection was shown as 'ROI position example' in the figure. Weak red area indicates the manually selected ROI.

4 Discussion

From the results of MC simulation, the vessel radius and the magnetic susceptibility is uniquely determined simultaneously from the combination of the SRSE-DWI and TRSE-DWI signal intensity under the assumption of uniform tissue diffusion coefficient. This suggests that the vessel radius can be estimated by acquiring SRSE-DWI and

TRSE-DWI. By performing MC simulation for each ADC with several vessel radius and magnetic susceptibility in advance, the vessel radius can be determined from SRSE-DWI and TRSE-DWI.

In the clinical image, the lower value in TRSE-DWI than SRSE-DWI may due to the twice number of motion probing gradient in TRSE-DWI. Because the motion probing gradient should decrease the signal. ADC-Diff was large in the tissue with relatively large susceptibility. It is reasonable that the difference image reflects the change in magnetic susceptibility. In addition, ADC-Diff shows a high value in the periphery of cerebral infarction in Fig. 3 and more clearly shown in Fig. 5. Identification of ischemic core was an ADC under 0.62×10^{-3} mm^2/s in previous literature [7]. The biggest difference in ADC-Diff was shown at ADC around 0.6×10^{-3} mm^2/s in Fig. 5. It may reflect the change in magnetic susceptibility associated with vasodilation. DWI is indispensable for clinical diagnosis and obtain the two DWI images in a short time. Therefore ADC-Diff is considered to be clinically useful as an image reflecting the tissue magnetic susceptibility.

5 Conclusion

The simulation results and clinical images suggested that vessel radius might be estimated by acquiring SRSE-DWI and TRSE-DWI. It could be useful for the clinical diagnosis.

References

1. Tropres, I., Grimault, S., Vaeth, A., Grillon, E., Julien, C., Payen, J.F., Lamalle, L., Decorps, M.: Vessel size imaging. Magn. Reson. Med. **45**(3), 397–408 (2001)
2. Pampel, A., Jochimsen, T.H., Moller, H.E.: BOLD background gradient contributions in diffusion-weighted fMRI—comparison of spin-echo and twice-refocused spin-echo sequences. NMR Biomed. **23**, 610–618 (2010)
3. Zhong, J., Kennan, R.P., Gore, J.C.: Gore, Effects of susceptibility variations on NMR measurements of diffusion. J. Magn. Reson. **95**(2), 267–280 (1991)
4. Galvosas, P., Stallmach, F., Kärger, K.: Background gradient suppression in stimulated echo NMR diffusion studies using magic pulsed field gradient ratios. J. Magn. Reson. **166**(2), 164–173 (2004)
5. Boxerman, J.L., Hamberg, L.M., Rosen, B.R., Weisskoff, R.M.: MR contrast due to intravascular magnetic susceptibility perturbations. Magn. Reson. Med. **34**, 555–566 (1995)
6. Fischl, B., et al.: Whole brain segmentation: automated labeling of neuroanatomical structures in the human brain. Neuron **33**, 341–355 (2002)
7. Purushotham, A., et al.: Apparent diffusion coefficient threshold for delineation of ischemic core. Int. J. Stroke **10**(3), 348–353 (2015)

Medical Devices

Surgical-Tools Tracking Based on Convolutional Neural Network and Long Short-Term Memory

Yu Yang and Zijian Zhao(✉)

School of Control Science and Engineering, Shandong University, Jinan, China
zhaozijian@sdu.edu.cn

Abstract. Real-time surgical tool tracking is a critical component of computer-assisted surgery, because it is highly instrumental to analyze and understand the surgical activities. Nowadays, many deep learning methods take fully advantage of very deep networks and track by detection. Although these methods work well, but they take up a significant amount of time and computational resources. To address this problem, we propose a new network which use the cascade of refined convolutional neural network and long short-term memory for real-time single tool tracking based on the Real-time Recurrent Regression Networks (Re^3). Our method is tested on the publicly available standard dataset from UCL (University College London). The experimental result show that our method achieves better performance than state-of-the-art tracking methods in terms of accuracy and speed.

Keywords: Surgical tool tracking · Convolutional Neural Network · Long Short-Term Memory

1 Introduction

Laparoscopic surgery is a type of minimally invasive surgery and is used in a wide variety of surgical procedures because it causes less pain and allows a faster recovery rate when compared to open surgery [1]. With the development of computer technology, deep learning and modern medical technology, robot-assisted minimally invasive surgery (RMIS) systems, like the daVinci Surgical System, have gained more and more attention in recent years. Hence, in this paper we focus on real-time and accurate tracking of 2D coordinates of surgical tools in endoscopic surgery videos.

In this paper, we proposed a refined real-time tracking framework based on the Real-time Recurrent Regression Networks (Re^3) [2], which consists of convolutional neural network and two blocks of long short-term memory (LSTM) networks [3]. The former network is used to learn the appearance features of surgical tool and we refined the Realtime Recurrent Regression Networks by remove some features input to the first LSTM. In other words, we don't use as many skip connections as Re^3. The first LSTM block learns the motion features and remember appearance and motion information [2] and the second LSTM block is used to regress the position of surgical tool. The features extracted by convolutional neural network are input to the first LSTM. To evaluate the performance of our method, we evaluate it on the publicly available standard dataset

Y. Shiraishi et al. (Eds.): APCMBE 2020, IFMBE Proceedings 82, pp. 19–24, 2021.
https://doi.org/10.1007/978-3-030-66169-4_3

from UCL (University College London). Our method obtains a better performance than three state-of-the-art tracking methods in terms of the accuracy and speed.

2 Methods

The overall design of our surgical tools tracking model is shown in Fig. 1. This section will describe the architecture of each sub-network and our refinement based on Re^3 in more detail [2]. In the proposed framework, the convolutional neural network (Sect. 2.1) takes the RGB image frames as input, and outputs two different levels of features and these features of two adjacent frames are connected in parallel, and input to the first LSTM (Sect. 2.2) after passed through a fully connected layer. The second LSTM takes the features extracted from convolutional neural network concatenated the first LSTM's output as input. The second LSTM's outputs are fed into a fully connected layer and outputs four numbers for surgical tool which encodes the bounding-box position in the input image.

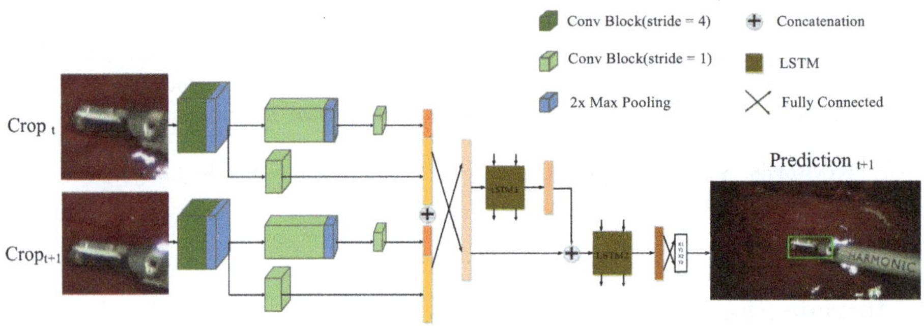

Fig. 1. Network structure

2.1 Convolutional Neural Network

Network Inputs. The input images are cropped from the video stream. By feeding a pair of crops, the network can directly compare differences between frame_t and frame_{t+1} and learn how motion affects the image pixels. Both crops are centered around the objects location in the previous frame, padded to two times the width and height of the object [4], then be resized to 227 * 227. In other words, our search area is twice the size of the tool bounding box in the previous frame (this multiple can be adjusted according to the extreme mechanical movement speed of the surgical tool), we can find the location of the tool in this area definitely. Because we use this method to get the input images, we need to give the bounding box of the surgical tool in the first frame of the endoscopic surgery videos.

Convolutional Stride. In order to reduce the size of the feature maps quickly, we use convolutional kernels (11 * 11 * 3) to convolve the input image with big convolutional stride 4 at the beginning and following the convolutional layer is rectified Linear unit (ReLU). Other convolutional layer's stride is 1 and followed by a 2x max pooling layer except last convolutional layers, and these convolutional layers reduce the dimensionality of the layers with higher spatial resolutions to keep computational cost low [2].

Skip Connection. Here is the biggest difference between our proposed network and the Re_3 Network [2]. As we know, the hierarchical structure [5] of convolutional network extracts different levels of information from different layers. But for different tasks, the level of features extracted by CNN is not the more the better. For example, our experiments prove that if the convolutional neural network extracts four levels of features and then feeds them into the fully connected layer, the experimental results are not ideal. After 30 frames or so, the network will not track the surgical tool. We analyze the reason is that the LSTM network may be more suitable for learning and memorizing the appearance and motion features from similar levels. But we still use skip connections to give the network a richer appearance model [1] when spatial resolution decreases and a level of features is really too little for identifying surgical tools, so we concatenate the pair of images features at the end of convolutional layers to allow the network to separate out the differences quickly between the images and allow LSTM to learn more useful information.

2.2 Long Short-Term Memory

LSTM network is designed to solve the long-term dependence of the general RNN (Recurrent Neural Network). Following the prior work Greff et al. [6], we choose to use two layers of LSTM, because the two layers of LSTM are able to remember longer term relationships between frame and frame such as the laws of surgical tool's movement and the change of shape. LSTM uses C_t to store this information for a long time and these information will not be affected by the current frame obviously. But it is difficult to regress the coordinates of surgical tool just rely on these laws, so we combine these laws with the features extracted from convolutional neural network and fully connected layer to regress the coordinates. We use an L1 loss on the outputs to encourage exact matches the ground truth.

3 Results

To evaluate the performance of our method, we apply the method to single tool dataset from UCL (University College London). We compare the method with three other state-of-the-art tracking methods in terms of tracking accuracy and speed (Fig. 2).

The standard dataset from UCL (University College London) consists 5725 frames of 1920 * 1080. Because input images are cropped in our method, once the target is lost, we can't crop any image for the input. Therefore we chose successive 1232 frames in which the surgical tool has been in the field of view. We separated these frames into a training set (1000 frames) and a test set (232 frames). The image frames fed into our model are all resized to 640 * 480 pixels. We train our network using Adam with initial

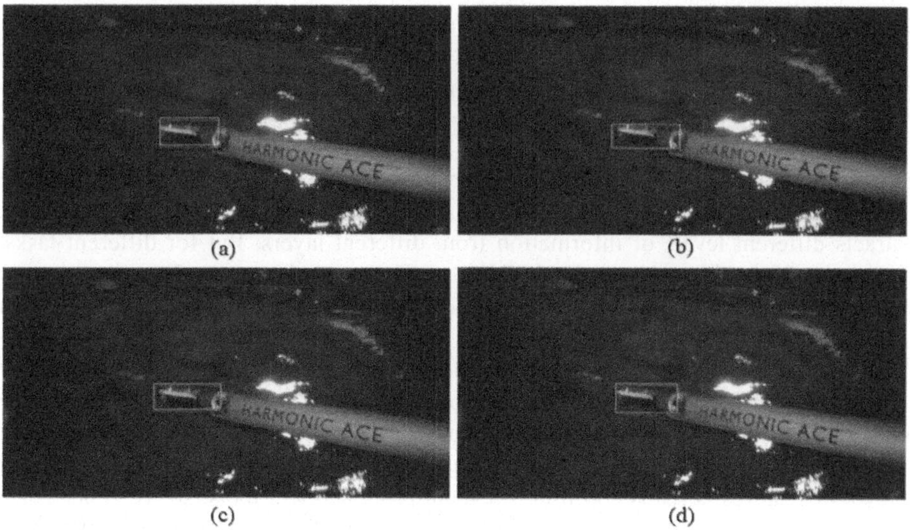

Fig. 2. Tracking examples for the standard dataset from UCL, the ground-truth is in red and tracking methods are in green. (a) BDF (b) ASMS (c) Staple (d) Our method.

learning rate of $1 * 10^{-5}$, and weight decay of $5 * 10^{-4}$. After 150 epochs, our network has achieved excellent results on the test dataset. We implement the proposed tracking method and the compared methods [7–9] on Pytorch 0.4.1, Ubuntu 16.04 LTS using an NVIDIA GeForce GTX TITAN X GPU accelerator.

To evaluate the accuracy of our tracking method, we use the following evaluation method: if the intersection over union (IoU) of the predicted bounding-box and the ground-truth is bigger than 0.5, we consider the instrument to be successfully tracked in this frame. As shown in Table 1, our method achieves a mean Average Precision (mAP) of 100.00% for the standard dataset from UCL, a mean computation time of 0.015 s for locating the surgical tool in each image frame.

Table 1. Compare accuracy and speed on the dataset from UCL.

Methods	mAP	Time (per frame)
Staple	100.0%	0.105 s
ASMS	100.0%	0.023 s
BDF	91.7%	0.017 s
Ours	100.0%	0.015 s

In order to evaluate these methods more accurately, we also evaluate these methods based on a distance evaluation approach, as shown in Fig. 3. If the distance between the center of the predicted bounding-box and the center of ground-truth bounding-box is smaller than a threshold in the image coordinates, the surgical instrument is considered to be correctly tracked. Staple achieves better performance than our approach for the standard dataset from UCL in terms of accuracy but at a cost of 6× lower tracking speed. This demonstrates that our proposed method achieves better performance than the other three methods for the standard dataset from a comprehensive perspective.

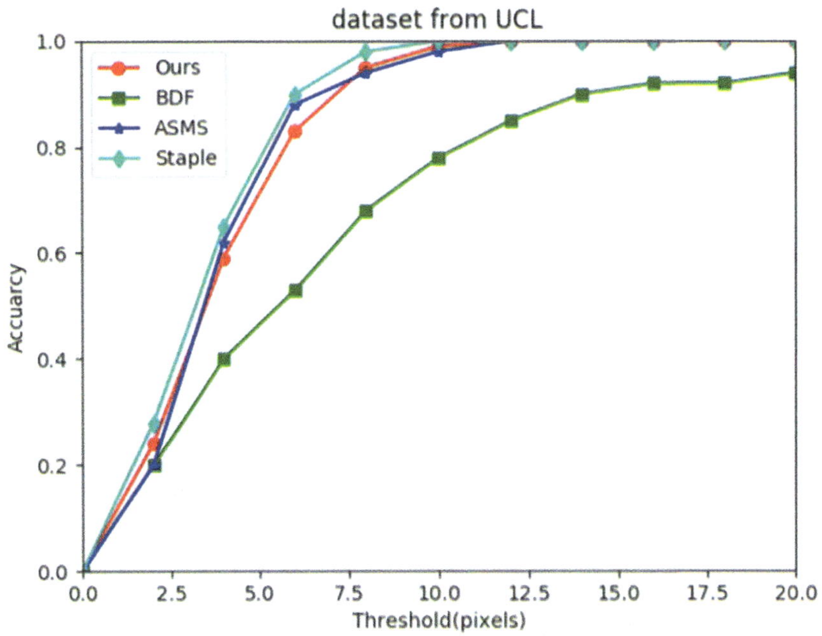

Fig. 3. Tracking accuracy of surgical instrument tips for the standard dataset from UCL

4 Conclusion

In this paper, we refine the Real-time Recurrent Regression Networks (Re^3) which using the cascade of convolutional neural network and long-short term memory for real-time single tool tracking. The convolutional neural network is used to learn the appearance features of surgical tool. The first LSTM learns the motion features and remember appearance and motion information [4] and the second LSTM is used to regress the position of surgical tool. The biggest difference between our method and Re^3 is that we don't use a lot skip connections in the features extraction stage. Our tracking model is validated on the standard dataset from UCL and the experimental results show that our method achieves a better tradeoff between tracking accuracy and speed than the other considered state-of-the-art methods. The most prominent feature of our network is that it achieves such high tracking accuracy and speed without occupying a lot of computing resources because our network can remember and take advantage of the relationship between frames. However, the images input to the network in our method are cropped, so once the surgical tool is lost in view, it will not be tracked by our network. Therefore, our network should be used in conjunction with some fast and lightweight detection networks to relocate the tool.

Acknowledgement. This research was funded by the National Key Research and Development Program of China (2019YFB1311302), and the National Natural Science Foundation of China (grant no. 81401543, 61273277).

References

1. Choi, F., Bareum, S.: Surgical-tools detection based on convolutional neural network in laparoscopic robot-assisted surgery. In: International Conference of the IEEE Engineering in Medicine & Biology Society (2017)
2. Gordon, F., Farhadi, A., Fox, D.: Re3: real-time recurrent regression networks for visual tracking of generic objects. IEEE Robot. Autom. Lett. **3**(2), 788–795 (2018)
3. Hochreiter, F., Schmidhuber, S.: Long short-term memory. Neural Comput. **9**(8), 1735–1780 (1997)
4. Garcia-Peraza-Herrera, L., Li, W., Gruijthuijsen, C., Devreker, A., Attilakos, G., Deprest, J., Vander Poorten, E., Stoyanov, D., Vercauteren, T., Ourselin, S.: Real-time segmentation of nonrigid surgical tools based on deep learning and tracking. In: CARE Workshop (MICCAI 2016), pp. 84–95(2016)
5. Lin, T.-Y., Dollár, P., Girshick, R., He, K., Hariharan, B., Belongie, S.: Feature pyramid networks for object detection (2016)
6. Greff, K., Srivastava, R.K., Koutník, J., Steunebrink, B.S., Schmidhuber, J.: LSTM: a search space Odyssey. IEEE Trans. Neural Netw. Learn. Syst. **28**(10), 2222–2232 (2016)
7. Bertinetto, L., Valmadre, J., Golodetz, S., Miksik, O., Torr, P.: Staple: complementary learners for real-time tracking. In: The IEEE Conference on Computer Vision and Pattern Recognition (CVPR), pp. 1401–1409 (2016)
8. Maresca, M.E., Petrosino, A.: Clustering local motion estimates for robust and efficient object tracking. In: European Conference on Computer Vision, pp. 244–253. Springer (2014)
9. Vojir, T., Noskova J., Matas, J.: Robust scale-adaptive mean-shift for tracking. In: Image Analysis, pp. 652–663. Springer, Heidelberg (2013)

Development of a Novel Laparoscopic Palpation System Using a Wearable Motion-Sensing Armband

Kerwin Caballas[1(✉)], Harold Jay Bolingot[2], Ingrid Salud[1], Lawrence Ibarrientos[1], Lea Cristina Macaraig[1], Nathaniel Joseph Libatique[1], and Gregory Tangonan[1]

[1] Ateneo Innovation Center, Ateneo de Manila University, 1108 Quezon City, Philippines
kerwin.caballas@gmail.com

[2] Kyushu Institute of Technology, Wakamatsu Ward, Kitakyushu City, Fukuoka Prefecture 808-0196, Japan

Abstract. This study proposes the development of a novel palpation system for use in the laparoscopic surgery setting, which relies on a wearable motion-sensing armband worn on the palpating arm. The system is tested by prodding the laparoscopic tool on a simulated human tissue with an anomaly underneath and measuring the effects of the forces experienced by the palpating arm by recording the accelerometer data embedded in the armband. Results show that the proposed system could detect the anomalies embedded in the gelatin phantom based on the change of the amplitude of the accelerometer data. A decrease in the average peak-to-peak amplitude of accelerometer signal by as much as 32.29% was observed for the single point prodding experiment, while a decrease of up to 22.93% was observed for the raster scan experiment. The results from this feasibility experiment confirmed the potential role of the accelerometer signal in the motion-sensing armband in being able to assist laparoscopic palpation.

Keywords: Laparoscopic palpation · Myo armband · Motion-based

1 Introduction

1.1 Laparoscopic Palpation

Palpation is one of the more common techniques being used when detecting tissue anomalies. During palpation, haptic information about the tissue being palpated can be obtained, such as its size, shape, and texture. Palpation is relatively easier and relatively faster to perform compared to other anomaly detection techniques, only needing the surgeons to examine the tissue for anomalies. As seen in Fig. 1a, Culmer et al. summarizes the crucial steps that take place during palpation. The entire process of palpation is performed until conclusions can be drawn from the obtained information, enabling surgeons to make more informed decisions.

Palpation, however, has its own set of limitations; namely the reliance on the surgeon's expertise and the need for proper contact between the surgeon's hand and the

Y. Shiraishi et al. (Eds.): APCMBE 2020, IFMBE Proceedings 82, pp. 25–30, 2021.
https://doi.org/10.1007/978-3-030-66169-4_4

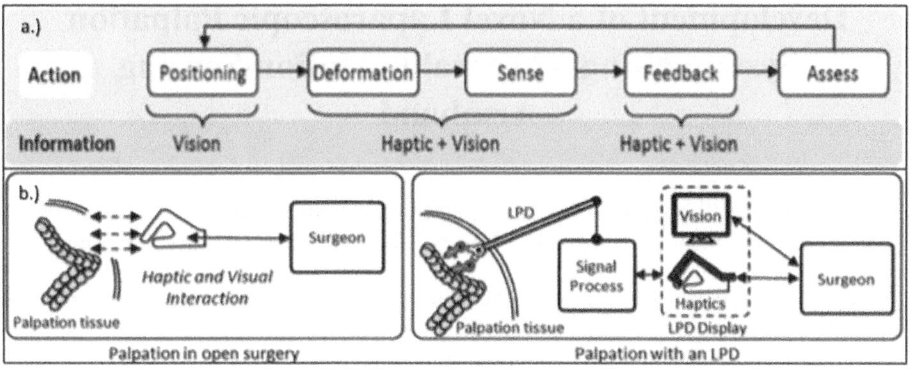

Fig. 1. (a) An overview of the iterative process of palpation and (b) a comparison of how palpation is performed in open surgery versus how it is performed with a laparoscopic palpation device (LPD) [1].

palpated tissue. The latter is not a problem in open surgery; but in laparoscopic surgery, there are issues with haptic and visual feedback [2, 3]. Due to these hindrances, there is clearly a need for laparoscopic palpation to be refined further using systems which can easily provide both haptic and visual information from palpated tissue in real time, shown in Fig. 1b. Designing devices like these would allow surgeons to detect anomalies quickly during operations.

Current Techniques of Laparoscopic Palpation. At present, there are numerous types of systems which aid in laparoscopic palpation. Talasaz and Patel mounted tactile sensors and the end of a probe which relied on tactile feedback to produce a force distribution map for detecting anomalies in robot-assisted laparoscopic surgery [4]. Kim, Choi, and Chung developed a palpation probe for laparoscopic surgery which detected the difference between the normal force and the surface normal as it rolled along the surface of the palpated tissue. Their method eliminates the factor of the contact conditions between the probe and the palpated tissue [5].

This study proposes a novel laparoscopic palpation system which makes use of a motion-sensing wearable armband to detect anomalies. The proposed system can be used with any laparoscopic tool, since the key device of the system is the armband. As seen in Fig. 2, it involves the surgeon prodding the area of interest while wearing the armband that can detect the forces experienced by the palpating arm.

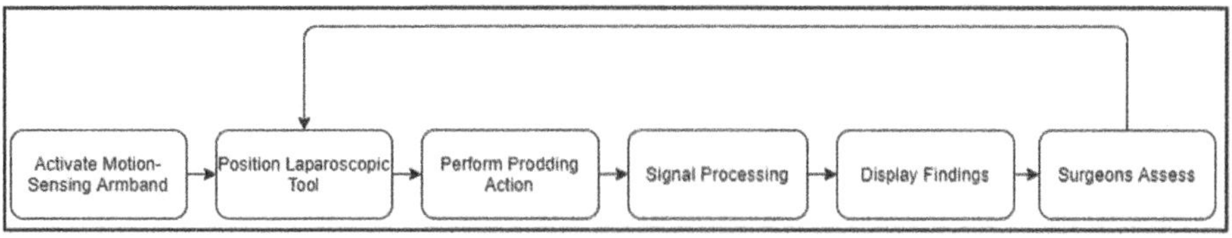

Fig. 2. System diagram of the proposed method.

2 Methodology

Fabrication of Anomaly and Simulated Human Tissue. In order to determine the efficacy of the proposed palpation method, simulated human tissue (SHT) was made from gelatin, with 10% gelatin powder and 90% water [6], and was molded into circular blocks which all had a diameter of 14.5 cm and a height of 5.5 cm. The proposed method was tested on gelatin blocks with anomalies embedded within them at different depths (d < 0.5 cm, d = 1 cm, and d = 2 cm), seen in Fig. 3a.

Fig. 3. (a) Gelatin blocks were used as SHT. A coin, serving as the anomaly, was embedded beneath the surface of each gelatin block. Two experiments were performed in the study: (b) the palpation of a single point marked on the SHT, and (c) a raster scan of the SHT.

Two different types of experiments were performed to test the proposed method. The first type of experiment was the palpation of a single point marked on the SHT. It was prodded with a laparoscopic grasper 30 times in each trial for three trials. It was ensured that the length of the grasper was normal to the surface of the SHT, seen in Fig. 3b. The second type of experiment involved the raster scan of an area of each SHT block marked with a 3×11 matrix of points, shown in Fig. 3c. The subject prodded each point 10 times until all points were prodded for three trials. The raster scan was performed while the grasper was inserted in a makeshift laparoscopy training box. It was ensured that the angle between the surface of the SHT and the length of the grasper was at 45°. Both types of experiments were repeated for the different SHTs with anomalies at different depths (d < 0 cm, d = 1 cm, and d = 2 cm).

Motion Telemetry. The use of a motion-sensing wearable armband, the Myo Armband, for the proposed system was inspired by a previous study by Bolingot et al. where it was used to monitor the performance and the fatigue experienced by laparoscopic surgeons-in-training using its surface electromyographic (sEMG) and inertial measurement unit (IMU) sensors [7]. For this study, the armband was the device which obtained data from the palpating arm. This armband collected data with a sampling rate of 50 Hz for the IMU sensors and 200 Hz for its EMG sensors. The data from the armband was collected using a C++ program developed by Thalmic Labs, which streamed data from the armband to a laptop through a Bluetooth dongle. The collected acceleration and EMG data was processed and visualized using MATLAB.

Motion Signal Measurement and Analysis. The back-and-forth prodding motion of the palpating arm in the proposed method produces a wave signal in terms of the accelerometer motion sensing. The peak-to-peak amplitude of the accelerometer signal is measured to detect the instance when the anomaly is present during the prodding action, and its average is calculated using Eq. 1. In the equation, the product of the frequency of the accelerometer signal f and the time duration of a prodding series t_{final} is equal to the total number of periods (i.e. prodding actions) recorded in a trial. The maximum and minimum values of the accelerometer data in a period are represented by a_i^+ and a_i^-, respectively, and i denotes the order of a single accelerometer wave period from start to finish of a prodding series. The average peak-to-peak amplitude (APPA) of the signal corresponds to the extent of capturing the changes of altitude in the different media involved in the experiment, representing different scenarios of organ anomaly.

$$APPA = \left(\frac{1}{[[f \times t_{final}]]}\right)\left(\sum_{i=0}^{i=[[f \times t_{final}]]} (a_i^+ - a_i^-)\right) \quad (1)$$

Figure 4 shows the correspondence between the accelerometer signal peaks and the presence of an anomaly during prodding. The change in the accelerometer peaks corresponds to the change in altitude of the surface due to the anomaly. As the end-effector approaches the anomaly, the altitude of the surface increases, affecting the peaks of the accelerometer signal due to a change in the height and motion of the palpating arm. As it leaves the anomalous area, the altitude of the surface decreases, translating into a change in the APPA of the accelerometer signal. The negative peak corresponds to the moment when the grasper pauses in mid-air before another prod, while the positive peak corresponds to the moment when the end-effector touches the SHT surface.

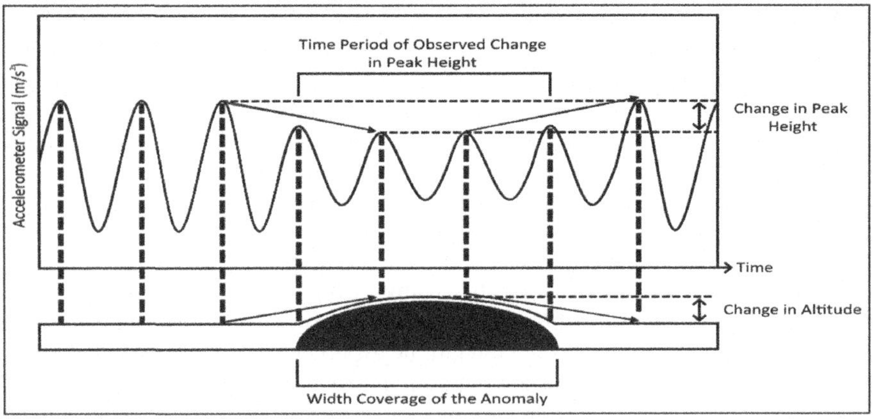

Fig. 4. Analysis of the accelerometer signals. The positive peaks indicate the end-effector hitting the SHT surface while the negative peaks show the moment the grasper is paused mid-air after pulling it away. The signal's peak-to-peak amplitude decreases as the palpation approaches the anomaly due to it elevating the tissue surface, lowering the distance between it and the end-effector. Thus, less acceleration occurs before the end-effector touches the SHT surface.

3 Results

When prodding a single point marked on the SHT, results showed that there were decreases of 32.29%, 18.67%, and 12.41% in the APPA when the anomalies were embedded at depths of less than 0.5 cm, of 1 cm, and of 2 cm, respectively.

Data from the raster scan experiment, shown in Fig. 5b, was presented per row of points prodded. When an anomaly was embedded at d < 0.5 cm, the value of the APPA for the middle row decreased by 22.93% when the points directly above the anomaly were prodded. A decrease in APPA was observed from the top and bottom rows as well, with decreases of 7.06% and 19.33% respectively, as seen in Fig. 5c. Embedded at d = 1 cm, there were decreases of 5.23% and 6.15% observed for the middle and bottom rows, respectively, as illustrated by Fig. 5d. However, for the top row, there was an increase of 1.59%, caused by problems with uneven curing of the SHT block. When the anomaly was buried at d = 2 cm, there were decreases of 2.44%, 7%, and 3.26% observed for the top, middle, and bottom rows of points, respectively.

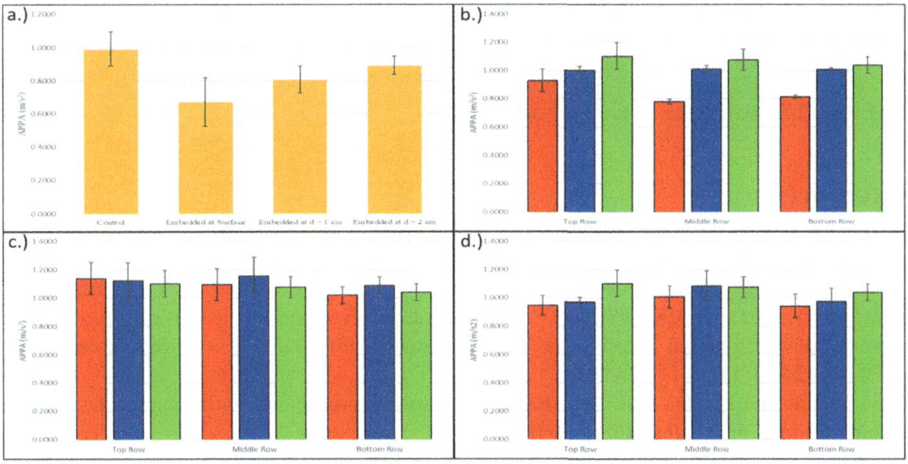

Fig. 5. Graphs showing the APPA values of the accelerometer data collected from the two types of experiments performed: (a) prodding of a single marked point, and the raster scan of SHT with anomalies embedded at (b) d < 0.5 cm, (c) d = 1 cm, and (d) d = 2 cm (red: SHT area with embedded anomaly, blue: SHT area without embedded anomaly, green: control SHT block).

4 Discussion

We present the comparisons of our proposed system with previously published solutions to the laparoscopic palpation problem, as well as the advantages and disadvantages of our system. One main advantage the proposed system has is its unobtrusiveness in accomplishing palpation. In other proposed laparoscopic palpation systems [4, 7, 8], sensors are attached to the laparoscopic tool handle and end-effector to capture meaningful signals that describe the tissue anomaly. Our proposed system instead relies on the subtle motion of the forearm of the surgeon during laparoscopy. The potential sources of disruption to the integrity of motion sensing in our system include irregular and inconsistent prodding of the end-effector, as well as organ movement during laparoscopy. Our proposed system does not necessarily contribute to a reconfiguration of the operation procedure, unlike in other methods which require a separate tool, as the laparoscopic tool in use is a nonfactor. However, our proposed system must be tested on broad cases of palpation to further validate its efficacy.

5 Conclusion

A novel palpation system that used a wearable motion-sensing armband was developed. It involved prodding the tissue to be palpated with a laparoscopic tool, as the surgeon wears the armband. The armband collected from the accelerometer of the armband, detecting changes in the acceleration of the arm as the tool collided with the surface of the palpated tissue. The method was effective when an anomaly was present down to a depth of 2 cm. Though the proposed method is simple and cost-efficient, improvements are needed on the regulation of force the surgeon applies during palpation.

Conflict of Interest. The authors declare that they have no conflict of interest.

References

1. Culmer, P., et al.: Reviewing the technological challenges associated with the development of a laparoscopic palpation device. Int. J. Med. Robot. Comput. Assist. Surg. **2**(8), 146–159 (2012)
2. van der Meijden, O.A.J., Schijven, M.P.: The value of haptic feedback in conventional and robot-assisted minimal invasive surgery and virtual reality training: a current review. Surg. Endosc. **23**(6), 1180–1190 (2009)
3. Berguer, R., Forkey, D., Smith, W.: Ergonomic problems associated with laparoscopic surgery. Surg. Endosc. **5**(13), 466–468 (1999)
4. Talasaz, A., Patel, R.: Telerobotic palpation for tumor localization with depth estimation. In: Proceedings of the 2013 IEEE/RSJ International Conference on Intelligent Robots and Systems, pp. 463–468. IEEE, Tokyo (2013)
5. Kim, H., Choi, S., Chung W.K.: Feasibility of a novel indicator for lump detection using contact pressure distribution. In: Proceedings of the 2015 IEEE/RSJ International Conference on Intelligent Robots and Systems (IROS), pp. 2111–2117. IEEE, Hamburg (2015)
6. Instructions for making "home-made" ballistic gelatin. https://www.customcartridge.com/pdfs/BallisticGel.pdf. Accessed 20 Jan 2020
7. Bolingot, H.J.M., et al.: Simulated laparoscopic training and measurement systems based on a low-cost sEMG and IMU Armband. In: Proceedings of the 39th Annual International Conference of the IEEE Engineering in Medicine and Biology Society. IEEE, Jeju Island (2017)
8. Illanes, A., et al.: Novel clinical device tracking and tissue event characterization using proximally placed audio signal acquisition and processing. Sci. Rep. **8**, 12070 (2018)

Surface-Based Electromyography Gesture Profiling of Laparoscopic Tools Using a Wearable Sensor

Ingrid Salud[1(✉)], Harold Jay Bolingot[2], Lea Cristina Macaraig[1], Nathaniel Libatique[1], and Gregory Tangonan[1]

[1] Ateneo Innovation Center, Ateneo de Manila University, 1108 Quezon City, Philippines
ingrid.salud@obf.ateneo.edu
[2] Kyushu Institute of Technology, Wakamatsu Ward, Kitakyushu City, Fukuoka Prefecture 808-0196, Japan

Abstract. This study explores surface-based electromyography (sEMG) to evaluate the effects of tool handling and fatigue on the muscle activation of laparoscopy surgeons. Specifically, the Myo Armband is used because it is a wearable, unobtrusive, and wireless sensor. Collected EMG signatures showed that more complex gestures have more active muscle groups. For the fatigued state, analysis of the signatures using the RMS feature showed that for more complex gestures, RMS increased because of muscle compensation due to fatigue. Accuracy exercise findings showed that there is a decrease in %accuracy when using an endoscope compared to when looking directly and there is a slight decrease in %accuracy once fatigue is induced. 100% accuracy cannot be reached due to awkward tool handling. Through various experiments, this study presented possible methods to evaluate the effects of tool handling and fatigue on the muscle activation of surgeons through use of sEMG. Analysis of these signatures may lead to better understanding on how specific tool design affects muscle activity and muscle fatigue, and to utilize this as a basis for a more ergonomic laparoscopic set-up and tool design.

Keywords: Laparoscopic surgery · Ergonomics · Surface-based electromyography · Myo Armband · Motion-based

1 Introduction

Laparoscopic surgery, also known as minimal access surgery (MAS) is a modern surgical technique that involves opening small incisions and uses specialized tools to perform the operation while a camera is used to monitor the surgery on a screen [1]. Compared to traditional open surgery, laparoscopic surgery avoids making large incisions that exposes internal organs to the environment, reducing the risk of infection and allows for faster wound healing and recovery. While laparoscopic surgery holds multiple advantages over open surgery, it also comes with disadvantages on the side of the surgeon. Due to the awkward man-machine interface of the laparoscopic set-up, surgeons encounter

Y. Shiraishi et al. (Eds.): APCMBE 2020, IFMBE Proceedings 82, pp. 31–37, 2021.
https://doi.org/10.1007/978-3-030-66169-4_5

more difficulties compared to open surgery. These difficulties are due to the limitations to two-dimensional viewing of a three-dimensional field and its corresponding loss of depth perception, the fewer degrees of freedom, and the loss of direct tactile feedback compared to open surgery [2, 3]. Because of limiting and poor ergonomic design of the laparoscopic set-up, surgeons are reported to sustain surgery-related injuries in a spectrum described by Reyes, Tang and Cuschieri [1] as MAS-related morbidity syndromes. Injuries included here are overuse syndrome or overuse of muscle groups, surgical fatigue syndrome and possible loss of visual acuity, among other injuries. These injuries, especially when accumulated, have long-term consequences affecting the surgeon's surgical skills and their daily lives. As more MAS-related injuries are being reported, there is an increasing need to study the ergonomics of laparoscopic tools and interface to improve safety, efficiency and comfort of the surgeon.

Surface electromyography (sEMG) has been used to evaluate various biomechanical characteristics including muscle activity and fatigue [4]. Several studies have demonstrated the use of sEMG as a screening tool for ergonomic purposes. Susic, Lulic and Veljovic [5] compared the muscle activity index (MAI) of two tasks. Correlating MAI with the work done by the muscles, they concluded that the task with the lower MAI average is the less demanding, therefore more ergonomic, of the two tasks. In the context of laparoscopic surgery, Berguer, Forkey and Smith [2] compared the EMG signatures of surgeons doing an exercise using a hemostat for open surgery and a laparoscopic grasper. Based on their results, EMG signals from the target muscle groups had a higher magnitude when using the laparoscopic grasper compared to the hemostat for open surgery. They concluded that laparoscopic instruments are less efficient mechanically and that the configuration of the grasper, causing excessive flexion of the wrist, leads to a decrease in the transmission of muscular force across the wrist.

The objective of this study is to make use of the Myo Armband, a commercial gesture control device, to collect EMG signatures for gesture analysis and profiling of simple manipulations of a commercial laparoscopic tool for the purpose of understanding muscle activation during tool handling. The accuracy of tool handling is also measured to evaluate the laparoscopic set-up.

2 Methodology

The Myo Armband has previously been used in the context of laparoscopic surgery to evaluate the performance of training surgeons in laparoscopy simulator training using sEMG. Their results show that the signals or the profiles can be used to differentiate trained and experienced surgeons. They also reported a decrease in signal amplitude when fatigue is induced as well as incoherent muscle firing and signal delay [6].

The Myo Armband is a commercial wireless gesture control device developed by Thalmic Labs Inc. It has 8 electromyography (EMG) sensors, each detecting muscle activation of different muscle groups as shown in Fig. 1, streaming EMG data at a sampling rate of 200 Hz. It also has a built-in accelerometer, gyroscope and magnetometer, streaming IMU data at a sampling rate of 50 Hz. In this study, the Myo Armband is used purely as a recording device, and the default features of detecting certain gestures were ignored and turned off.

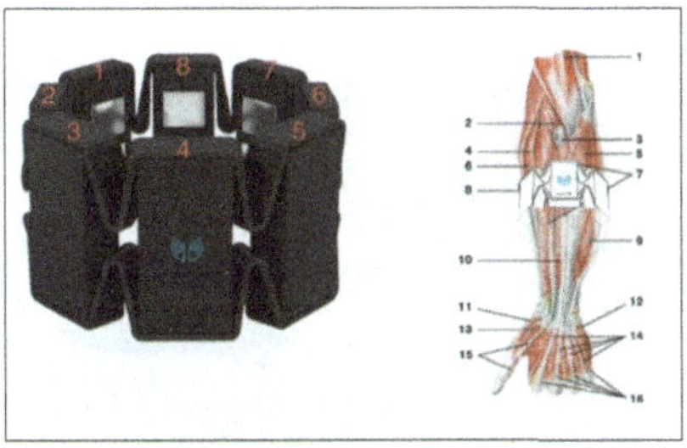

Fig. 1. The 8 sEMG channels of the Myo armband and its corresponding muscle groups

For gesture profiling and fatigue studies, the specific laparoscopic tool evaluated is the Ethicon Enseal™ G2 Articulating Tissue Sealer. The EMG signals of five gestures related to the Enseal™ tool were collected from four participants, with each participant having done three trials. This experiment was done with the Myo Armband worn on the participant's right arm, the same arm that will be handling the tool. The participants were then asked to do five gestures as illustrated in Fig. 2. To test the effects of short-term fatigue on muscle activation, the participants were asked to do hand grip exercises for one minute after which, they were asked to repeat the five gestures.

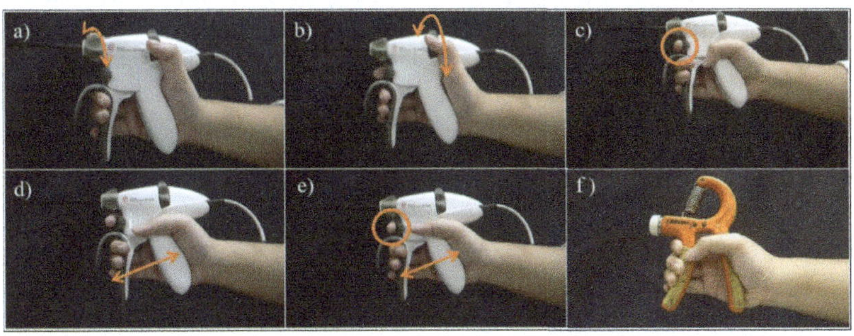

Fig. 2. The five gestures in the manipulation of the Ethicon Enseal™ Laparoscopic tool: (a) rotation by the right index finger (b) rotation by the right thumb (c) pressing of the right index finger (d) gripping of the index-middle-pinky finger (e) combination of gestures a and b and (f) the hand gripper for inducing fatigue

For the accuracy exercise, the participant must hit a set of targets (Fig. 3a) using the Ethicon Endopath™ Grasper laparoscopic tool while wearing the Myo Armband. Accuracy was measured using Eq. 1, where R is the arm length/radius of the target and L is the length between the center of the target and the marked point and took the average of all 25 targets per trial.

$$\%\boldsymbol{Accuracy} = \frac{R - L}{R} \times 100 \tag{1}$$

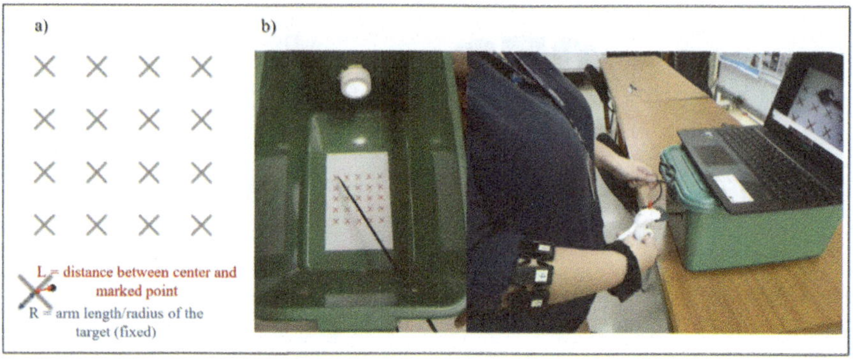

Fig. 3. (a) the set of targets used in the accuracy exercise and the variables measured to calculate for the accuracy, and (b) the two set-ups for the accuracy exercise experiments: looking directly and looking through an endoscope

The exercise was repeated after inducing short-term fatigue. There were 2 different set-ups for this experiment: (1) looking directly at the targets and (2) using an endoscope to view the targets, recreating a laparoscopic surgery set-up, as illustrated in Fig. 3b.

The Myo Armband was connected to a laptop via Bluetooth. In order to collect EMG data from the device, a C++ program developed by Thalmic Labs, called Myo Data Capture, was used. This program had a sampling rate of 200 Hz for EMG data and 50 Hz for IMU data, similar to the Myo Armband. The EMG (in Muscle Activation Units) and x, y, z-acceleration (in *g*) data obtained from the Myo Data Capture was visualized and processed using MATLAB. In both exercises, the RMS of the EMG signals was also calculated to obtain the average power of the signal.

3 Results

Each gesture yields a unique EMG signature across the 8 channels that correspond to a particular muscle group. Using the same data, the muscle groups most active for a gesture is also determined and is presented in Table 1.

Table 1. sEMG Channels of the most active muscle groups per gesture based on the visualized EMG signal

Gesture		Active muscle groups (sEMG Channel)
1	Rotation by the right index finger	4, 5
2	Rotation by the right thumb	3, 4, 5
3	Pressing of the right index finger	4
4	Gripping of the index-middle-pinky finger	5, 7, 8
5	Combination of gestures 1 and 2	2, 3, 4, 5, 6, 8

Based on the results, more complex gestures have more active muscle groups as seen in the case of gesture 5 which has 6 active muscle groups compared to gesture 3 which only has 1 active muscle group.

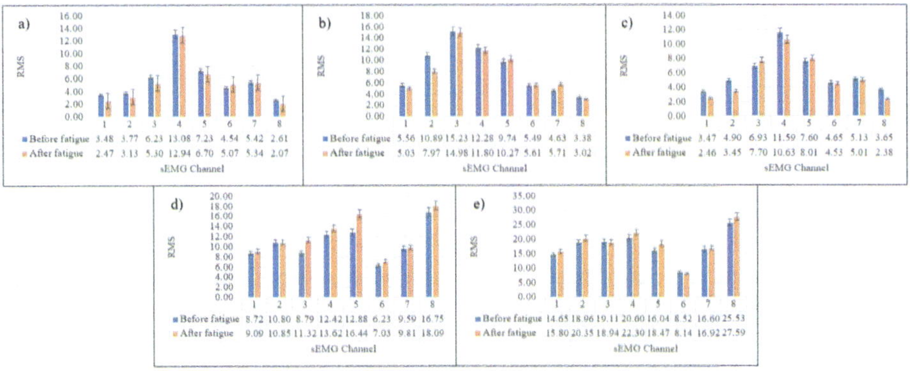

Fig. 4. Graphs of the average RMS per channel for (a) Gesture 1 (b) Gesture 2 (c) Gesture 3 (d) Gesture 4, and (e) Gesture 5

To determine the effects of short-term fatigue on the different gestures, comparison of the overall RMS per channel per gesture was done. Graphs of the results, presented in Fig. 4, show that in gesture 1, there is a 40.9% to 1.08% decrease in the RMS of 7 muscle groups and a 10.45% increase in 1 muscle group. In gesture 2, there is a 36.64% to 1.67% decrease in the RMS of 5 muscle groups and a 18.91% to 2.14% increase in 3 groups. In gesture 3, there is a 42.02% to 2.40% decrease in the RMS of 6 muscle groups and a 10% to 5.12% increase in 2 groups. In gesture 4, there is a 21.65% to 0.46% increase in the RMS of all muscle groups. In gesture 5, there is 4.67% to 9% decrease in the RMS of 2 muscle groups and a 13.16% to 1.89% increase in 6 groups. After comparing the RMS of the EMG signals, results show that the more complex the gesture is, the more sEMG channels show an increase in RMS for induced fatigue.

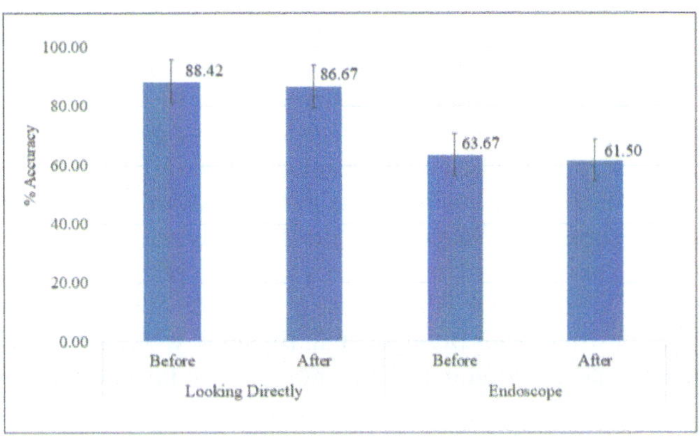

Fig. 5. Graph of the average %accuracy of the accuracy exercise

Results of the accuracy exercises, presented in Fig. 5, show that even when looking directly, the accuracy does not reach 100%. For the exercise where the participant is looking directly, there is a slight decrease in the accuracy before and after fatigue. Trying to recreate a laparoscopic set-up where an endoscope was used to look at a screen to see the targets, results show a decrease in %accuracy. There is also a slight decrease in accuracy before and after fatigue.

4 Discussion

The Myo Armband is a low-cost commercial gesture control device that can effectively measure the EMG signatures of the arms. In this study, the Myo Armband was used to collect EMG data for two experiments exploring two problems associated with laparoscopic surgery: the poor ergonomic design of laparoscopic tools and the awkward interface of the laparoscopic set-up.

The first experiment demonstrates the possible use of sEMG as ergonomic analysis for the Ethicon Enseal™ laparoscopic tool through a series of gesture profiling and fatigue studies experiments. Through EMG signal visualization, this study presents the unique signatures related to the tool and determines the most active muscle groups for each gesture. Based on the results, more complex gestures activate more muscle groups. Fatigue studies were also conducted by measuring the RMS of the EMG signals of the gestures done before and after inducing short-term fatigue. Results show that more complex gestures have more muscle groups experiencing an increase in RMS after induced fatigue. This result is consistent with other studies that have used RMS as one of the features they used in the EMG and fatigue studies. They reported that at muscle fatigue there is a synchronization of motor units firing, a decrease in muscle fiber conduction velocity, drop in motoneuron excitability with sustained muscle activity and decrease firing rates of active motor units. Because complex gestures require higher muscle contraction and muscle activation firing, and with the decrease in firing rates in muscle fatigue, the muscle must compensate. As such, this increase in RMS implies that more and more effort is needed as muscle fatigue intensifies and can lead to long-term fatigue [2, 7, 8]. Complex laparoscopic operations have longer durations and may induce long-term fatigue due to the poor ergonomic interface and the need to use specialized tools. Based on the results with the increased RMS, certain gestures related to the tool require a lot of extra effort to sustain that gesture.

The second experiment evaluates the effects of fatigue and the laparoscopic set-up on tool positioning accuracy. Based on the results, a 100% accuracy cannot be obtained even while looking directly at the targets due to the awkward tool handling. There is a slight decrease in accuracy after induced fatigue. However, the decrease in accuracy can be attributed more to the awkward interface and loss of depth perception when using an endoscope. For both set-ups, there is a decrease in the RMS of the obtained EMG signal. These results are consistent with Bolingot et al. [6] study where they saw a decrease in the RMS of the EMG signals after induced fatigue while doing laparoscopic peg transfer exercises.

5 Conclusion

Due to the advantages of laparoscopic surgery compared to open surgery, advancements to laparoscopic surgery have led to its use in more complex operations with longer execution times. Because of this, more MAS-related injuries have been reported but are still not fully understood [1]. This study made use of a wearable, unobtrusive, and wireless sensor to collect the unique EMG signatures related to tool handling of the Ethicon Enseal™ laparoscopic tool and analyzed the effects of fatigue on these signatures. Analysis of the signatures using the RMS feature showed that for more complex gestures, RMS increased because of muscle compensation due to fatigue. For the accuracy exercise, there is a slight decrease in EMG and the decrease in %accuracy can be attributed more to the awkward interface of the laparoscopic set-up. Through the designed experiments, this study presented possible methods to evaluate the effects of tool handling and fatigue on the muscle activation of surgeons using sEMG. Analysis of these signatures may lead to better understanding on how specific tool design affects muscle activity and muscle fatigue, and to utilize this as a basis for a more ergonomic laparoscopic set-up and tool design. In future studies, the designed experiments in this study may be coupled with machine learning techniques to further study tool ergonomics and effects of fatigue during laparoscopic surgery. Furthermore, further analysis of these unique signatures may be used to design a robotic assist system that is controlled by gestures or one that can aid surgeons once fatigue is detected.

Conflict of Interest. The authors declare that they have no conflict of interest.

References

1. Reyes, D.A.G., Tang, B., Cuschieri, A.: Minimal access surgery (MAS)-related surgeon morbidity syndromes. Surg. Endosc. **20**, 1–13 (2006)
2. Berguer, R., Forkey, D.L., Smith, W.D.: Ergonomic problems associated with laparoscopic surgery. Surg. Endosc. **13**, 466–468 (1998)
3. Supe, A., Kulkarni, G., Supe, P.: Ergonomics in laparoscopic surgery. J. Min. Access Surg. **6**(2), 31–36 (2010)
4. Lee, K., Jung, M.: Ergonomic evaluation of biomechanical hand function. Saf. Health Work **6**, 9–17 (2015)
5. Susic, A., Lulic, T., Veljovic, F.: Ergonomic evaluation of task execution: surface electromyography in muscular activity screening. Periodicum Biologorum **112**(1), 33–38 (2010)
6. Bolingot, H., Abrajano, G., Libatique, N., Salvana, M., Reyes, D., Tangonan, G., Ngeo, J., Shibata, T.: Simulated laparoscopic training and measurement systems based on a low-cost sEMG and IMU Armband. In: 39th Annual International Conference of the IEEE Engineering in Medicine and Biology Society (2017)
7. Chang, J., Chablat, D., Bennis, F., Ma, L.: Estimating the EMG response exclusively to fatigue during sustained static maximum voluntary contraction. In: Goonetilleke, R., Karwowski, W. (eds.) Advances in Physical Ergonomics and Human Factors. Advances in Intelligent Systems and Computing, vol. 489, pp. 29–39. Springer, Cham (2016)
8. Cao, L., Wang, Y., Hao, D., Rong, Y., Yang, L., Zhang, S., Zheng, D.: Effects of force load, muscle fatigue and magnetic stimulation on surface electromyography during side arm lateral raise task: a preliminary study with healthy subjects. Biomed. Res. Int. **2017**(4), 1–9 (2017)

Development of Rehabilitation System with Bilateral Upper Extremity Training Robot and Virtual Reality for Stroke Patient

Thitikorn Kaewlee(✉) and Panrasee Ritthipravat

Department of Biomedical Faculty, Faculty of Engineering, Mahidol University, Nakhon Pathom, Thailand
thitikorn.kae@student.mahidol.ac.th

Abstract. Recent trends in the development of rehabilitation for stroke patients suggest that the integration of motor and cognitive training is one of the promising techniques. Gyro-roller rehabilitation system has been developed based on the integration of robotics and virtual reality game for bilateral upper extremity training. In this study, the feasibility of the combined training is investigated with ten healthy participants and a chronic stroke patient. Amplitude analysis of electromyography in both arms in relation to training levels is carried out to investigate the effects of virtual reality based bilateral training on motor activity. Also, the integrated system with short term memory test has been found to induce the result metrics as digit span assessment and stimulate muscle activity. Overall, results suggest that combined cognitive and motor training with Gyro-roller rehabilitation system is possible for stroke rehabilitation.

Keywords: Stroke · Rehabilitation · Robotic · Virtual reality · Bilateral · Digit span · Electromyography

1 Introduction

Stroke or cardiovascular disease (CVD) is one of the leading causes of death among non-communicable diseases [1]. Post-stroke survival patients usually suffer from loss of motor function and cognitive and perceptive dysfunction depending on affected brain areas. After medical treatment, the patients require physical and occupational rehabilitation to improve and recover their affected functions. Bilateral movement training is one of traditional rehabilitations that provide efficiency in extensive severity level of stroke patients [2]. However, traditional repetitive rehabilitation requires long intensive training periods. Recently, robotic technology and virtual reality have been integrated with bilateral training for motor or cognitive function independently [3]. Although several studies demonstrate strong relationship between motor and cognitive development in stroke patients, few researches study the effectiveness of combination training

Y. Shiraishi et al. (Eds.): APCMBE 2020, IFMBE Proceedings 82, pp. 38–44, 2021.
https://doi.org/10.1007/978-3-030-66169-4_6

[4,5]. Gyro-roller bilateral rehabilitation system, which was developed by Mahidol University, Thailand, is also an integration of robotic training with virtual reality. Results from a pilot study with healthy subjects indicate that gyroscopic effect can stimulate muscle activities as observed by electromyography (EMG) analysis [6]. However, the efficiency of combined motor and cognitive training with Gyro-roller rehabilitation system has not been studied. In this study, novel upper extremity rehabilitation system based on integration of motor and cognitive bilateral training has been designed for a pilot study with healthy participants and a stroke patient.

2 Materials and Methods

2.1 Gyro-Roller Rehabilitation Device

Gyro-Roller rehabilitation system is developed based on bilateral training of upper extremity for stroke patients [6]. It provides reaching movement in two degrees of freedom along longitudinal axis (roll) and vertical axis (yaw). The fundamental principle in this robot is the utilization of gyroscopic effect generated from spinning wheel inside spherical cover. The gyroscopic effect is adjustable as a disturbing force to stimulate motor activities of upper extremities. Virtual reality game based on bilateral training is designed to motivate engagement of participants to perform rehabilitation task. Performance of a training session is recorded in the system for statistical analysis.

2.2 Experiment Design

This study propose to investigate the feasibility of combined cognitive and motor training using Gyro-Roller rehabilitation system. Ten healthy subjects including five males and five females are involved. All participants are undergraduate and graduate students from Faculty of Engineering, Mahidol University and right-hand dominant. Average ages are 27.27 ± 5.47 years old. No one has been injured or has muscle fatigue before experiments. Moreover, one chronic stroke patient participate in this study. A 70 years old male with right hand dominant had

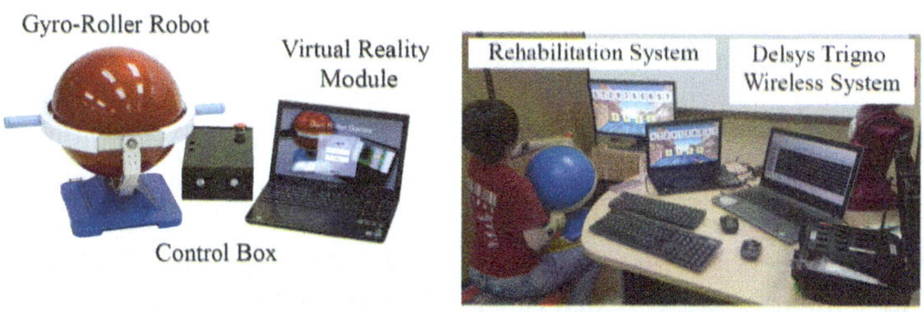

Device components Experiment Setup

Fig. 1. Gyro-roller rehabilitation system components and experiment setup

ischemic stroke on the right side scoring 40 on Montreal Cognitive Assessment (MOCA) and 43 on Fugl-Meyer Assessment for upper extremity (FMA-UE) (Fig. 1).

Fig. 2. Virtual reality game based on bilateral training (Upper: Journey game) and Short term memory test (Lower: Memory game)

Virtual reality games are developed by using Unity® game engine software following bilateral training and short term memory test literature. The underlying theme of game design is endless adventure along the river by paddling the boat to perform rehabilitation tasks as shown in Fig. 2. Firstly, participants have to register basic information to the system and calibrate baseline range of movement with gyro-roller robot. To study effect of the training on motor activity while performing experiments, surface EMG of triceps brachii and extensor carpi ulnaris muscles on both arms are recorded by Delsys Trigno™ wireless system and smart sensors. Journey game which is designed as a target reaching task is selected for this experiment. Participants will perform 3 levels of the game which different in the time of appearing target and obstacle. The memory game is designed as a combination of target reaching task with short term memory test [7] since the objective is to study efficiency of combined training in memory and motor activity. Participants will perform 4 tasks, namely forward and backward digit span (DS) and forward and backward letter span (LS). Response Speed, score per minute and percent of correctness are collected after finish each task.

Data were analyzed using the SciPy package, a Python-based open-source software. Because most distributions deviate from normality, non-parametric statistical tests were used. For EMG amplitude analysis [8], Maximum Voluntary Contraction (MVC) of each muscle of participants is recorded while attempting to oppose the resistance force of the researcher. To remove noises, artifacts including baseline shifts, bandpass filter with the range 20 Hz 500 Hz were applied. Root mean square (RMS) of the filtered EMGs were then calculated and normalized with peak RMS value of the MVC trial. Motor activities were displayed as a percentage of MVC (%MVC) value. Training results were compared to percent of MVC for studying the relationship between motor activities. For assessing the effect of different levels, a Friedman test was used. In order to study the effective memory task with bilateral training, memory results were analyzed by the same metrics as the digit span assessment [7]. Parametric statistical tests were used in the metrics to compare with digit span assessment.

3 Results

For experiments to study motor activities in both arms with the Journey Game and Memory Game, Table 1 shows the medians and interquartile ranges of percent of MVC on each muscles and training results in each task for healthy subjects, while percent of MVC on each muscles and training results played by the stroke patient indicate below as gray label. The results show that all muscles are stimulated during training tasks in both groups. For Journey Game, the percent of MVC of extensor muscles across game levels in healthy subjects, which according to the Friedman's test, shows significantly increasing trends on both arms. In stroke patient, The percent of MVC of almost all muscles show an upward

Table 1. Medians and correlations of amplitude analysis of EMG and training results in each task

Task	Amplitude Analysis (%MVC)				Training Results		
	Left		Right		CP	SPM	Speed
	Extensor	Triceps	Extensor	Triceps			
Level							
1	3.50(2.87)	5.04(2.08)	3.82(3.44)	3.50(2.59)	97.18(4.45)	28.67(1.00)	0.52(0.11)
	6.81	9.07	3.44	7.67	97.72	28.67	0.54
2	4.81(4.22)	4.81(1.85)	4.51(7.52)	3.81(1.69)	98.01(1.43)	57.50(1.00)	1.13(0.19)
	5.33	10.05	4.67	9.20	98.29	57.67	1.24
3	6.37(3.90)	5.40(2.31)	6.91(7.44)	4.07(2.57)	94.01(4.45)	91.50(4.33)	1.83(0.18)
	5.85	11.40	5.29	10.96	88.69	86.33	1.81
p	**≤.001**	.061	**≤.001**	.149	-	-	-
Digit							
FS	3.53(2.73)	5.49(3.05)	4.01(1.67)	3.75(1.98)	64.29(5.36)	66.12(7.25)	1.19(0.05)
	4.01	7.28	3.37	7.68	62.50	59.33	1.18
BS	3.41(2.37)	5.24(5.76)	3.59(1.41)	3.81(2.02)	53.57(15.80)	62.25(11.87)	1.29(0.14)
	3.54	7.08	3.15	8.09	55.56	57.99	1.20
Letter							
FS	3.60(1.94)	3.87(3.76)	3.59(1.04)	3.37(1.79)	57.14(7.14)	66.34(7.23)	1.21(0.15)
	2.82	6.97	3.05	7.94	55.56	50.99	1.42
BS	2.81(1.54)	4.54(5.61)	3.76(1.81)	3.46(2.01)	53.57(7.14)	64.88(10.81)	1.21(0.18)
	3.06	6.89	3.14	8.62	40.00	41.33	1.27
p	**.023**	.948	.101	.948	-	-	-

Notes. Result report as Median(Interquartile Range); p value = compared between training level; **bold** = p ≤ .05; EMG = electromyography; CP = percent of correctness; SPM = score per minute; FS = forward span; BS = backward span

Table 2. Means of digit span metric and training results in each memory task of healthy participants and stroke patient

Task	Healthy Patient				Stroke Subject			
	TE-ML	TE-TT	ML	MS	TE-ML	TE-TT	ML	MS
Digit								
FS	7.50±2.17	6.40±2.32	9.10±1.20	8.35±1.27	6.00	5.00	6.00	6.30
BS	6.90±2.18	5.90±2.56	8.00±2.26	7.54±2.04	4.00	2.00	6.00	5.58
Letter								
FS	7.60±2.32	6.80±2.86	8.70±1.95	7.87±1.96	5.00	3.00	5.00	5.50
BS	7.30±2.31	6.40±3.24	8.10±1.85	7.16±1.85	4.00	2.00	4.00	4.07

Notes. FS = forward span; BS = backward span. TE-ML = two-error maximum length; TE-TT = two-error total trials; ML = maximum length over all trials; MS = mean span over all trials

trend, except the left extensor muscle. Motor activities in each muscles on left side is higher than right side because of his stroke affected side. For each level, percent of correctness (CP) is greater than 80%. Score per minute (SPM) and movement speed are also increasing with the game level. For the results of the Memory Game on motor activities in healthy group, Only the percent of MVC of left extensor muscles is significantly different on each tasks. The percent of MVC of left triceps muscles is highest in all tasks. For stroke subject, the percent of MVC of triceps brachii muscles on both arms are higher than extensor carpi ulnaris muscles in all tasks. CP is less than 80% for both groups. CP and SPM of forward span (FS) are higher than backward span (BS) in both DS and LS. Response time in digit FS is faster than digit BS.

For the analysis of cognitive training with memory game, Table 2 shows the means and standard deviations of the digit span assessment metrics in each tasks of healthy participants and stroke. For healthy participants, means and standard variations of two-error maximum length (TE-ML) and two-error total trials (TE-TT) are less than maximum length (ML) and mean span (MS) in almost all of tasks. In stroke patient, ML is also greater than TE-ML and TE-TT.

4 Discussion

Our study has shown that novel upper extremity rehabilitation system with Gyro-Roller robot and virtual reality can stimulate motor and cognitive functions in healthy subjects and stroke patient. Percent of Maximum Voluntary Contraction (MVC) of triceps brachii and extensor carpi ulnaris muscles on both arms in each training reflected the motor activities during the training. With increasing difficulties in each level of journey game significantly affect the percent of MVC in extensor carpi ulnaris muscles on both arms of healthy subjects. In stroke patient, progressive levels also affect muscle activities, especially on triceps brachii. Furthermore, percent of MVC on left arm in each levels are more than right arm related to affected side of stroke. The results suggest that the feasibility for using the rehabilitation system to stimulate muscle activities in stroke training. For training result, CP, SPM, and movement speed of game character can show the performance of healthy and stroke participants. CP which indicates engagement to perform the task are higher than 80% on all subjects for the journey game. The effect of stroke rehabilitation also depends on engagement of participants. So, this training can motivate and challenge stroke patients. SPM and movement speed in the game are relative to the level that measures how quickly participants response during performing tasks.

For combined motor and cognitive training, we developed virtual reality game based on short term memory test and bilateral training. Our system can also stimulate the motor activities of healthy participant during training session, especially in left triceps brachii muscle, while stroke patient was stimulated dominantly on triceps brachii muscle in both arms same as reported in previous study [6]. Triceps muscle was induced on bilateral training movement with gyroscopic interfere. In training result on normal and patient subjects, CP and SPM are less than 80% in both groups because of difficulty of the tasks that is related to memory capacity of participants. Normal subject performed in each tasks better than stroke patient. Digit span metrics from the experiment have similar relationship as digit span assessment of short-term verbal memory [7], even though our experiment is using visual memory and answer selection by reaching movement. In our study, mean values of ML and MS are higher than mean of digit span assessment metrics because of education level of subjects. Stroke patient have mild cognitive impairment and older than normal subjects, hence the memory result is lower [9]. This results suggest that the integrated system have validity for cognitive and motor function training in stroke rehabilitation.

5 Conclusion

Upper extremity rehabilitation system based on bilateral training robot and virtual reality have effects on motor activities and cognitive function. The study in ten healthy participants and a chronic stroke patient suggests that the training system can stimulate percent of MVC of triceps brachii and extensor carpi ulnaris muscles on both arms. Moreover, participants can still perform short term memory task similarly as digit span assessment. Further research in stroke patients is essential in order to be supportive of the viability of the system.

References

1. Thai Non-communicable Diseases Information. http://www.thaincd.com/document/file/info/non-communicable-disease/
2. Cauraugh, J.H., Lodha, N., Naik, S.K., Summers, J.J.: Bilateral movement training and stroke motor recovery progress: a structured review and meta-analysis. Hum. Mov. Sci. **29**, 853–870 (2010). https://doi.org/10.1016/j.humov.2009.09.004
3. Van Delden, A.E.Q., Peper, C.E., Kwakkel, G., Beek, P.J.: A systematic review of bilateral upper limb training devices for poststroke rehabilitation. Stroke Res. Treat. **2012** (2012). https://doi.org/10.1155/2012/972069

4. Lee, S., Kim, Y., Lee, B.-H.: Effect of virtual reality-based bilateral upper extremity training on upper extremity function after stroke: a randomized controlled clinical trial. Occup. Ther. Int. **23**, 357–368 (2016). https://doi.org/10.1002/oti.1437
5. Faria, A.L., Cameirão, M.S., Couras, J.F., Aguiar, J.R.O., Costa, G.M., Bermúdez i Badia, S.: Combined cognitive-motor rehabilitation in virtual reality improves motor outcomes in chronic stroke - a pilot study. Front. Psychol. **9**, 1–13 (2018). https://doi.org/10.3389/fpsyg.2018.00854
6. Ruangrong, T., Ritthipravat, P.: EMG analysis on Gyro-roller rehabilitation device for stroke patients. Int. Conf. Biomed. Eng. Technol. **7**, 153–161 (2017). https://doi.org/10.17706/ijbbb.2017.7.3.153-161
7. Woods, D.L., Kishiyama, M.M., Yund, E.W., Herron, T.J., Edwards, B., Poliva, O., Hink, R.F., Reed, B.: Improving digit span assessment of short-term verbal memory. J. Clin. Exp. Neuropsychol. **33**, 101–111 (2011). https://doi.org/10.1080/13803395.2010.493149
8. Amplitude Analysis: Normalization of EMG to Maximum Voluntary Contraction (MVC). https://www.delsys.com/emgworks-analysis-techniques-using-emgscript/
9. Woods, D.L., Wyma, J.M., William Yund, E., Herron, T.J., Reed, B.: Age-related slowing of response selection and production in a visual choice reaction time task. Front. Hum. Neurosci. **9**, 1–12 (2015). https://doi.org/10.3389/fnhum.2015.00193

Therapeutic Medical Engineering

Numerical Analysis of Aortic Valve Stress for Phantom Experiments

Takuya Watanabe[1(✉)], Asuka Hatano[1], Satoshi Izumi[1], Naoki Tomii[1], Eriko Maeda[2], Haruo Yamauchi[2], Kan Nawata[2], Shu Takagi[1], Ichiro Sakuma[1], Minoru Ono[2], and Hiroshi Tsukihara[2]

[1] University of Tokyo, Hongo, Tokyo, Japan
watanabe.takuya@fml.t.u-tokyo.ac.jp
[2] University of Tokyo Hospital, Hongo, Tokyo, Japan

Abstract. In order to analyze stress and flow field around the aortic valve, various numerical method have been developed. The validity of the results has been confirmed from the point of hemo dynamical, fluid dynamical, view. However, it has not been evaluated from the point of view of structural mechanics. The purpose of this study is to construct a stress measurement experiment using a phantom model and to establish a numerical analysis method for it. The phantom model of the aortic valve was created with reference to the valve shape used for aortic valve reconstruction surgery. The displacement field of the phantom model is measured using Digital Image Correlation (DIC), which measures the three-dimensional displacement field on object surface using a stereo camera. In the numerical analysis, a finite element model simulating a phantom experiment is created, and quasi-static analysis is performed by an explicit method using LS-Dyna. Though the experimental results have not been obtained yet, according to the numerical analysis, max principle strain will increase at the fixed commissure and center of the coaptation.

Keywords: Aortic valve · Digital Image Correlation (DIC) · LS-Dyna

1 Introduction

In order to drive blood efficiently throughout the body, an aortic valve is placed at outlet of the left ventricle. Aortic valve dysfunction is due to mechanical stress on the valve. Therefore, various studies of the disease progression have been performed both in vivo and vitro [1], but it is still difficult to measure stress and detailed blood flow distribution by each of the methods. In recent years, a fluid structure interaction (FSI) simulator using numerical analysis has been developed in order to know more detailed stress and flow around the valve and, using this simulator, studies have been conducted on the effects of valve and sinus Valsalva shapes on hemodynamics and stress states [1, 2]. However, the validity of these simulators has been confirmed based only on the blood flow measured by in vitro experiments, but it has not been confirmed based on the stress. This is because it was difficult to measure the stress of membranes over a wide range, but digital image

Y. Shiraishi et al. (Eds.): APCMBE 2020, IFMBE Proceedings 82, pp. 47–51, 2021.
https://doi.org/10.1007/978-3-030-66169-4_7

correlation (DIC) that measures strain by recognizing random patterns on the object surface has been realized and, it made it easier to measure the strain field on an object surface. In this study, the purpose of this study is to construct an experiment to measure the displacement field of a phantom model of an aortic valve by DIC and to establish a numerical analysis method.

2 Material and Method

2.1 Aortic Valve Phantom Model

In this study, we made phantom aortic valve model, referencing aortic valve reconstruction [3]. In this surgical operation, aortic valve re-construct from glutaraldehyde-treated autologous pericardium. There were 444 male and 406 female patients in Japan and, this shape has been demonstrated to the function as an aortic valve. Silicon rubber (t = 0.5) was used as leaflets.

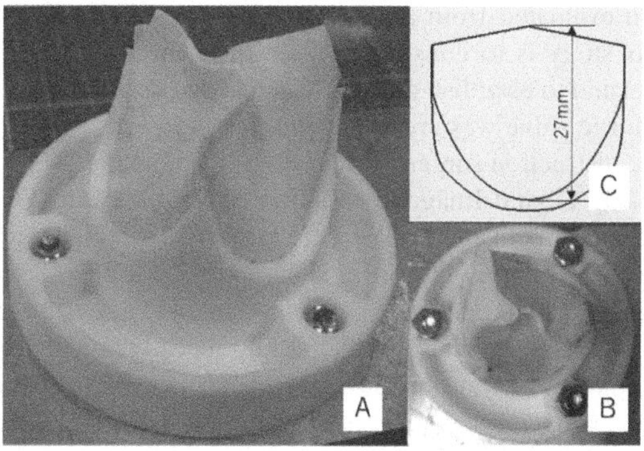

Fig. 1. A: Phantom aortic valve model from an oblique (A) and top (B) point. The Coaptation was twisted. C: Leaflet shape made with reference to Ozaki's model [3].

2.2 Digital Image Correlation

Digital image correlation (DIC) method measures displacement of the object surface by comparing random patterns drawn on the surface before and after deformation [4]. The strain can be calculated by differentiating the displacement field or apply the displacement field to nodes of a finite element model. If one camera is used, a 2D displacement field can be measured, but if two, a 3D displacement field can be measured (Fig. 2).

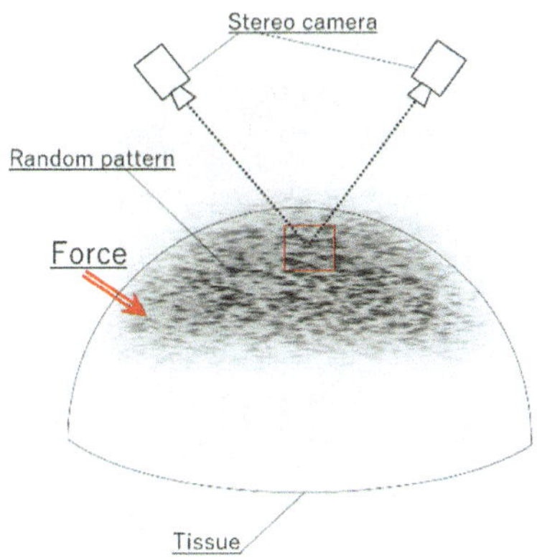

Fig. 2. Matching random patterns before and after deformation in DIC [4]

2.3 Experiment

By applying a negative pressure of 16 kPa to the left ventricle, the closed state was reproduced. The strain on the leaflets in this state was measured by the DIC method, and the contact pressure of the valve was measured by a pressure sensor (Fig. 3).

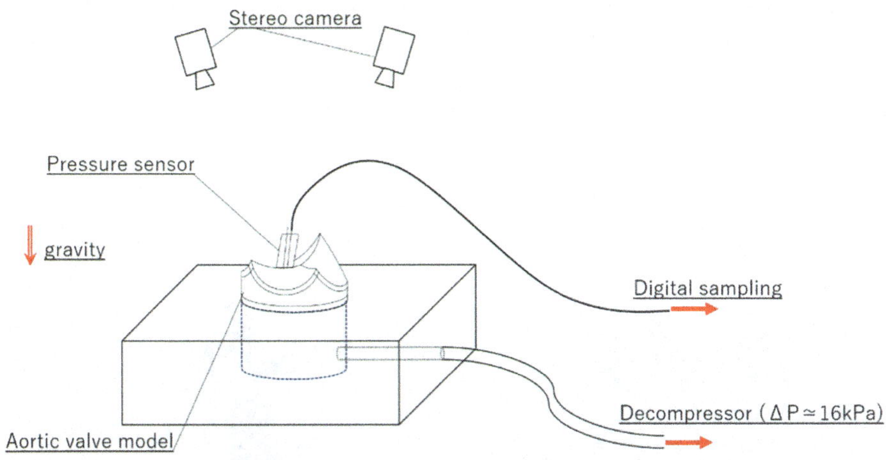

Fig. 3. Overview of the Phantom experiment

2.4 Numerical Analysis

We also performed numerical analysis that simulate the previous experiment. In this study, numerical analysis was performed on an Intel core i7-4770 3.4 GHz workspace, using explicit finite element software LS-Dyna. The valve leaflets were discretized into 2266 4 and 3-noded Belytschko-Tsay Shell elements. Material properties of the leaflets was same as silicon rubber. Density was equal to 9.5×10^{10}ton/mm^3, Young's modules was equal to 1.0MPa and Poisson's ratio was 0.45. Commissure was fixed completely, and constant pressure (16 kPa) was applied to the surface of the leaflets (Fig. 4).

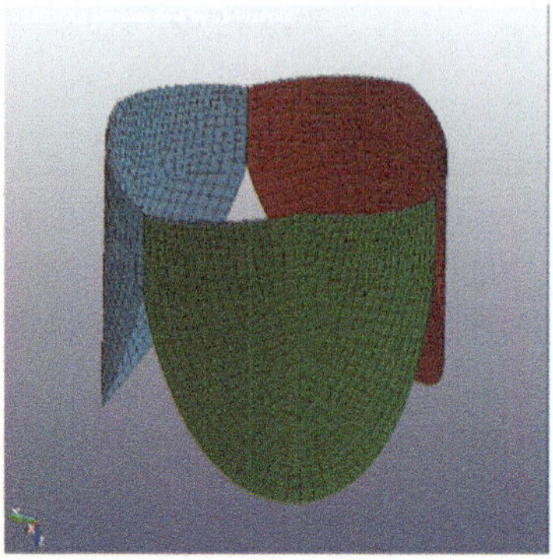

Fig. 4. Initial state mesh shape of the aortic valve model

3 Result and Discussion

3.1 Numerical Analysis

As a result of the analysis, the distribution of Mises stress and the first principal strain was as shown in Fig. 1. It is observed that the stress increases at the fixed commissures and center of coaptation. At present, there are no results of phantom experiments, but it is expected that the strain distribution shown in the left of Fig. 1 will be obtained.

The twist of coaptation, which was observed in the phantom model, was also observed in numerical model. This shows that the twisted shape of the coaptation has a stable state when valve is closed(Fig. 5).

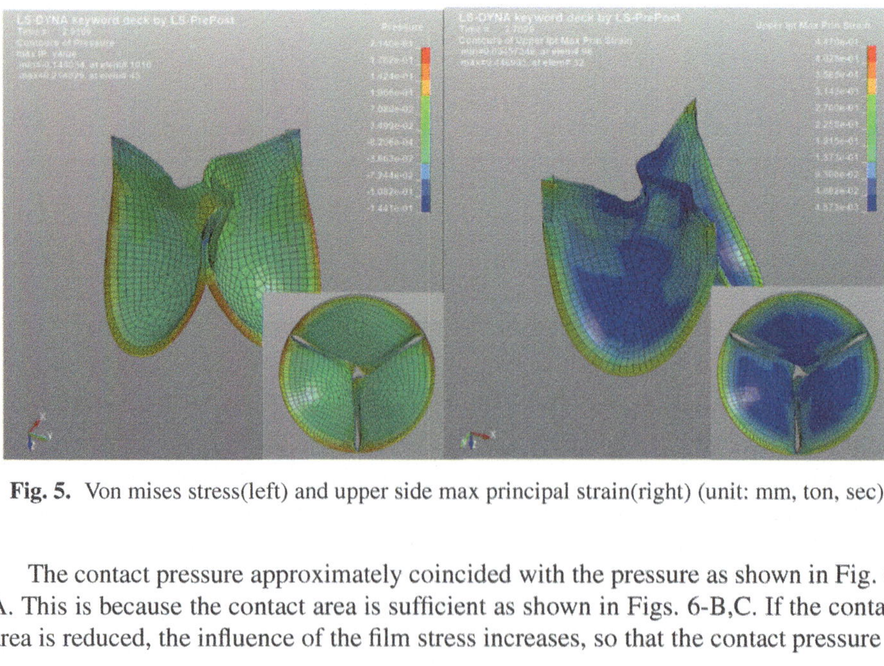

Fig. 5. Von mises stress(left) and upper side max principal strain(right) (unit: mm, ton, sec)

The contact pressure approximately coincided with the pressure as shown in Fig. 6-A. This is because the contact area is sufficient as shown in Figs. 6-B,C. If the contact area is reduced, the influence of the film stress increases, so that the contact pressure is also expected to decrease. Since the contact pressure and the contact area have a great effect on regurgitation, it will be a future task to evaluate the this relationship with the result of the phantom experiment.

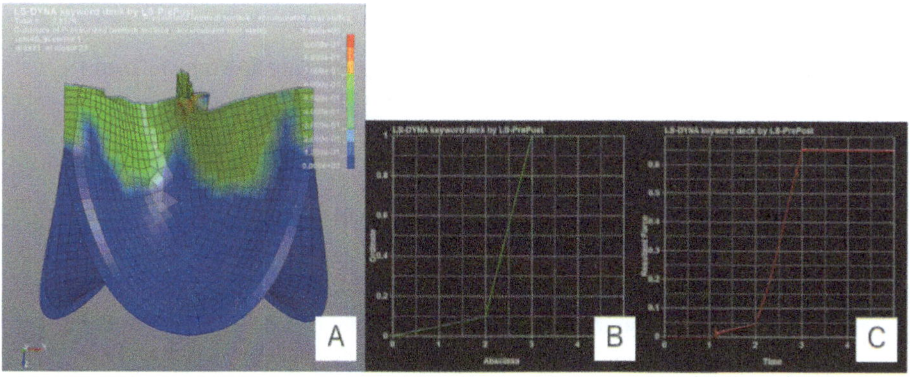

Fig. 6. A: Contact Pressure. Green area is equal to coaptation. B: Applied pressure. C: Contact pressure. Contact starts in 1.2 s, and the contact pressure increases as the applied pressure increases

Acknowledgement. This work is supported by Canon Medical.

References

1. Tango, A., Salmonsmith, J., et al.: Validation and extension of a fluid–structure interaction model of the healthy aortic valve. Cardiovasc. Eng. Technol. **9**(4), 739–751 (2018)
2. Katayama, S., Umetani, N., et al.: Bicuspid aortic valves undergo excessive strain during opening: a simulation study. J. Thorac. Cardiovasc. Surg. **147**(4), 1439–1440 (2014)
3. Ozaki, S., et al.: Aortic valve reconstruction using self-developed aortic valve plasty system in aortic valve disease. Interact. CardioVasc. Thorac. Surg. **12**, 550–553 (2011)
4. Yoneyama, S., MUrasawa, G.: Digital Image Correlation. In: Experimental Mechanics

Development of an Ultrastructural Model of Cardiomyocytes Based on Electron Microscopy

Makoto Someya[1(✉)], Hiroaki Tanaka[1], Asuka Hatano[1], Satoshi Izumi[1], Masahiko Hoshijima[2], and Andrew D. McCulloch[2]

[1] Tokyo University, Hongo 7-3-1, Bunkyo Ward, Tokyo, Japan
someya.makoto@fml.t.u-tokyo.ac.jp
[2] University of California San Diego, 9500 Gilman Drive, La Jolla, CA, USA

Abstract. How ultrastructural alterations in cardiomyocytes lead to heart disease remains unclear. Therefore, there is interest in simulate cardiac myocytes with realistic 3-D models of myocyte shape and structure. Using deep learning and image processing, we developed an efficient method to generate 3-D models of cardiac myocytes without the need for manual segmentation. Image processing was used to segment borders between mitochondria in two series of serial-block-face scanning electron microscopy (SBF-SEM) images of healthy mouse ventricular cardiomyocytes. Computed mitochondrial size and morphologies were in good agreement with published measurements. These new automated methods will facilitate more realistic simulation studies of cardiac myocytes and enable new investigations of the links between altered mitochondrial morphology and heart diseases.

Keywords: Cardiac ultrastructure · Image segmentation · Serial-block-face imaging

1 Introduction

To date, medicine has evolved primarily through empirical population-based medical knowledge and experience. Since the cellular and molecular mechanisms of many diseases have not been fully elucidated, a current goal is to develop mechanistic computational models of cells and organs.

Hearty disease frequently disrupts normal cardiomyocyte ultrastructure but the mechanisms remain poorly understood. Recent advancements in 3-D electron microscopy technologies are creating new scientific insights. Rog-Zielinska *et al.* [1] reported the effects of contraction on the mitochondrial morphology by analyzing main axis orientation and eccentricity of reconstructed mitochondria from 3-D electron micrographs.

In order to investigate mitochondrial morphology further, large-scale segmentations data are needed to account for individual differences between animals and cells. Hussain *et al.* [2] presented an algorithm to automatically segment SBF-SEM images using

Y. Shiraishi et al. (Eds.): APCMBE 2020, IFMBE Proceedings 82, pp. 52–59, 2021.
https://doi.org/10.1007/978-3-030-66169-4_8

bimodal pixel intensities. Their algorithm improved segment accuracy, however mitochondria were segmented into aggregates without being divided individually. Therefore, this algorithm is not well suited for creating models for simulation.

The purpose of this study was to develop a new algorithm for easily segmenting the whole slices of electron micrograph images.

2 Methods

2.1 3-D Modeling

Segmentation. We prepared as training data 70 images obtained by extracting the contours of mitochondria in SEM Images. 50 images were from one further the (cell A), and 20 were from a second further the (cell B) (Fig. 1).

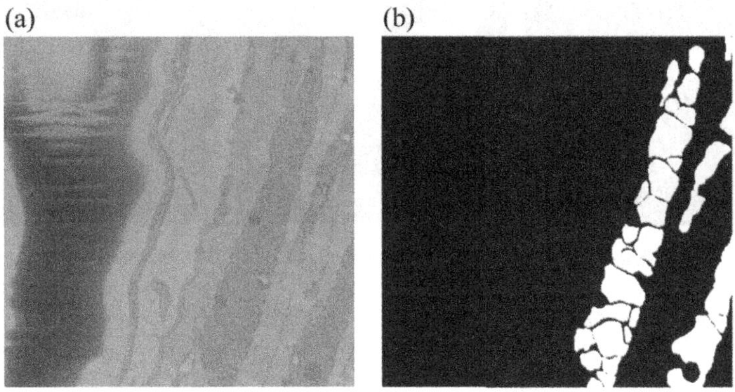

Fig. 1. Example of dataset for training. (a) Raw image for input, (b) Corresponding label image.

Five learning models were created, based on a total of 50, 55, 60, 65, and 70 datasets by adding datasets five at a time to 50 training sets from cell A. In this study, CDeep3M [3], a cloud-based deep convolutional network specialized for segmentation of electron microscope images, was used for deep learning. The number of training data sets was reduced by transfer learning. For the transfer learning, we used a model in which 80 mitochondrial datasets were pre-trained with 30,000 iterations. Input images were 400 cardiomyocyte images from cell A, and 690 images from cell B.

For the 70 data set case, the intensities of input images were adjusted, and we tested whether accuracy was improved or not. Since mitochondria and myofibrils form two peaks in the intensity distribution, the intensity was adjusted by linear transformation so that the peaks of the input images coincided with the average of the peaks of input images.

Separation. The division process comprised five steps: (a) rough image segmentation, (b) nucleus extraction, (c) nucleus allocation, (d) nucleus expansion, and (e) three-dimensional processing. we created 3-D models (cells A and B) from the divided images (Fig. 2).

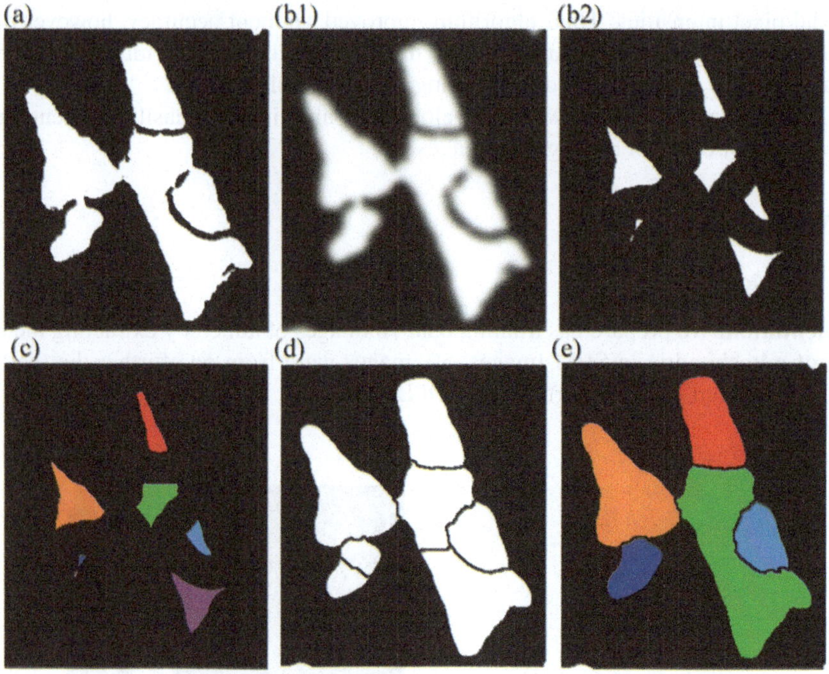

Fig. 2. Division processing. (a) Subtract contour result from the whole object; (b1) Gaussian blur; (b2) shrink objects and extract nucleus; (c) allocate nucleus; (d) expand nucleus on top of the segmentation; (e) extend these processes to 3-D to correct the extra division.

2.2 Validation

Labeled images not used as training data were used for validation.

Pixel-Based Validation. Labeled images not used as training data were used for validation.

Pixel-based classification was based on the following contingency Table 1.

Table 1. Result classification

	Ground truth positive	Ground truth negative
Segmentation positive	TP: True Positive	FP: False Positive
Segmentation false	FN: False Negative	TN: True Negative

Based on this classification, 3 evaluation indices, precision, recall, and F-score, were set. Each index is calculated as follows. Precision and recall are in a trade-off relationship. Precision ignores FN, while recall ignores FP. To compensate for this, the

F-score is derived by taking the harmonic mean of two indices.

$$\text{Precision} = \text{TP}/(\text{TP} + \text{FP}) \quad (2.1)$$

$$\text{Recall} = \text{TP}/(\text{TP} + \text{FN}) \quad (2.2)$$

$$\text{F} - \text{score} = 2\text{TP}/(2\text{TP} + \text{FP} + \text{FN}) \quad (2.3)$$

Model Evaluation. We evaluated 3-D structure of the segmented mitochondria in the volumes. quantitative values were calculated and compared with literature data. Specifically, the number and volume of objects in the model, the eccentricity when mitochondria were approximated as a prolate spheroid, and the orientation of the principal axis to the direction of myofibrils were evaluated. for the spheroid approximation, a least-squares method was used.

3 Results and Discussion

3.1 Pixel-Based Result

The result of binarizing the segmentation result with the threshold value that maximizes each F-value is the following Table 2 (Fig. 3).

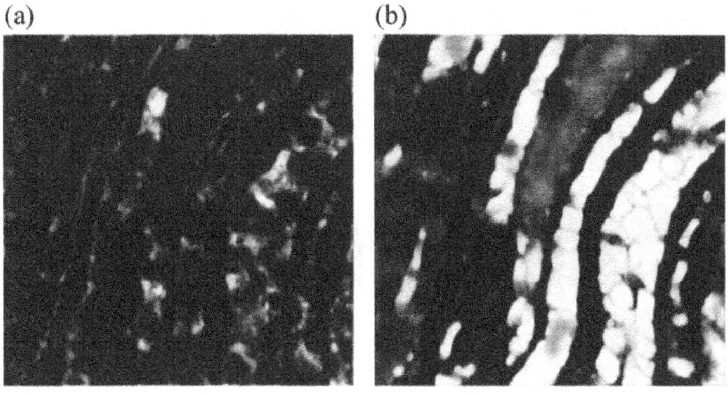

Fig. 3. Result of segmentation of myocyte B by CDeep3M(a) with 50 datasets; (b) with 70 datasets.

Accuracy for cell B was slightly lower than that for cell A. The main factor was the difference in recall, which indicates that there were some missed in myocyte B. Since myocyte A has been able to extract with approximately the same accuracy as Table 2 in 20 datasets, it can be said that the number required for learning has not decreased. In order to create a versatile deep learning model, it is important to create datasets from many different cardiomyocytes incrementally rather than from only a few cells. Therefore, it

Table 2. Accuracy at the maximum F-score

		Precision	Recall	F-Score
Cell A	50 sets	0.922	0.967	0.948
Cell B	50 sets	0.466	0.625	0.534
	55 sets	0.414	0.635	0.501
	60 sets	0.533	0.739	0.620
	65 sets	0.884	0.921	0.902
	70 sets	0.912	0.932	0.922

is necessary to create an efficient learning model that can reduce the number of datasets needed per cell.

The effect of intensity adjustment is noticeable in Fig. 4. Several slices near the slice in Fig. 4 with higher overall intensity than other slices were initially rejected by binarization and could have not be used. However, with adjustment, mitochondria could be extracted with high probability, as shown in Fig. 4 (b). As seen in Table 3, the F-value of the validation also increased, albeit slightly.

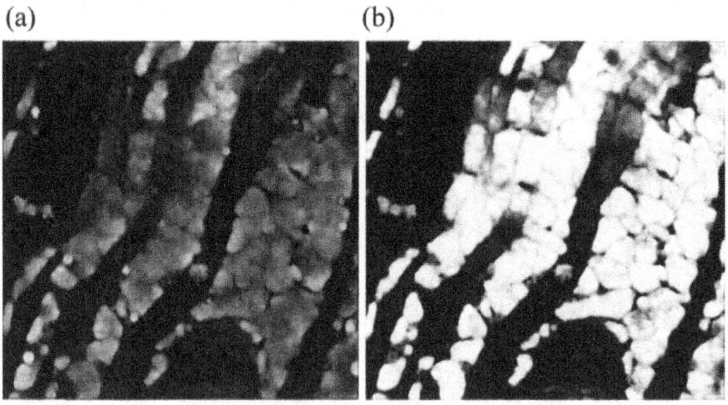

Fig. 4. Effect of intensity adjustment. (a) Before adjustment; (b) After adjustment.

Table 3. Accuracy before and after adjustment

	Precision	Recall	F-Score
Before adjustment	0.912	0.932	0.922
After adjustment	0.912	0.935	0.923

This method assumes that SEM images are taken in an imaging environment where the intensity peaks of mitochondria and myofibrils are separated. Among the images used in this study, there were slices in which myofibrils and mitochondria were captured at similar intensities. Results from before and after intensity adjustment are shown in Fig. 5 (a & b). From Fig. 5, it is appreciated that not only the difference in intensity but also the subtlety of intensity and macro shape are learned and inferred, and it is impossible to make use of them with adjustment.

As a countermeasure, by adjusting the intensity of all images first according to the intensity distribution of the data used for transfer learning and then using learning, it is expected that improvement of the segmentation can be achieved.

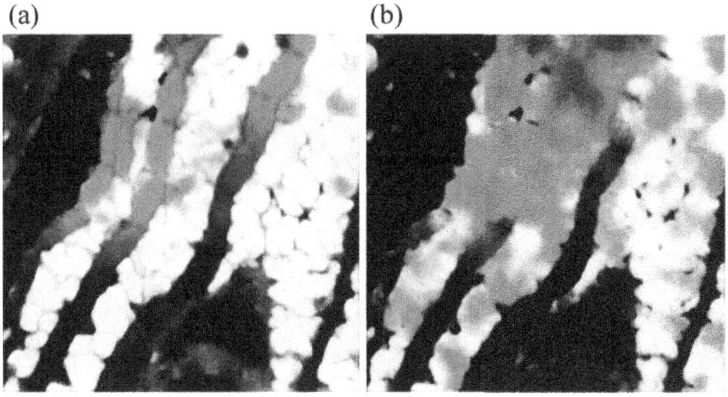

Fig. 5. Segmentation result. (a) Myofibrils are in low probability before adjustment. (b) Mitochondria and myofibrils are at almost the same probability after adjustment.

3.2 Model-Based Result

Table 4 and Fig. 5 shows the validity of the models. From (c), (d1), and (d2), it is likely that mitochondria who orientation did not parallel the myofibrils were detected because mitochondria were cut off during the division process. It is necessary to determine an appropriate threshold for contour data by taking feedback from the number of individuals in the label (Fig. 6).

Table 4. Mitochondrial measurements

	Model size (μm^3)	Quantity	Volume avg. (μm^3)	Volume ratio (%)
Cell A	14 × 14 × 28	1177	0.9367	20.04
Cell B	14.3 × 14.3 × 48.4	2801	0.9472	26.68

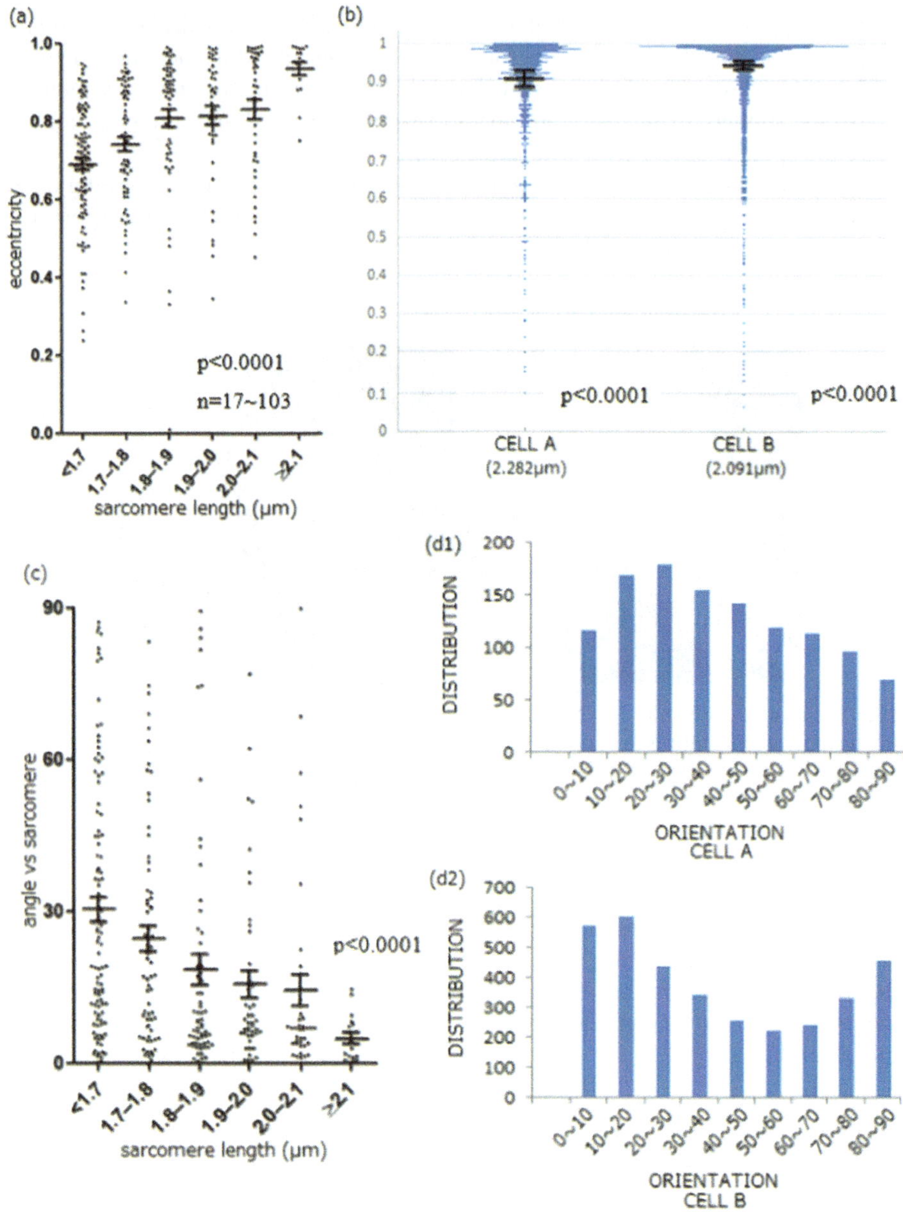

Fig. 6. Quantitative Measurements: (a) eccentricity distribution from manual segmentations [1]; (b) eccentricity by machine learning in this study. (a) and (b) are almost the same for the same sarcomere length. (c) orientation of mitochondria from manual segmentations [1]. (d1) (d2) are obtained from cells A and B. Both cells A and B have many mitochondria at angles greater than 50°. (Copyright permission is provided by American Association of Anatomists.)

4 Conclusion

The reliability of this new automatic mitochondrial extraction method enables the creation of a large-scale models from SBF-SEM images. We confirmed that this method is suitable for 3D model generation by increasing division accuracy.

References

1. Rog-Zielinska, E.A., O'Toole, E.T., Hoenger, A., Kohl, P.: Mitochondrial deformation during the cardiac mechanical cycle. Anat. Rec. **302**(1), 146–152 (2019)
2. Hussain, A., et al.: An automated workflow for segmenting single adult cardiac cells from large volume serial block-face scanning electron microscopy data. J. Struct. Biol. **202**, 275–285 (2018)
3. Haberl, M.G., et al.: CDeep3M—Plug-and-Play cloud-based deep learning for image segmentation. Nat. Methods **15**(9), 677–680 (2018)

Fluid-Structure Interaction Analysis of 3D Human Aortic Valve Model Constructed from CT Images

Sakurako Miyazaki[1](✉), Asuka Hatano[1], Satoshi Izumi[1], Yusuke Morishita[1], Bowen Fan[1], Shu Takagi[1], Naoki Tomii[2], Hiroyuki Tsukihara[3], Eriko Maeda[3], Haruo Yamaguchi[3], Kan Nawata[3], Ichiro Sakuma[1], and Minoru Ono[3]

[1] Graduate School of Engineering, The University of Tokyo, Tokyo, Japan
miyazaki.sakurako@fml.t.u-tokyo.ac.jp
[2] Center for Disease Biology and Integrative Medicine, The University of Tokyo, Tokyo, Japan
[3] The University of Tokyo Hospital, Tokyo, Japan

Abstract. The aortic valve is always under great mechanical stress. It is known that stress is greatly involved in the progression of the disease. For treatment, information such as valve stress, strain, and blood flow are helpful. Until now, the relationship between mechanical stress and disease progression has been studied in clinical research, phantom experiments, and numerical simulations.

The purpose of this study is to show a new diagnostic indicator of the aortic valve based on numerical simulation. Based on the three-dimensional shape of the aortic valve reconstructed based on CT images, a personalized mesh was constructed. NURBS is used to construct personalized mesh. Using the mesh, we perform numerical simulations and obtain useful information for diagnosis.

Deep learning is used to reconstruct three-dimensional shapes from CT images. As an analysis method, ALE fluid-structure interaction analysis is used. The Navier-Stokes equation is weakly formalized, finite elements discrete, linearized, and conformed with structural analysis formulas. SUPG and PSPG stabilization are used for fluid analysis.

Three mesh based on CT images of patients(patient1–3) were created and analyzed. The stress condition of the valve, the flow field of the blood, and the pressure field was visualized and compared. There was a significant difference in the analysis results for each personalized mesh. The length of the perimeter, area, and Young's modulus greatly affected the way the valve opened.

Keywords: Aortic valve · Fluid-structure interaction analysis · CT images

1 Introduction

Typical aortic valve diseases include stenosis and regurgitation. The aortic valve is exposed to a flow rate of as much as 1 m/s during the systolic period of the heart, and the diastolic period is always subject to a large mechanical load because it supports a pressure difference of as much as 10 kPa. It is known that stress is greatly involved in

Y. Shiraishi et al. (Eds.): APCMBE 2020, IFMBE Proceedings 82, pp. 60–66, 2021.
https://doi.org/10.1007/978-3-030-66169-4_9

the progression of the disease. For treatment, information such as valve stress, strain, and blood flow are helpful.

Many attempts have been made to clarify the relationship between the progression of aortic valve diseases and mechanical stress. In the area ranging from animal studies to clinical studies, the stress on the valve during normal and disease were compared. The movement of the valve, blood pressure in the ventricle and aorta, and strain and stress of the valve when opening and closing were measured. Also, since the measurement using the living body is accompanied by a large restriction, experiments using artificial flow path and pump have been carried out [1].

Numerical simulation was applied to the heart valve as a method that was neither a theory nor an experiment [2]. The application of numerical simulation to clinical research is progressing gradually.

The goal of this study is to perform a more detailed diagnosis of the aortic valve based on the three-dimensional shape of the aortic valve obtained by the noninvasive testing method. Based on the three-dimensional shape reconstructed from CT images, an personalized mesh is constructed and an ALE fluid-structure coupled analysis is applied.

2 Methods

2.1 Fluid-Structure Interaction Analysis

The ALE (Arbitrary Lagrangian-Eulerian) fluid-structure interaction analysis program is provided by Katayama [3]. 4/4c elements were used for fluid analysis. DKT shell elements are used for structural analysis. As an analysis method, the Navier-Stokes equation is weakly formalized, finite elements discrete, linearized and conformed with structural analysis formulas. SUPG and PSPG stabilization are used for fluid analysis. Remeshing is performed at each time step.

2.2 Personalized Mesh Construction

The personalized mesh was constructed based on the three-dimensional shape of the aortic valve which Fan [4] reconstructed from CT images. The aortic valve consists of three leaflets and a bulge in the vascular wall, the sinus of Valsalva. The three-dimensional shape of each leaflet and sinus of Valsalva was created individually. We created a personalized mesh based on the three-dimensional shape reconstructed by Fan. HyperMesh and Evolve, software from Altair, were used to create the mesh. To eliminate the noise generated when reconstructing a three-dimensional shape, we simplified the three-dimensional shape and made it into NURBS and created the geometry. A mesh was created from the geometry. We bound leaflets and the wall and created the upper and lower flow paths.

Three mesh were constructed based on the three-dimensional shape of each patient (patient1–3).

2.3 Analysis Conditions

The Navier–Stokes equations of incompressible flows were solved using 4/4c elements. Non-Newtonian behavior of blood is known negligible to simulate aortic flow condition [5]. The density is $\rho = 998.2$ [kg/m2] and the viscosity coefficient is $\mu = 0.038$ [Pas]. The leaflet is a DKT shell element of a linear elastic body, and it is assumed that it is an isotropic material. The modulus of longitudinal elasticity was $E = 0.7$ [MPa], and the Poisson's ratio was $\nu = 0.45$, and the modulus of transverse elasticity was set to $G = E/2(1 + \nu)$ from the assumption of isotropic material. The boundary condition of the blood vessel was a fixed wall. As a boundary condition, the fluid inlet pressure is specified (Fig. 1), and the exit side is connected to the three-element Windkessel model. At the entrance, we assume the Poiseuille flow which deformed along the boundary surface. The inlet velocity is degenerated by the profile function.

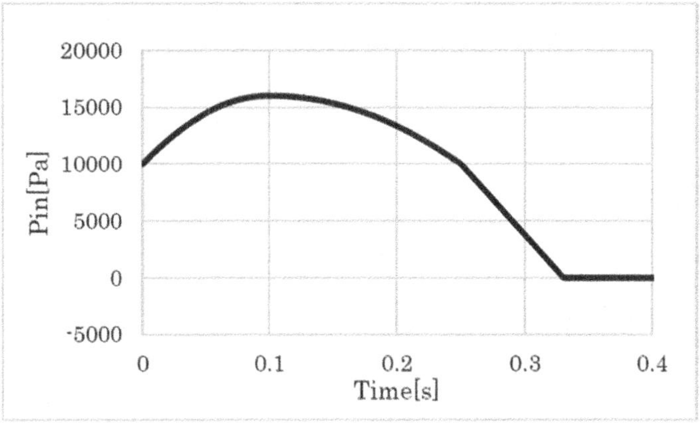

Fig. 1. Inlet pressure (0 s < t < 4s)

3 Results and Discussion

3.1 Comparison of Analysis Results by Patient

We conducted the analysis in patient1, 2, and 3 personalized mesh, and compared the results of the analysis (Fig. 2).

Figure 3, 4 and 5 shows the primary principal stress at 0.1 s. At 0.1 s the valve opened the most. The unit of the first principal stress is Pa. Because bending stress is dominant than tensile stress, the part where the fold occurs has higher stress. When the valve is opened, the valve becomes hard to open because the force that the valve tries to return to the original condition is in the direction of closing the valve.

In patient1, the NCC (Non-Coronary Cusp) and the LCC (Left Coronary Cusp) have folds near the center of the valve, and the valve is not as open as the RCC (Write Coronary Cusp) (Fig. 3). The RCC has a fold at the base of the valve, but the valve is open. In patient2, a fold occurs near the center of all valves, and all leaflets are almost equally

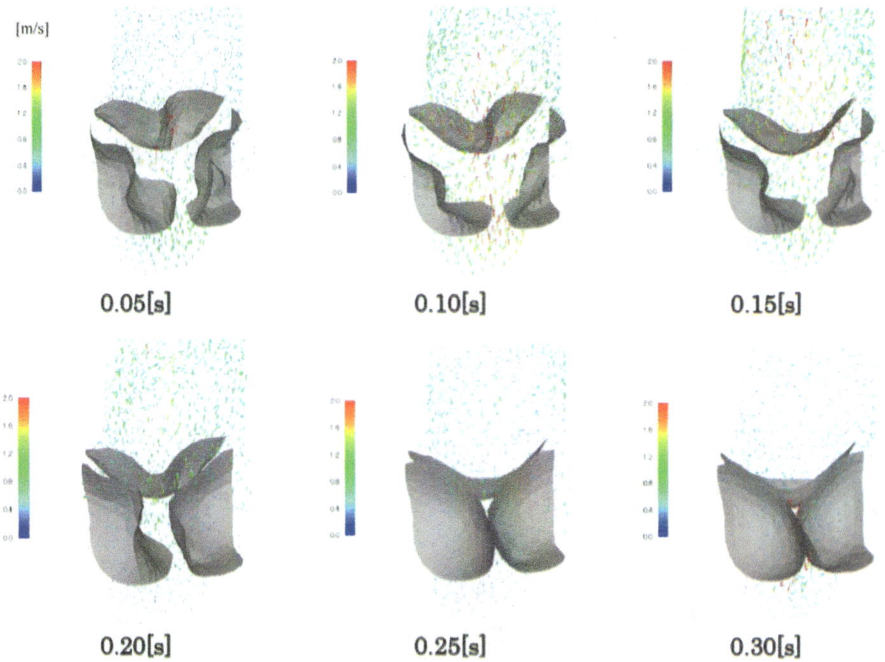

Fig. 2. Analysis result of patient2

open (Fig. 4). In patient3, the folds occur only at the base, but the opening area is narrow (Fig. 5). This is because the area is small relative to the outer periphery of the valve, and the valve does not follow the wall even though the valve is opened.

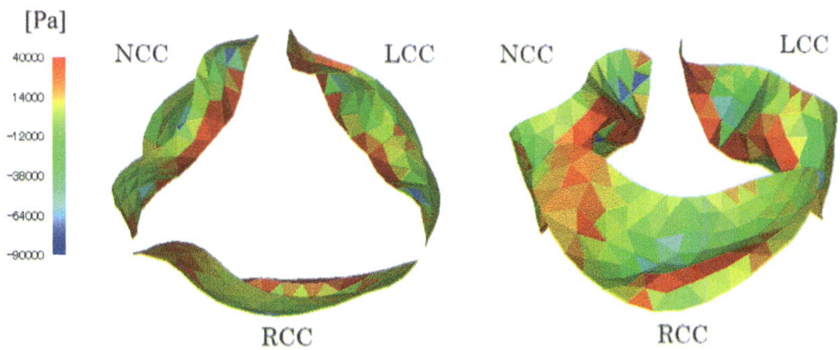

Fig. 3. First principal stress of patient1 (0.1 s)

The more the ratio of the opening area of the valve to the inlet area smaller was, the inlet velocity smaller was (Fig. 6). The outlet velocity was proportional to the inlet velocity. Proportional constants are considered to depend on the flow rate and the area ratio of the inlet boundary surface and the outlet boundary surface.

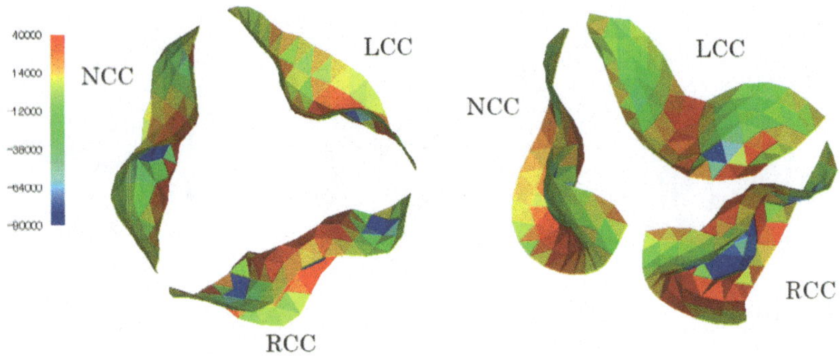

Fig. 4. First principal stress of patient2 (0.1 s)

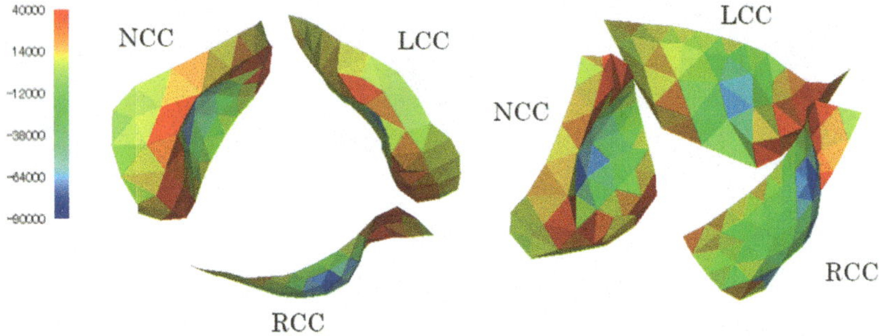

Fig. 5. First principal stress of patient3 (0.1 s)

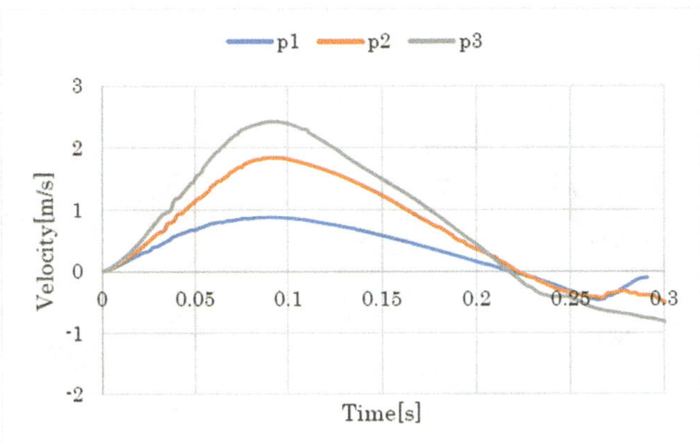

Fig. 6. Time history of the inlet velocity

3.2 Evaluating Mesh Sizes

We evaluated the mesh size using three personalized mesh of 0.8, 1.2, and 1.6 mm of patient2 mesh size (Fig. 7).

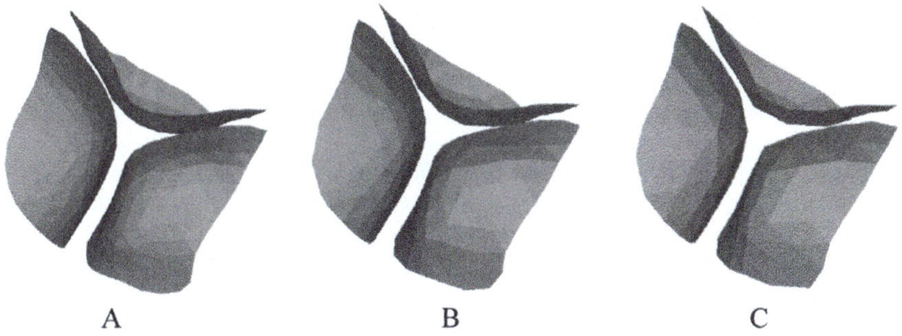

Fig. 7. (A) meshsize 0.8 mm, (B) meshsize 1.2 mm, (C) meshsize 1.6 mm

Here, an index called aperture ratio is introduced to represent the performance of the aortic valve. The aperture ratio is an indicator of how much the valve is open. When the leaflets are observed from the axial direction of the blood vessel, the difference between the area of leaflets when the valve is completely closed and the area of leaflets at each time is shown as a percentage.

We analyzed the personalized mesh with mesh sizes 0.8, 1.2, and 1.6 mm, and calculated the aperture ratio at each time and compared them. A graph of the time change of the aperture ratio is shown in Fig. 8.

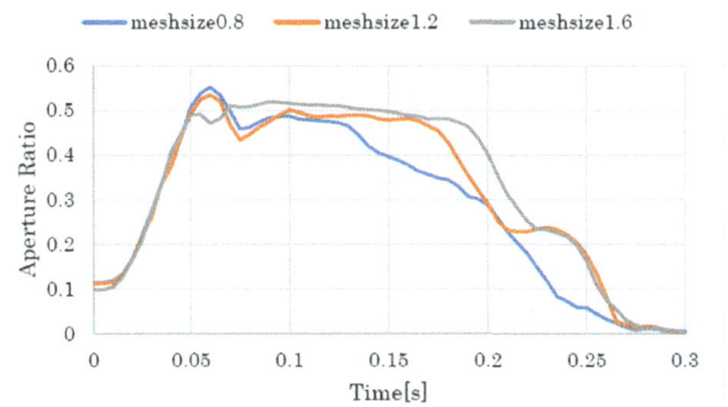

Fig. 8. Time history of the aperture ratio

From Fig. 8, it can be seen that the speed at which the valve opens is about the same regardless of the mesh size, but the speed at which the valve is closed is faster when mesh size is smaller.

The mesh sizes of 0.8 mm and 1.2 mm are well-matched until the valve begins to close after the valve's aperture ratio is maximized, but the difference in the aperture ratio of the valve closes has increased since then. Therefore, it is not possible to say that the analysis is fully converged with a mesh size of 0.8 mm, and it is necessary to perform the analysis with a smaller mesh size in the future.

Acknowledgement. This work is supported by Canon Medical Systems.

References

1. Toninato, R., et al.: Physiological vortices in the sinuses of Valsalva: an in vitro approach. J. Biomech. **49**, 2635–2643 (2016)
2. Tango, A.M., et al.: Validation and extension of a fluid-structure interaction model of the healthy aortic valve. Cardiovasc. Eng. Technol. **9**(4), 739–751 (2018)
3. Katayama, S., et al.: The sinus of Valsalva relieves abnormal stress on aortic valve leaflets by facilitating smooth closure. J. Thoracic Cardiovasc. Surg. **136**(6), 1528–1535 (2008)
4. Fan, B., et al.: Attention-guided decoder in dilated residual network for accurate aortic valve segmentation in 3D CT scans. In: MLMECH 2019/CVII-STENT 2019, LNCS 11794, pp. 121–129 (2019)
5. Hsu, M.C., et al.: Fluid–structure interaction analysis of bioprosthetic heart valves: Significance of arterial wall deformation. Comput. Mech. **54**(4), 1055–1071 (2014)

Experiment on Flow Estimation of Simulated Urination in a Western Style Toilet with Built-in Image Sensor

Yusuke Niwa[1(✉)], Haruki Kawanaka[1], Eiichi Watanabe[2], and Koji Oguri[1]

[1] Graduate School of Information Science and Technology, Aichi Prefectural University, 1522-3 Ibaragabasama, Nagakute-city, Aichi, Japan
y.niwa@bme.ist.aichi-pu.ac.jp

[2] Fujita Health University, 1-98 Dengakugakubo, Kutsukake-cho, Toyoake, Aichi, Japan

Abstract. The multiple cylindrical model to estimate urinary volume using image sensors without collecting urine in to the cup has been proposed. However, it was not supposing that the measurement is performed on a western style toilet. So, it is necessary to devise a way to embed an image sensor into a western style toilet where it can keep the geometric parallelism of the multiple cylindrical model. In this study, we constructed an environment where a multiple cylindrical model can be applied by using a raising toilet seat in a western style toilet. And we confirmed the accuracy of estimation of liquid flow rate using simulated urination.

Keywords: Urine volume · Flow estimation · Image processing · Multiple cylinder model

1 Introduction

When we measure our urinary volume, the urine is put into a measuring container such as a cup, and it takes time and effort to clean the surroundings and the used cup due to urine droplets. In addition, measurement of urine flow rate has a problem that urination is not ordinary because they are a little nervous when urinating in a special environment. It is said that measuring natural urinations in daily life is important for healthcare.

In the previous studies, various methods has been proposed to estimate the urinary volume without contact and not collecting urine in to the cup; temperature sensor [1], doppler sensor [2], image sensor [3]. In particular, Isomura et al. proposed the multiple cylindrical model for estimating urination volume using images [4]. Then, Honda et al. Improved it to get higher accuracy from the viewpoint of image processing [5]. Urination volume estimation using an image sensor is relatively accurate and stable. The image of urination has the potential to measure not only the amount and flow rate of urination but also the torsion form and color of urination. Therefore, urination image is expected to be widely applied.

Y. Shiraishi et al. (Eds.): APCMBE 2020, IFMBE Proceedings 82, pp. 67–73, 2021.
https://doi.org/10.1007/978-3-030-66169-4_10

In this study, we construct an environment for estimating urination volume in a western style toilet, and perform a flow estimation experiment for simulated urination in order to confirm the applicability of the multiple cylindrical model. In consideration of the geometric parallelism condition of the multiple cylindrical model, we propose a method of incorporating an image sensor using a raising toilet seat.

2 Estimation of Flow Rate Using Multiple Cylindrical Model

In the previous research [5], a flow estimation experiment has performed using an experimental device as shown in Fig. 1. The initial radiation velocity and diameter of the liquid are obtained from the extraction of the liquid area and the parabolic approximation of the streak line (Fig. 2(b)) for each image (Fig. 2(a)) of urine in the air taken in this environment. From Eq. (1), the urination volume in each frame is obtained by applying the multiple cylindrical model. The total urination volume is derived by summing individual volume from the start to the end of urination.

$$V = \pi I t \sum_{i=1}^{n} \left\{ \left(\frac{d_i}{2} \right)^2 v_{i0} \right\} \tag{1}$$

where, let n be the total number of frames, d_i be the diameter of the ith frame, v_{i0} be the initial velocity in the ith frame image, t be the frame time, I be a constant to convert from pixel to mL.

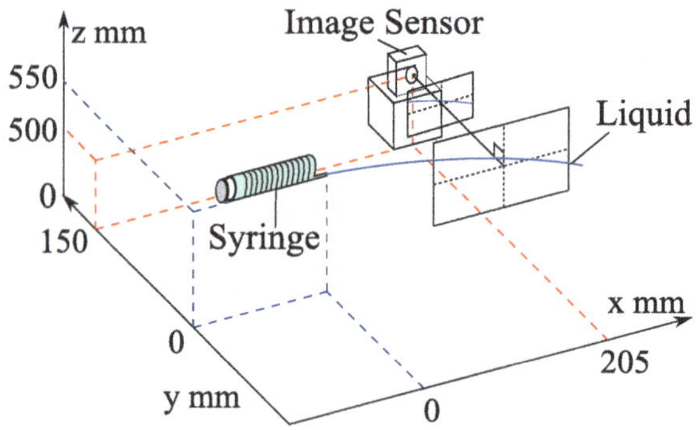

Fig. 1. Experimental environment for a previous study [5]

A method applying background subtraction is used for the extraction of the liquid area, and the parabola is robustly approximated using RANSAC [6]. Please refer to the previous studies' paper [4,5] for details of the multiple cylindrical model.

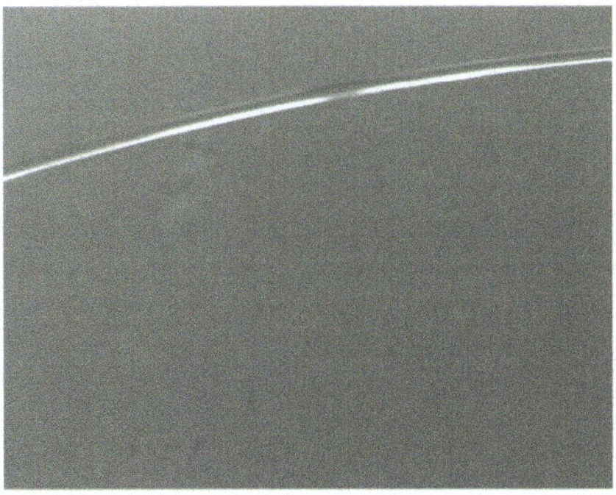

(a) Input image

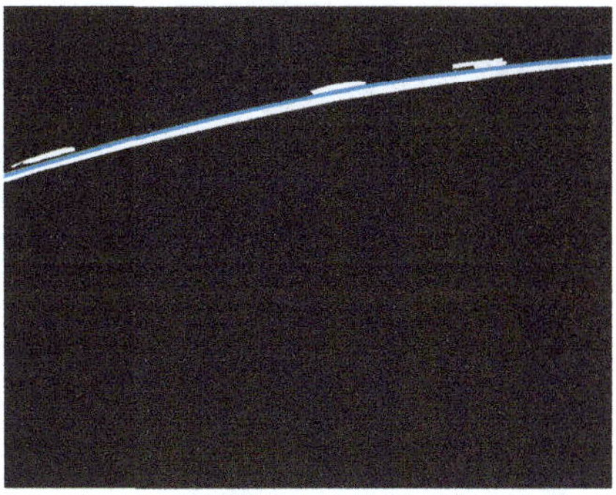

(b) Parabolic approximation

Fig. 2. Previous study's an input image and the parabolic approximation

A prerequisite for the multiple cylindrical model is geometric parallelism. This is a constraint on the positional relationship between the image sensor and urine that the plane of motion of the discharged liquid is parallel to the projection plane of the camera. Because of this, the initial velocity of the liquid is obtained from the parabolic approximation of the streak line in the image. This condition must be satisfied to incorporate a photographing device into the toilet.

3 Construction of Photographing Environment for Western Style Toilet

For example, even if an image sensor is built on the outside of the urinal for men, there is no problem because the clearance (movement space of urine) for taking urine in the air is wide. While, when a person is sitting on the western style toilet, it is not possible to shoot the movement space of urine from the outside of the western style toilet. An image sensor must be built inside the toilet seat to capture the movement space which is spread from the urethral opening to the toilet bowl surface.

In this study, from the viewpoint of flexibility of image sensor and wiring management, an image sensor is built in the raising toilet seat, because it is difficult to embed an image sensor in pottery toilet bowl. As a side effect, a larger movement space of urine can be obtained. The raising toilet seat is one of the toilet aids and nursing care tools shown in the Fig. 3(a). The higher seat reduces the burden on the knees and makes it easier to sit on the toilet seat and stand up.

As shown in the Fig. 4, an image sensor is installed on the side where liquid can be taken in parallel from inside the raising toilet seat. And it must be installed at a position where the area from the position of the male genitalia when sitting in the toilet to the front of toilet bowl can be photographed. A hole where the size is 20×10 mm was drilled inside the raised toilet seat for

(a) Raising toilet seat

(b) Image sensor

Fig. 3. Raising toilet seat and image sensor

photography. The image sensor built in the western style toilet was Raspberry Pi Camera Module V2, the resolution was 1920 × 1080 pix, and the frame rate was 30 fps. The position of the urethral orifice when an adult male of average physique sits on the toilet seat was 140 mm distance from the front of the western style toilet bowl and 90 mm height from the deepest part of the toilet bowl. As shown in Fig. 4, under the assumption that all urine would be emitted from this range, the optical center and optical axis of the sensor were adjusted to shoot the movement space of urine.

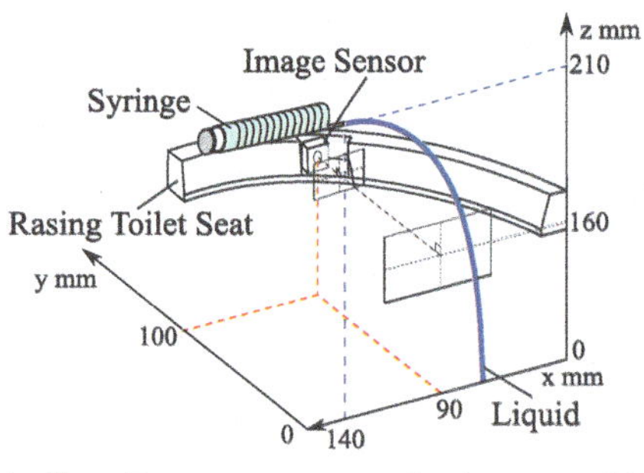

(a) Corrdinate system and photographing environment

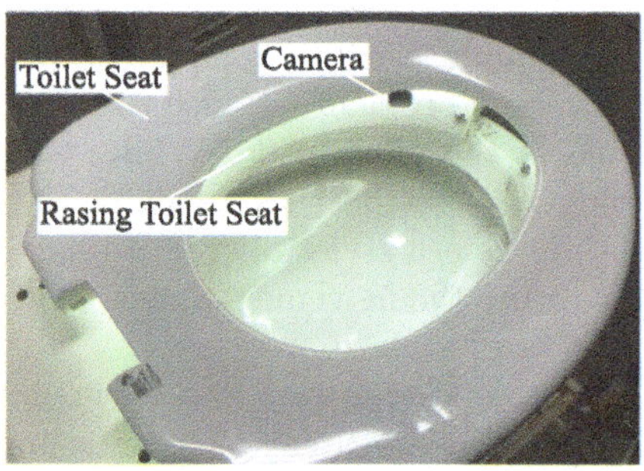

(b) Appearance picture

Fig. 4. Built-in an image sensor into a western style toilet

The inside of a western style toilet becomes dark when a person sits on the toilet seat. Then, the extraction of only a liquid region from a captured image is difficult because the exposure of the captured image is too low. Therefore, a tube-type LED was attached on the lower surface of the toilet seat in order to keep the brightness inside the western style toilet.

4 Experiment of Liquid Volume Estimation by Simulated Urination

We performed an evaluation experiment of the estimation accuracy by simulated urination in order to evaluate the validity of liquid volume estimation when an image sensor was built in a western style toilet. Urination was simulated using a syringe with a capacity of 50 mL and a diameter of 4 mm. The liquid to be discharged had a uniform momentum roughly because the cylinder was kept pressed with a constant force. Two types of lighting is used: fluorescent lights in the room and LEDs mounted on the lower surface of the toilet seat.

Five trials about 50 mL liquid were performed. The data from the first trial used to calculate the calibration of the multiple cylinder model. The accuracy was evaluated in the second to fifth trials. The absolute error E of the flow volume estimated for each trial is calculated.

Figure 5 shows a captured image and the result of processing the image. From this figure, it can be seen that the liquid area was extracted against the background of the raising toilet seat and the parabolic approximation was performed. Table 1 shows the estimated liquid volume V, the average initial velocity $\overline{v_0}$, and the average diameter $\overline{d}$ obtained by image processing for each trial. The average absolute error from #2 to #5 was 14.2 ± 5.2%. The absolute error of #3 was as small as 5.1%, while #2, #4 and #5 had errors exceeding 15%.

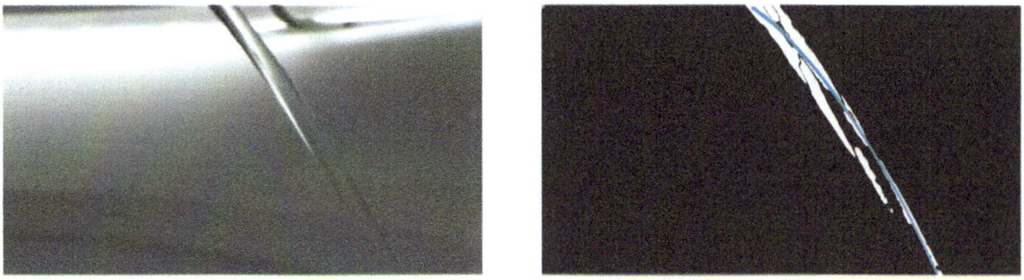

(a) Captured image (b) Parabolic approximation

Fig. 5. Our original study's an image captured in western style toilet and the processing parabolic approximation

Table 1. Accuracy of flow estimation of simulated urination

#	V [mL]	E [%]	$\overline{v_0} \pm SD$ [pix/s]	$\overline{d} \pm SD$ [pix]
2	41.9	16.2	1578.2 ± 860.2	112.3 ± 13.2
3	47.4	5.1	1849.2 ± 755.1	110.6 ± 16.9
4	58.8	17.6	1886.6 ± 638.1	124.5 ± 25.2
5	58.9	17.7	1940.8 ± 739.0	128.5 ± 15.7

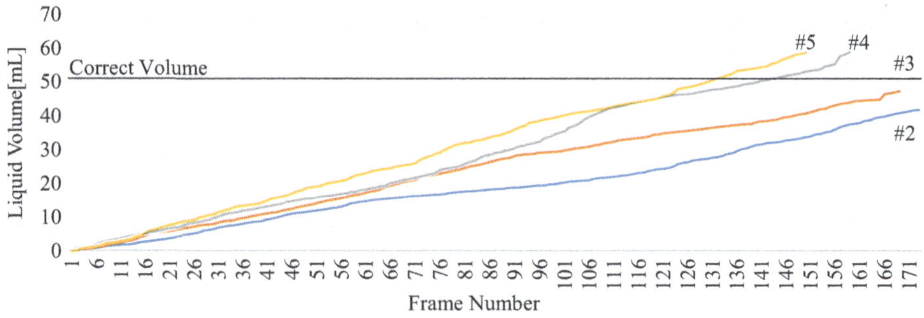

Fig. 6. Total volume of liquid up to each frame

Figure 6 shows the change in the total volume of liquid up to each frame. From this result, some frames of #4 and #5 were greatly affected by the parabolic approximation error. Especially in #4, the error continued to increase from the 95th to 110th frames due to the estimation error in the radius of the cylinder. In #2, the initial velocity of the liquid was estimated to be low between 60th and 120th frames.

The causes of these errors are the accuracy of the parabolic approximation and the diameter calculation. As shown in Fig. 7, as the liquid is discharged from the urethral opening and approaches the toilet bowl surface, the velocity of the liquid increases in the vertical direction under the influence of gravity. Then, in the image, the liquid streak line becomes a silhouette close to a straight line. As a result, the estimated liquid initial velocity becomes unstable, and causes in a large error.

In an ideal experimental environment of previous studies, it is easy to approximate with a parabola because the images could be captured immediately after discharging. However, in this study, due to the position of the image sensor, the image can only be captured after strongly affected by gravity as shown in Fig. 6. Therefore, it is necessary to improve the approximation accuracy of the parabola by expanding the photographing range of the image sensor so that the liquid area before being greatly affected by gravity can be captured.

(a) Captured image

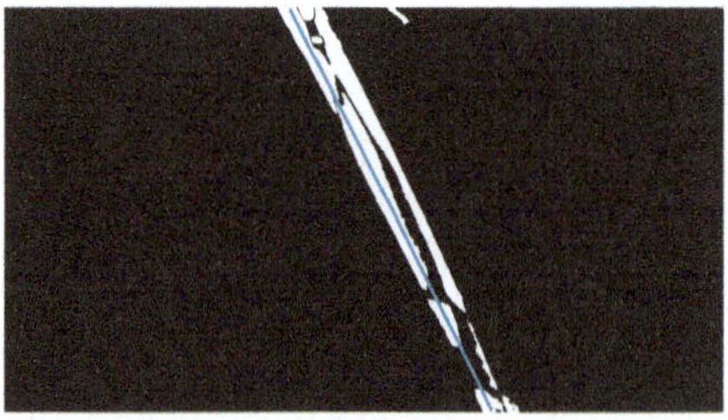

(b) Parabolic Aapproximation

Fig. 7. Large error occurred image after strongly affected by gravity and its parabolic approximation

Nevertheless, the estimated accuracy of the simulated urination in the previous study [5] was 15.4 ± 12.4%, and the accuracy resulted in our experiment was almost same with it.

5 Conclusions

An image sensor was installed at a position that satisfies the prerequisite of the multiple cylindrical model inside the raising toilet seat. As a result, in the flow rate estimation experiment using simulated urination was obtained an average absolute error of 14.2%. Although a large error occurred in several trials, the performance was almost the same as the accuracy of the previous study which was an ideal environment.

The future work is to improve the estimation accuracy of the quadratic equation by making it possible to photograph the range from liquid enters the toilet vowel area to it hits the toilet vowel surface to measure the movement space of urine without oversight. And, the experiment with actual urination is the next step.

References

1. Fujita, K., Kanayama, Y., Kim, J., Nakajima, K.: A preliminary study on a voided volume measuring method using noncontact temperature sensors under the toilet seat. Adv. Biomed. Eng. **8**, 1–6 (2019)
2. Matsumoto, S., Takeuchi, Y., Kakizaki, H.: Development and utility evaluation of new uroflowmetry device using wearable airborne ultrasound Doppler system. J. Urol. **189**(4), 806 (2013)
3. Weins, K., Green, S., Grecov, D.: Novel optical uroflowmeter using image processing techniques. Measurement **47**, 314–320 (2014)
4. Isomura, A., Bhuiyan, Md.S., Kawanaka, H., Watanabe, E., Oguri, K.: Estimating flow rate and volume of simulated urine flow noninvasively from a monocular camera. In: The 37th Annual International Conference of the IEEE Engineering in Medicine and Biology Society (2015)
5. Honda, C., Bhuiyan, Md.S., Kawanaka, H., Watanabe, E., Oguri, K.: Robust estimation of simulated urinary volume from camera images under bathroom illumination. In: The 38th Annual International Conference of the IEEE Engineering in Medicine and Biology Society, pp. 415–418 (2016)
6. Fischler, M.A., Bolles, R.C.: Random sample consensus: a paradigm for model fitting with applications to image analysis and automated cartography. Commun. ACM **24**(6), 381–395 (1981)

Diagonstic Evaluation

Effect of a Television Commercial on the Autonomic Nervous System

Tomoyuki Yambe[1(✉)], Yasuyuki Shiraishi[1], Yusuke Inoue[1], Akihiro Yamada[1], and Makoto Yoshizawa[2]

[1] Tohoku University, Sendai, Japan
yambe@tohoku.ac.jp
[2] Cyber Science Center, Tohoku University, Sendai, Japan

Abstract. As shown in cases of photosensitive epilepsy, audio-visual stimuli can have various effects on the human body. However, until now, there have been no reports investigating the effect of television commercials on the human body. In this study, for which we obtained ethical approval, we are the first ones to investigate the effect of television commercials on the human cardiovascular and autonomic nervous systems. Cardiovascular and autonomic nervous functions were evaluated using the human face blood flow analyzing system invented in Tohoku University (Jpn.Pt. 4789203, 5161671, 5390851, 5408751). No epilepsy and no patho-physiological reactions were observed after viewing of soft TV commercials including commercials from the advertising awards. Autonomic nerve responses were shown to be altered depending on the interests and preferences of the viewer. Our results suggest that assessing the response of the human body to TV commercials will become a key aspect of the process of creating TV commercials.

Keywords: Television commercial · Autonomic nervous system · Human face blood flow analyzing system · Heart rate variability · Sympathetic tone

1 Introduction

1.1 Importance of Autonomic Nervous System for Audio-Visual Stimulation

As shown in cases of photosensitive epilepsy [1, 2], audio-visual stimuli can have various effects on the human body [3, 4]. Commercials at the cinema, on television, and on the internet can be highly stimulatory for the human body. It should be recognized that television commercials (TVCMs) can have a strong effect on the autonomic nervous system. However, until now, there have been no reports studying the effect of TVCMs on the human body.

No one knows which level of stimulation is safe and which level is dangerous.

1.2 Application to the Television Commercial

This study is the first one in the world to evaluate the effect of TVCMs on the human cardiovascular and autonomic nervous systems [5–8].

Y. Shiraishi et al. (Eds.): APCMBE 2020, IFMBE Proceedings 82, pp. 77–82, 2021.
https://doi.org/10.1007/978-3-030-66169-4_11

This study received ethical approval. Cardiovascular and autonomic nervous functions were evaluated using the human face blood flow analyzing system invented in Tohoku University (Jpn.Pt. 4789203, 5161671, 5390851, 5408751). No epilepsy and no patho-physiological reactions were observed following viewing of soft TVCMs including commercials with content from the advertising awards. Autonomic nerve responses were shown to be altered depending on the viewers interests and preferences. Our results suggest that assessing the response of the human body to TV commercials will become a key aspect of the process of creating TV commercials.

2 Material and Methods

The purpose of TVCMs is to influence the viewers to buy products. An increasing effect of commercials on the human body, also increases the health risks they carry. Thus, the influence of TVCMs on the human body should be evaluated before they are broadcasted.

Until now, no previous reports have investigated the risks of TVCMs and, because of this, there is also no established method to evaluate the risks of TVCMs.

In this paper, a newly invented system that can evaluate responses in the human body, was applied in order to investigate the effects of TVCMs on the cardiovascular and autonomic nervous systems. This method is a multi-channel human face blood flow analyzing system. The multi-channel system means that both heart rate variability and pulse wave velocity could be evaluated (Fig. 1).

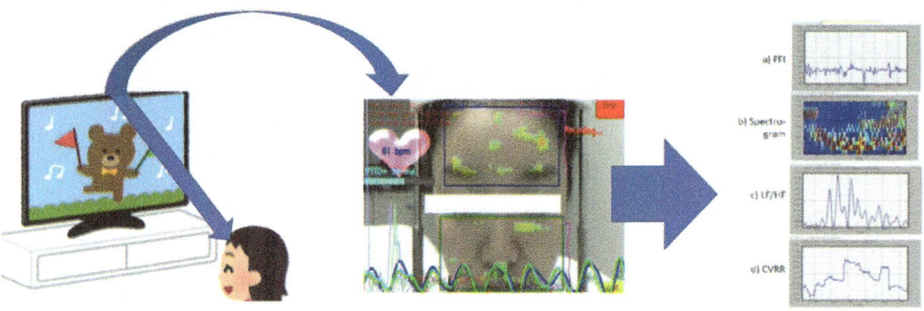

Fig. 1. Schematic illustration of the newly invented quantitative diagnosis system for the autonomic nerve activity during TV-CM immersion

As shown with cases of photosensitive epilepsy, audio-visual stimuli can have various effects on the human body. However, until now, there has been no research investigating the effects of TVCMs on the human body.

In order to evaluate human reactions when watching TVCMs, it is desirable to be able to evaluate indices reflecting autonomic nervous system function, such as LF/HF or CVRR, and cardiovascular function indices, such as variability in blood pressure, with a non-contact method. This is because measurement methods relying on direct contact with the subject can give a feeling of invasiveness.

To avoid this, we used video plethysmography (VPG) in our experiments. VPG is a non-contact method that works by detecting pulse waves in a remote manner using video obtained from video cameras.

Replacement, Reduction and Refinement are important principles that should be followed before clinical evaluation. Experimental animal models are important for evaluating a system before using them for clinical application. Thus, once we obtained the appropriate ethical approval, we used animals to obtain an accurate time series data using highly sensitive sensors. We used the property that hemoglobin in the blood absorbs green light and obtained a time series of the averaged value of the green component of a color video signal taken from the skin portion.

The VPG used in these experiments has multi-channel pulse waves, and so, the phase difference between the two VPG signals, taken at the proximal and the distal parts of the body based on the position of the heart or the distortion time of the shape of the pulse wave were used to obtain information correlated with blood pressure and stiffness of an artery.

After testing the feasibility of the multi-channel pulse wave recording on animal models, we then tested it on humans to assess its potential for clinical application.

We used automatic techniques to select the target region of interest (ROI) and thus simplify the troublesome procedure. In this way we were also able to determine the validity of the proposed system for automatically detecting and separating the face and hand, and tracking the face in an online and real-time fashion.

The newly invented system uses the phase difference between two VPGs, taken at the proximal and distal parts of the body and based on the position of the heart to obtain the information correlated with blood pressure and tension level of an arterial wall.

3 Results and Discussion

Before investigating its clinical application, it was important to first test the feasibility of the study using animals models.Before the clinical application, fusibility study with animals were important (Fig. 2).

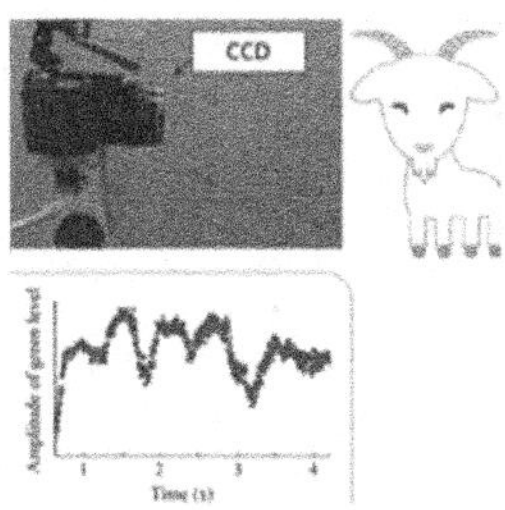

Fig. 2. Experimental model of the pulse wave detector using a goat

After carrying out the feasibility studies, we evaluated the effect of the television commercial on the cardiovascular and autonomic nervous system of human volunteers. Our study received the necessary ethical approval.

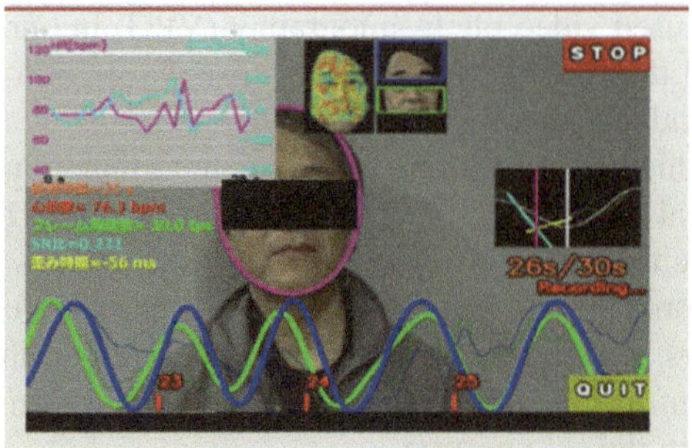

Fig. 3. Example of the video image pulse wave series recording

An example of the experiment, showing a volunteer during audio visual stimulation, is shown in Fig. 3.

During experiments multi-channel recording was initiated after the ROI was set to a partial skin area of the face.

Based on the successful incorporation of multi-channel pulse wave detection, clinical examination to diagnose the effects of TVCM viewing on the cardiovascular and autonomic nerve was initiated (Fig. 4).

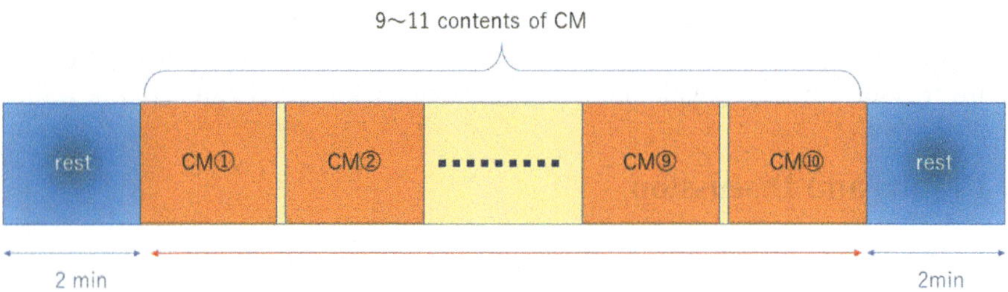

Fig. 4. Experimental protocol for TV-CM immersion

In this study, we used nine to eleven sets of TVCMs obtained from the internet. After obtaining informed consent, healthy normal volunteers viewed the TVCMs. The volunteers had resting time both before and after TVCM viewing. Cardiovascular and autonomic nervous function were evaluated with the human face blood flow analyzing system invented in Tohoku University.

Results showed that no epilepsy and no pathophysiological reactions in the cardiovascular systems were observed during viewing of a soft TV commercial which included contents from the advertising awards. Autonomic nerve responses were altered depending on the interests and preferences of the viewer.

The more the person is interested in the audio-visual content of the TVCM the more their body responses are influenced. If the response of the human body is too extreme, it could become a pathological response which could be harmful to the human body and particularly in elderly people and children.

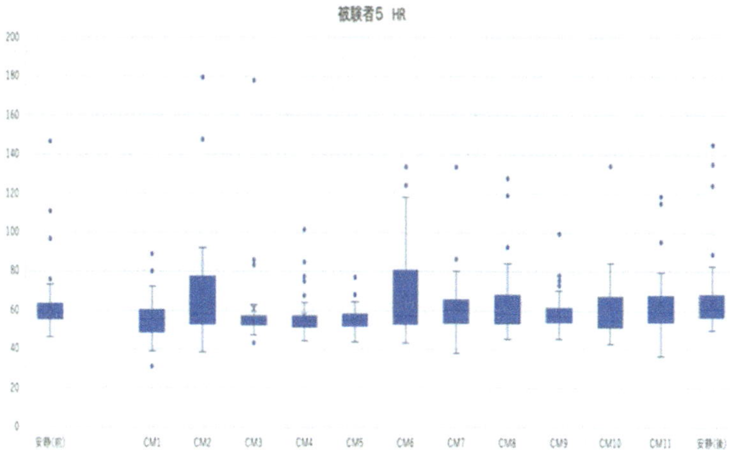

Fig. 5. Heart rate responses to TVCM viewing detected from the volunteer's faces

When considering autonomic nerve responses it is also important to evaluate heart response (HR). Figure 5 shows HRs for every TVCM. In some types of CM, HR was high. Unfortunately, pulse wave detection was sometimes missed. Almost tendency of the HR responses was diagnosed well when we consider the average.

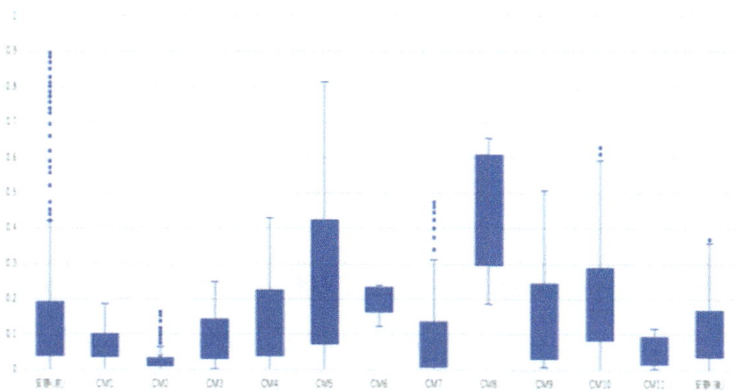

Fig. 6. LF/HF during various TVCM immersion

Previous studies have reported various methodologies to monitor the autonomic nervous system. Figure 6 shows LF/HF, which has been reported to reflect autonomic function. Various values were recorded with our method. Overall these findings suggest that it may be possible to use this technique to determine whether TVCMs are good or bad by analyzing medical responses.

In summary, we have been the first to evaluate the effect of TVCMs on the human cardiovascular and autonomic nervous systems. We did so by using the human face blood flow analyzing system invented in Tohoku University. No epilepsy and no pathophysiological reactions were observed following viewing of soft TV commercials including the contents from the advertising awards. Autonomic nerve responses were altered depending on the interests and preferences of the viewer. Our results suggest that assessing the response of the human body to TV commercials will become a key aspect of the process of creating TV commercials.

Acknowledgment. This work was supported in part from the Yoshda Hideo Memorial Foundation, Paroma Foundation, SCAT foundation, Miyagi Yobouigaku Association, Mitusi-Sumitomo Foundation and Japanese Foundation of promotion and research of Endoscopy.

References

1. Aso, K., Watanabe, K., Negoro, T., Takaesu, E., Furune, A., Takahashi, I., Yamamoto, N., Nomura, K.: Visual seizures in children.Epilepsy Res. **1**(4), 246–253 (1987)
2. Aso, K., Watanabe, K., Negoro, T., Haga, Y., Kito, M., Maeda, N., Ohki, T.: Photosensitive epilepsy in children. Seizure. **3**(1), 67–71 (1994)
3. Carruthers, M., Taggart, P., Somerville, W.: The heart's response to the portrayal of violence. Med Sci Law. **13**(4), 252–255 (1973)
4. Carruthers, M., Taggart, P.: Vagotonicity of violence: biochemical and cardiac responses to violent films and television programmes. Br. Med. J. **3**(5876), 384–389 (1973)
5. Sugita, N., Yoshizawa, M., Tanaka, A., Abe, K., Chiba, S., Yambe, T., Nitta, S.: Evaluation of the effect of visual stimulation on humans by simultaneous experiment with multiple subjects. In: Conference on Proceedings of the IEEE Engineering Medical Biology Social, vol. 5, pp. 5572-–5575 (2005)
6. Sugita, N., Yoshizawa, M., Tanaka, A., Abe, K., Chiba, S., Yambe, T., Nitta, S.: Quantitative evaluation of the effect of visually-induced motion sickness using causal coherence function between blood pressure and heart rate. In: Conference on Proceedings of the IEEE Engineering Medical Biology Social, vol. 4, pp. 2407–2410 (2004)
7. Yambe, T., Yoshizawa, M., Fukudo, S., Fukuda, H., Kawashima, R., Shizuka, K., Nanka, S., Tanaka, A., Abe, K., Shouji, T., Hongo, M., Tabayashi, K., Nitta, S.: Can personality traits predict pathological responses to audiovisual stimulation? Biomed Pharmacother. **57**(Suppl 1), 83s–86s (2003)
8. Yoshizawa, M., Sugita, N., Abe, M., Tanaka, A., Obara, K., Yamauchi, T., Homma, N., Yambe, T.: Blood perfusion display based on video pulse wave. In: Conference on Proceedings of the IEEE Engineering Medical Biology Social, pp. 4763–4767, August 2016

Modeling of Pupillometric Signals for Studying Children's Rare Diseases

Rachele Fabbri[1], Ernesto Iadanza[1(✉)], Monica Gherardelli[1], Paolo Melillo[2], Francesca Simonelli[2], and Leonardo Bocchi[1]

[1] Department of Information Engineering, University of Florence, Via S. Marta 3, 50139 Florence, Italy
{ernesto.iadanza,leonardo.bocchi}@unifi.it

[2] Eye Clinic, Multidisciplinary Department of Medical, Surgical and Dental Sciences, University of Campania Luigi Vanvitelli, 80100 Naples, Italy

Abstract. The use of chromatic pupil responses may be a novel way to diagnose and monitor diseases affecting either the outer or inner retina.

Here we developed a model-based approximation of the dynamics of pupillary diameter after an optical stimulation; we assume the response of the pupil is approximated as the output of a 2nd-order linear model. Model parameters are identified by using a least-squares fitting procedure, thus obtaining an optimal estimate of the activation time and of the shape of the pupil response to the stimulation.

Results indicate the model adequately represents the curve shape; despite the presence of artifacts, that hinder the fitting procedure, a significant difference in the time constants of the model of controls and of patients is present.

Keywords: Computational geometry · Graph theory · Hamilton cycles

1 Introduction

The work presented in this paper is framed within a research project aimed at using chromatic pupillometry for an early diagnosis and monitoring of rare genetic diseases in children. Inherited Retinal Diseases (IRDs) represent a significant cause of severe visual deficits in children [1] and can be divided in degenerations of photoreceptor (diseases of the outer retina) and retinal ganglion cell degenerations (diseases of the inner retina). Retinitis Pigmentosa (RP) falls in the former group. Because the clinical evaluation of IRDs is routinely implying the adoption of invasive clinical tests, sometimes requiring complex settings, the clinical diagnosis is not easy and requires specialized centers. The prolonged melanopsin-mediated constriction has been used in clinical protocols to assess inner-retina function [2–4]. Thus, the use of chromatic pupil responses may be a novel way to diagnose and monitor diseases affecting either the outer or inner retina. One of the results of the above-mentioned project, involving three

Y. Shiraishi et al. (Eds.): APCMBE 2020, IFMBE Proceedings 82, pp. 83–90, 2021.
https://doi.org/10.1007/978-3-030-66169-4_12

operative units in northern, central and southern Italy, was a novel Computerized Decision Support System (CDSS) for the diagnosis of RP in pediatric subjects. Artificial intelligence has been gaining a growing role in ophthalmology in the last years [5,6]. The unit at the Department of Information Engineering of the University of Florence, exploiting the experience gained in the recent years in designing RESTful [7] CDSSs integrating ML capabilities [8–11], has been in charge of designing and deploying the cloud-based CDSS for this project [12,13]. It is composed of an Electronic Medical Record (EMR) web-application [14] and of a Machine Learning (ML) custom set of Application Program Interfaces (APIs) performing signals' filtering and classification [15]. The communication between the EMR and the ML APIs is RESTful.

2 Materials and Methods

2.1 Pupillometric Data

The dataset used in this work comes from the database of the EMR developed to collect the ophthalmological and pupillometric data. It consists of 30 subjects affected by RP, labeled as *Patients* and 30 healthy subjects, labeled as *Controls*. The pupillometric analysis has been performed for each subject and the samples of the pupillary diameter trend over time have been output in a .DAT file. The protocol for the acquisition of these data is fully described in [12,14,15]. The response of the pupil to each light stimulation has been filtered, using a FIR filter as described in [15], in order to remove the noise overlapping the useful signal. For each of the 60 subjects, 20 acquisitions are available both for the right and left eye. A total amount of 2400 signals of the pupil contraction has been considered in this modeling analysis.

2.2 Signal Model

The exact dynamics of the pupil diameter, depending on the physiological connections and the nervous networks, are not fully described by a mathematical model. However, the experimental shape of the curve resembles the output of a linear model, suggesting that a linear approximation of the system could yield reasonable results. Thus, we adopted a simplified model that is, however, able to replicate the observed shape of the pupillary reaction to a stimulus; the most simple model that is able to realize such a shape is a second order linear time invariant model. The impulse response of this family of models, produced by an impulse input at time $t = t_0$, can be expressed as:

$$f(t) = \begin{cases} k - A\left(e^{-\frac{t-t_0}{\tau_d}} - e^{-\frac{t-t_0}{\tau_r}}\right) & \text{if } t > t_0 \\ k & \text{otherwise} \end{cases} \tag{1}$$

where the parameters of the model are: k, representing the pupillar diameter at equilibrium; t_0, representing the starting time of the response (possibly delayed with respect to the stimulation); A, related to the maximum diameter change; τ_d, the time constant of the contraction phase; and τ_r, the time constant of the dilation (return to resting diameter) phase.

2.3 Parameter Estimation

A simple but effective procedure [16], shown in Fig. 1, allows to estimate suitable values $(\hat{k}, \hat{t}_0, \hat{A}, \hat{\tau}_d, \hat{\tau}_r)$ of the model parameters from the signal $s(t)$ in order to obtain a good convergence of the Least Squares (LS) algorithm. The procedure starts by removing possible outliers (bottom right in figure); the first fiduciary point corresponds to the point with minimum diameter that is labelled as (t_m, s_m). The next fiduciary point is located in correspondence with the maximum pupillary diameter, occurring before t_m, that is labelled as (t_M, s_M). The maximum value s_M is assumed as initial estimate for the constant $\hat{k} = s_M$, corresponding to the rest value of the pupillary diameter, both before and after the stimulation. The difference $\hat{A} = s_M - s_m$ is used as an estimate of the amplitude A. The halving times t_d and t_r represent the time instants where $s(t) = s_m + (s_M - s_m)/2$, respectively during contraction and dilation phases. The halving times allow to estimate the two time-constants: $\hat{\tau}_d = ln(2) * (t_m - t_r)$ and $\hat{\tau}_r = ln(2)(t_r - t_m)$. Lastly, the start of the contraction $\hat{t}_0$ is estimated as the point where $s(t) = s_M - 0.2(s_M - s_m)$. As shown in the figure, in some cases several signal points satisfy the above requirements (in the example, several candidates exist for t_r); as the determination of the initial estimate is not critical, when several points are available, heuristics are used to perform a selection among points that satisfy the condition. In the example, the point closest to t_m is selected.

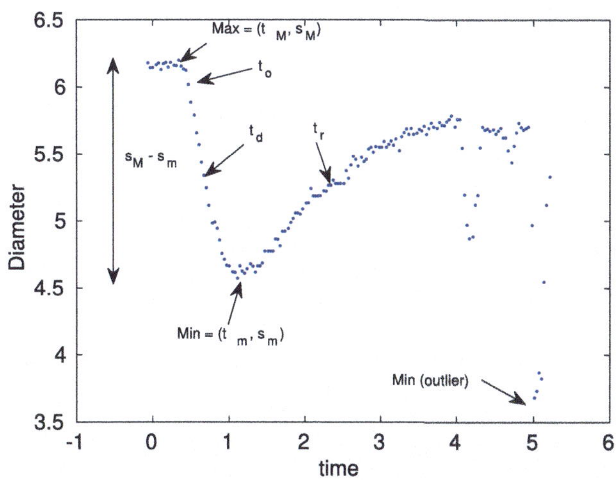

Fig. 1. Parameter estimation procedure.

2.4 Fitting and Optimization

The parameter estimation procedure, described in the previous section, provides a suitable starting point for further optimization, in order to obtain the optimal fitting. As shown in Fig. 2, the model output resulting from the initial estimate is reasonably close to the signal itself, but it does not follow exactly the curve shape. Thus, a fitting procedure optimizes the parameter values by matching the model output with the signal.

The procedure consists in a Least Squares optimization, that tunes the vector $\tilde{p}$ of model parameters to their optimal values ($\tilde{p} = \{\tilde{k}, \tilde{t}_0, \tilde{A}, \tilde{\tau}_d, \tilde{\tau}_r\}$).

The variability ranges imposed on the parameters are defined as follows: the range of the parameter t_0 equals to half of the signal duration, while the variability range of the other parameters spans from a tenth to ten times of the estimated value.

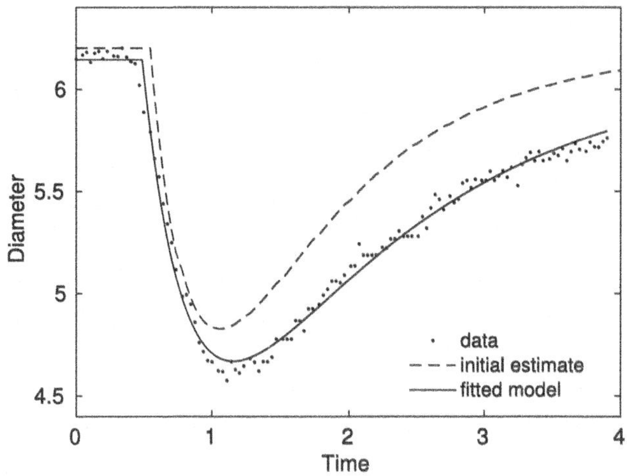

Fig. 2. Sample signal, model resulting from parameter estimation and after Least Squares fitting

The pupil response to each stimulation included in the previous-mentioned pupillometric protocol has been independently fitted with the model and the Root Mean Square Error (RMSE) has been analysed as a metric of the goodness of the fit.

The procedure allows to identify each stimulus-response pair with a vector $\tilde{p}$; therefore, from each measure, we obtained 40 vectors corresponding to the 20 stimulation for each eye.

3 Results

A preliminary validation allows to identify the cases where the fitted procedure fails to properly identify the model. Visual inspection of the failures indicates that they are related to noise in the input data. Indeed, since the paediatric patients are subject to discomfort during the whole acquisition process, both because of the several stimulations and of the duration of the procedure), the pupillary diameter signal shows several artefacts that can be visually identified as eye blinks, head movements, and similar patient-related effects.

In this preliminary study, we introduced a selection step in order to avoid including bad estimates in our analysis.

The selection step is based on the evaluation of the fitting parameters, and of the termination condition of the LS minimization. In detail, an analysis of the histogram of the model parameters, and of the RMSE after the minimization step suggested to use a threshold value of $e_t = 1.5$ for discriminate acceptable fitting results: all the signals fitting with a RMSE higher than the threshold have been excluded from the final analysis; in addition, we also eliminated those samples where one of the model parameters hit one of the boundaries of the acceptable domain. In particular, when an artifact is present, the estimated $\tilde{t}_0$ assumes unreasonable high values, as the model fits the eye-blinking pattern, that appears in the curve as a sudden decrease in the estimated pupillary diameter.

The selection process extracts about 1400 valid models from the full data set, thus we have a significant part of samples (evenly distributed between controls and patients) that were not correctly processed.

In this phase, we focused the analysis on the time constants characterizing the model, namely $t_d = 1/c$ (time constant of the contraction phase) and $t_r = 1/b$ (time constant of the relaxation phase). The distribution of both parameters, in form of box plots, is shown in Figs. 3 and 4, respectively. In both cases we observe a clear variation of the distribution; also, mainly in the second figure, a larger number of outliers is present. Note that the distribution contains some more outliers, with unreasonably large values, that are not visible with the selected scale.

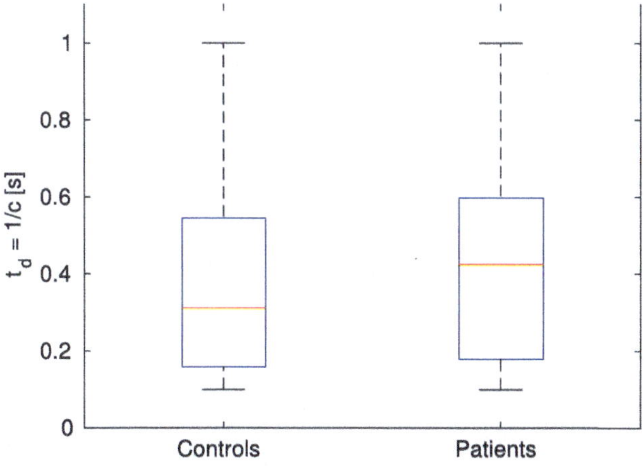

Fig. 3. Distribution of the parameter $t_d = 1/c$.

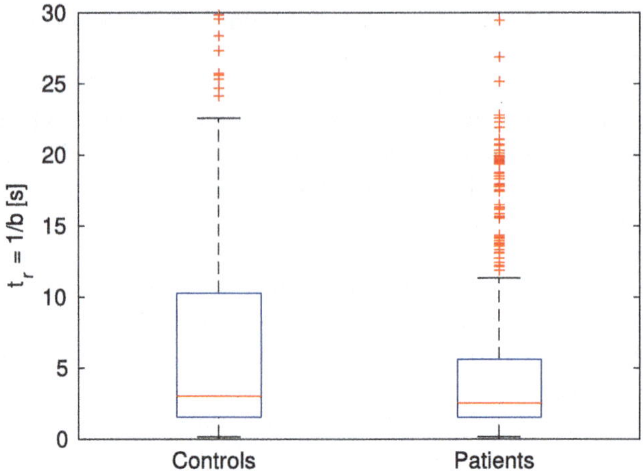

Fig. 4. Distribution of the parameter $t_s = 1/b$. A few more outliers are present for $t_s > 30$; the graphic has been rescaled for clarity

Visual inspection indicates that also in such cases the fitting is not satisfactory, although it satisfies our selection criterion.

The analysis of the statistical significance of the differences appearing in the plots confirms that a significant difference exists between values of t_d in controls and patients ($p < 0.0001$).

On the contrary, probably due to the presence of outliers, no statistically significant difference is present between t_s in patients and in controls.

4 Discussion and Conclusions

Results indicate the model adequately approximates the shape of the observed curve. However, more studies will be performed for testing different models that might further improve the quality of the fit. Also, the large number of fitting errors indicates the robustness of the procedure is not optimal. Therefore, it will be necessary to introduce a refinement step aimed at detecting noisy samples, in order to improve the quality of the fit in those cases.

Nevertheless, the procedure proves capable of detecting and describing, in terms of parameters with physiological meaning, the eye response to the stimulation. Also, the parameters values are in accordance with the expected physiological behavior.

References

1. Huang, X.F., Huang, F., Wu, K.C., Wu, J., Chen, J., Pang, C.P., Lu, F., Qu, J., Jin, Z.B.: Genotype–phenotype correlation and mutation spectrum in a large cohort of patients with inherited retinal dystrophy revealed by next-generation sequencing. Genet. Med. **17**(4), 271–278 (2015)
2. Kawasaki, A., Crippa, S.V., Kardon, R., Leon, L., Hamel, C.: Characterization of pupil responses to blue and red light stimuli in autosomal dominant retinitis pigmentosa due to NR2E3 mutation. Invest. Ophthalmol. Vis. Sci. **53**(9), 5562–5569 (2012). https://doi.org/10.1167/iovs.12-10230
3. Kawasaki, A., Munier, F.L., Leon, L., Kardon, R.H.: Pupillometric quantification of residual rod and cone activity in Leber congenital amaurosis. Arch. Ophthalmol. **130**(6), 798–800 (2012). https://doi.org/10.1001/archophthalmol.2011.1756
4. Kawasaki, A., Collomb, S., Léon, L., Münch, M.: Pupil responses derived from outer and inner retinal photoreception are normal in patients with hereditary optic neuropathy. Exp. Eye Res. **120**, 161–166 (2014). https://doi.org/10.1016/j.exer.2013.11.005
5. Grzybowski, A., Brona, P., Lim, G., Ruamviboonsuk, P., Tan, G., Abramoff, M., Ting, D.: Artificial intelligence for diabetic retinopathy screening: a review. Eye **34**, 1–10 (2019)
6. Hogarty, D.T., Mackey, D.A., Hewitt, A.W.: Current state and future prospects of artificial intelligence in ophthalmology: a review. Clin. Exp. Ophthalmol. **47**(1), 128–139 (2019)
7. Bianchi, L., Paganelli, F., Pettenati, M.C., Turchi, S., Ciofi, L., Iadanza, E., Giuli, D.: Design of a RESTful Web information system for drug prescription and administration. IEEE J. Biomed. Health Inform. **18**(3), 885–895 (2013). https://doi.org/10.1109/JBHI.2013.2282827
8. Guidi, G., Pollonini, L., Dacso, C., Iadanza, E.: A multi-layer monitoring system for clinical management of Congestive Heart Failure. BMC Med. Inform. Decis. Mak. **15**(3), 1–14 (2015). https://doi.org/10.1186/1472-6947-15-S3-S5
9. Guidi, G., Melillo, P., Pettenati, M.C., Milli, M., Iadanza, E.: Performance assessment of a clinical decision support system for analysis of heart failure, vol. 41, pp. 1354–1357 (2014). https://doi.org/10.1007/978-3-319-00846-2_335
10. Martini, L., Vannetti, F., Fabbri, L., Gerli, F., Mosca, I., Pazzi, S., Baglio, F., Bocchi, L., Sorbi, S., Macchi, C., Migliazza, T., Stoppini, A., Lucidi, G., Savazzi, F., Group, G.P.: GOAL (games for olders active life): a Web-application for cognitive impairment tele-rehabilitation. In: IFMBE Proceedings, vol. 68, pp. 177–182 (2019). https://doi.org/10.1007/978-981-10-9023-3_32
11. Bocchi, L., Orlandi, S., Manfredi, C., Puopolo, M., Guzzetta, A., Vicari, S., Scattoni, M.: Early diagnosis of autism spectrum disorders - design of the data acquisition and management system. In: IFMBE Proceedings, vol. 37, pp. 187–190 (2011). https://doi.org/10.1007/978-3-642-23508-5_50
12. Iadanza, E., Fabbri, R., Luschi, A., Gavazzi, F., Melillo, P., Simonelli, F., Gherardelli, M.: ORÁO: RESTful cloud-based ophthalmologic medical record for chromatic pupillometry. In: IFMBE Proceedings, vol. 73, pp. 713–720. Springer (2020)
13. Romero-Aroca, P., Valls, A., Moreno, A., Sagarra-Alamo, R., Basora-Gallisa, J., Saleh, E., Baget-Bernaldiz, M., Puig, D.: A clinical decision support system for diabetic retinopathy screening: creating a clinical support application. Telemed. e-Health **25**(1), 31–40 (2019)

14. Iadanza, E., Fabbri, R., Luschi, A., Melillo, P., Simonelli, F.: A collaborative RESTful cloud-based tool for management of chromatic pupillometry in a clinical trial. Health Technol. **10**, 1–14 (2019)
15. Iadanza, E., Goretti, F., Sorelli, M., Melillo, P., Pecchia, L., Simonelli, F., Gherardelli, M.: Automatic detection of genetic diseases in pediatric age using pupillometry. IEEE Access, 1 (2020, in press). https://doi.org/10.1109/ACCESS.2020.2973747
16. Tarchi, P., Bologna, F., Scortecci, C., Tiribilli, E., Pennati, D., Sacconi, L., Sorelli, M., Bocchi, L.: Modeling of voltage imaging for the study of action potential propagation. In: IFMBE Proceedings, vol. 73, pp. 309–314 (2020). https://doi.org/10.1007/978-3-030-17971-7_47

Current Technology in Prosthesis, Orthosis and Rehabilitation

Introduction of Prosthesis and Orthosis Evaluation System that Utilizes Human Models

Akihiko Hanafusa[1(✉)], Mohd Syahmi Jamaludin[1], Le Van Tuan[2], Hayato Ikebata[1], Ryogo Suzuki[1], Taiki Kawamura[1], Shin-ichiro Yamamoto[1], Yukio Agarie[3], Hiroshi Otsuka[4], and Kengo Ohnishi[5]

[1] Shibaura Institute of Technology, Saitama 337-8570, Japan
hanafusa@shibaura-it.ac.jp
[2] Ha Noi University of Science and Technology, Ha Noi, Viet Nam
[3] Niigata University of Health and Welfare, Niigata 950-3198, Japan
[4] Arts and Sciences, University of Human, Saitama 339-8555, Japan
[5] Tokyo Denki University, Saitama 350-0394, Japan

Abstract. The fit of prostheses and orthoses is important, because, if the fit is inadequate, patients may feel uncomfortable. A poor fit can cause difficulty in gait, and, in the worst cases, it can cause inflammation. An evaluation system for checking the fit that utilizes human models has been developed. Human models of the residual limb and lower thigh with bone, muscle, and fat were composed. These models were used to evaluate the fit of the transfemoral socket and ankle foot orthosis (AFO) by dynamic finite-element analysis (FEA). For the transfemoral socket, insertion simulation was performed, and the validity of the fat part and pressure distribution of the residual limb were evaluated using magnetic resonance imaging and the force measured by sensors built in the socket. For the AFO, gait simulation of the stance phase was performed, and the transition of strain was compared with the measured result by the 3D-printed AFO, on which strain gauges were pasted. In addition, by performing topological optimization, the weight of the AFO can be reduced by maintaining the stiffness of the AFO. The results indicate the qualitative validity of dynamic FEA for analyzing the prosthesis socket and AFO using a human leg model.

Keywords: Prosthesis · Orthosis · Human model · Dynamic finite element analysis

1 Introduction

Prostheses and orthoses should be fitted to individual persons. Otherwise, patients who use devices for the lower limb may feel discomfort or have difficulty in gait. In the worst cases, a poor fit can cause subsequent complications, such as inflammation. However, most of these devices are handmade by prosthetists, and the fit de-pends on their skill and experience. In recent years, new technologies, such as computer-aided design and computer-aided manufacturing systems and three-dimensional (3D) printers, have

Y. Shiraishi et al. (Eds.): APCMBE 2020, IFMBE Proceedings 82, pp. 93–101, 2021.
https://doi.org/10.1007/978-3-030-66169-4_13

been adopted for manufacturing. However, the function of the system is not sufficient, especially in the field of fitness evaluation.

In this research, a simulation system is being developed that can be used for evaluating how well prostheses and orthoses fit. In this system, human models with bones and soft tissues are composed in the computer, and how well the prosthesis or orthosis fits when the patient wears it is evaluated by the dynamic finite-element analysis (FEA) method. Currently, two systems are being developed for transfemoral prosthesis [1–3] and ankle foot orthosis (AFO) [4, 5].

2 Evaluation System for a Transfemoral Prosthesis

2.1 Method

The shape of the socket of a transfemoral prosthesis is very important, because it is an interface between the prosthesis and the human leg. The shape is made smaller than that of the residual limb, because it should be adsorbed. If the fit is inadequate, it may cause pain and, even worse, inflammation. The human models of the residual limb of the patients were obtained from magnetic resonance imaging (MRI) data. Each layer of the image every 5 mm was segmented into bone, muscle, fat, and parts. From the layers of data, 3D components were built and divided into finite tetrahedron elements. The socket was also divided into triangular shell elements. The bone model was treated as rigid; however, muscle and fat are defined as viscoelastic material, and fat is defined as the softest material. Because the socket is made of acrylic plastic material, the model is defined by the same material. The insertion simulation was performed using the dynamic FEA software LS-DYNA (LSTC Corp.) by applying 50% of the body weight to the top of the bone (Fig. 1).

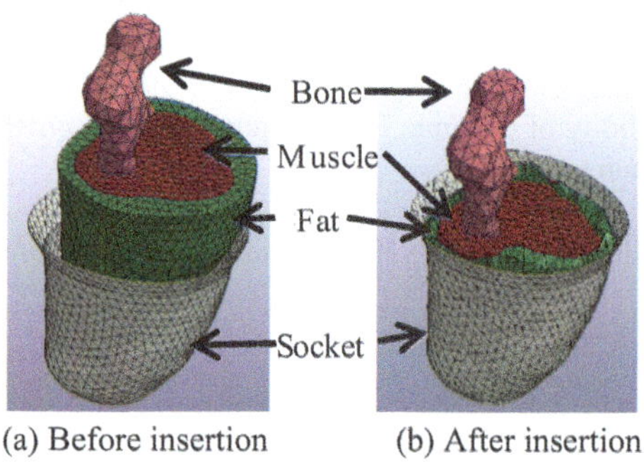

Fig. 1. Composed finite-element model of residual with bone, muscle, fat, and socket: (a) before insertion and (b) after insertion

2.2 Results

The inside distribution of fat and muscle can be estimated by simulation. The differences in the area of fat with the MRI image in each layer are shown in Fig. 2. The results before and after insertion are shown. The average errors before and after the insertion were 2.8% and 7.8%, respectively, and the error in the distal area was still large after insertion. This is because the boundaries of the muscle and skin are fixed and do not slip enough in the simulation.

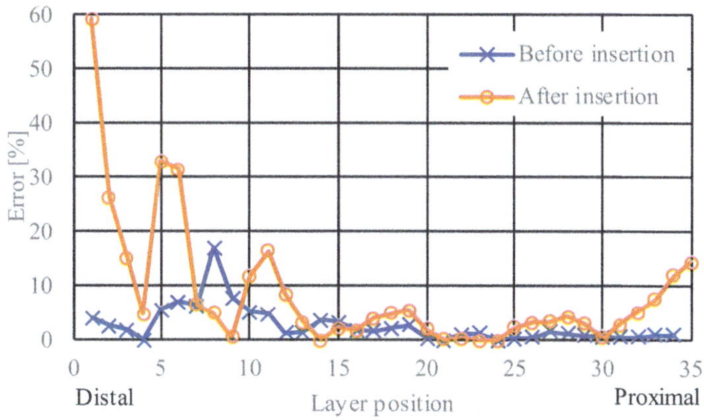

Fig. 2. Distribution of error of fat area in each layer before and after the insertion simulation.

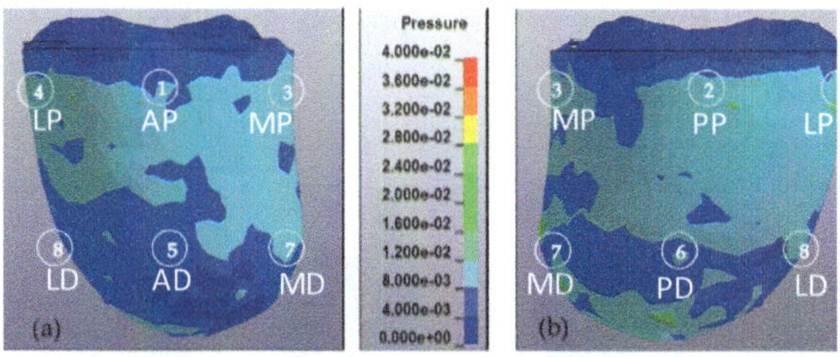

(a) Simulated pressure distribu-tion around the residual limb

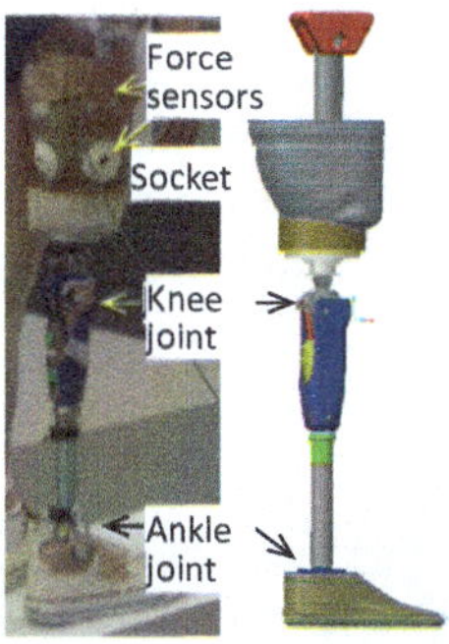

(b) Prosthetic socket with the contact force sensors

Fig. 3. Simulated distribution and measuring system of pressure around residual limb.

In addition, the internal pressure distribution could be simulated, and the result is shown in Fig. 3(a). A system was also developed that can measure the contact force between the residual limb and the socket on which eight triaxial force sensors (NITTA PD 3–32-05–015) were buried, as in Fig. 3(b). The positions of the sensors were the proximal part of the anterior (AP), posterior (PP), medial (MP), lateral (LP) and the distal part of the anterior (AD), posterior (PD), media l(MD), and lateral (LD) positions. The socket was worn by the subject, and the pressure applied to the sensor was measured while the subject was standing on both legs. Simulated results of the pressure on the

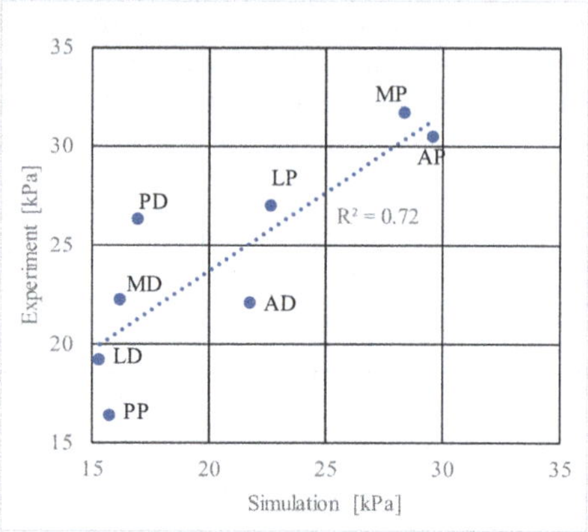

Fig. 4. Relationship between measured and simulated pressure at the position of sensors.

same position where the sensors were mounted were compared with the measured results (Fig. 4). The correlation ratio between the simulated and measured values was 0.72, and relatively high values were obtained.

2.3 Gait Simulation

The pressure distribution during gait is also important, and simulation with the transfemoral prosthesis was attempted. Currently, the evaluation system is under development, and the simulated floor reaction force is compared with the measured result [1]. To evaluate the pressure of the residual limb, the insertion process should be included in the gait simulation (Fig. 5).

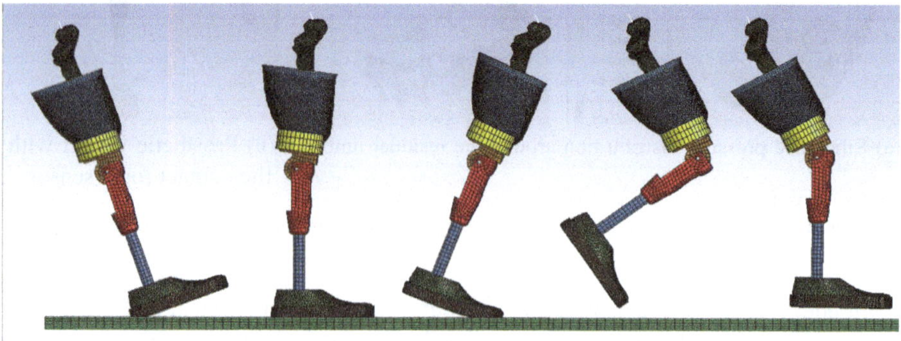

Fig. 5. Trajectory during the gait simulation with prosthesis.

3 Evaluation System for AFO

3.1 Method

The shape of the AFO is also important, because the fit affects the comfort and gate of the patients. In addition, new attempts to produce the AFO using a 3D printer, by which more-flexible shapes can be produced, are being considered. Therefore, evaluating the 3D shape is becoming essential. In this research, an attempt is being made to simulate the dynamic state of the lower thigh and AFO using dynamic FEA. LS-DYNA (LSTC Corp.) was used for the analysis.

A three-dimensional lower-thigh model was made from MRI data and was com-posed of bone, muscle, and fat data, as shown in Fig. 6(a). The AFO model with two belts was modeled by shell elements. The bone models were composed of 13 parts, as shown in Fig. 6(b). To realize the rotational movement of the ankle joint, five metacarpophalangeal joints were defined. To simulate the gait during the single-stance phase, human weight was applied to the position of the center of mass of the human body, and an initial walking speed of 0.85 m/s was also applied, as shown in Fig. 6(c). Currently, rotations around the Z- and X-axes are restricted. As a result, the model moves along the sagittal plane.

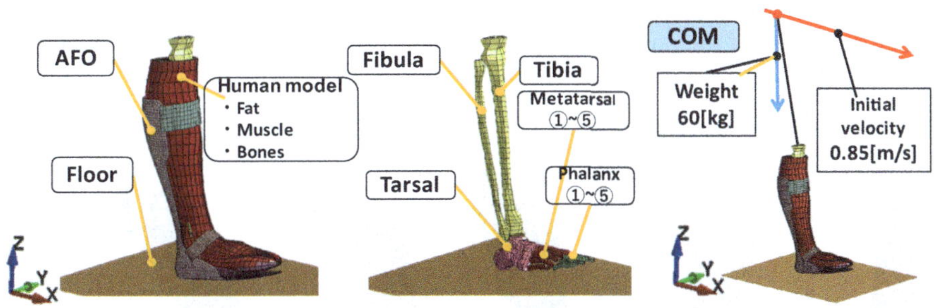

(a) Composed lower thigh and AFO model (b) Bone models (c) Initial condition of analysis

Fig. 6. Lower thigh and ankle foot models used for the analysis and initial condition of the analysis.

3.2 Results

An AFO produced by Konica Minolta Inc. Using a 3D printer with polypropylene material was used for the experiment, as shown in Fig. 7(a). The same AFO model with the same material was used for the simulation. Strain gauges were pasted on the posterior, medial, and lateral positions of the neck part of the AFO, as in Fig. 7(b), to measure the strain during gait. The gait experiment was performed by a healthy subject.

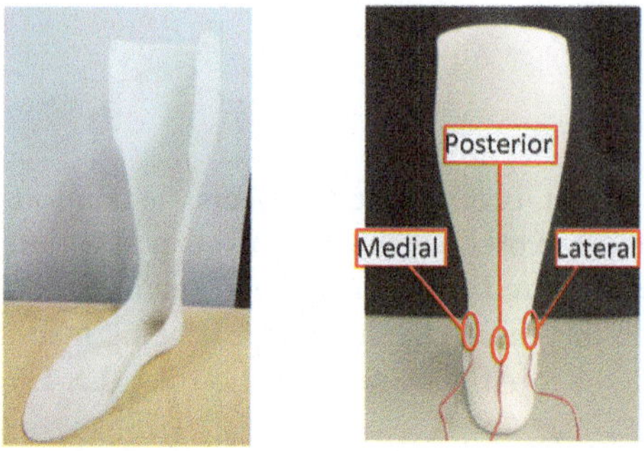

(a) Front view (b) Back view showing the position of strain gauges

Fig. 7. AFO produced by 3D printer with polypropylene material.

The results of the simulation are shown in Fig. 8. Planter flexion occurred until 0.1 s, the lower thigh was bent forward, and dorsal flexion occurred. The measured and simulated results of the strain transition during the stance phase where the strain gage is pasted are shown in Fig. 9. During the plantar flexion at the initial stage of the stance phase, the posterior position shrunk, and other positions were extended from the measured results. However, during the dorsal flexion at the midstage, the posterior position was extended, and other positions were shrunk. These trends coincided with the simulation results of the posterior and medial positions. However, the simulated result of the lateral side showed the reverse movement.

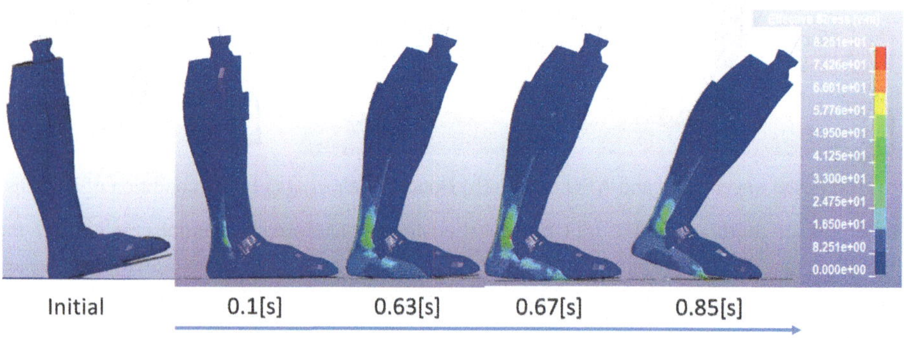

Fig. 8. Result of the gait simulation during a single-stance phase.

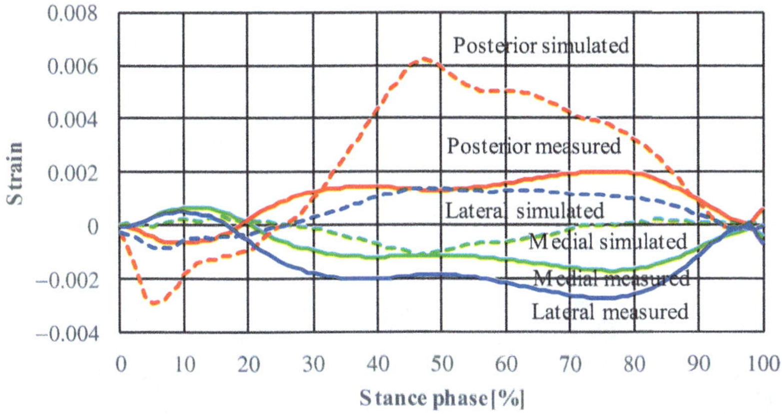

Fig. 9. Comparison of simulated and measured strains at the posterior, medial, and lateral positions.

3.3 Topological Optimization

The advantage of using the 3D printer is that a more flexible shape can be produced. In the first attempt, holes were made on the back side of the AFO to reduce the weight and prevent humidity. Topology optimization [6] is a method to reduce the volume while maintaining the rigidity of the structure. Altair Inspire (Altair Corp.) was used as the topology optimization software. The calf area of the AFO, shown in pink in Fig. 10(a), was defined as the optimized area, and the shapes of the heel, sole, and boundary area were not modified. For the boundary and force conditions, the sole was fixed, and 50-N force was applied to the left and right belt fixed area, each toward the forward, assuming dorsal flexion. The shape was optimized so that the volume of the optimized area was 40% of the total volume. The generated shape is shown in Fig. 10(b). The upper calf area was mainly processed, and holes were also made around the ankle part. Gait simulation

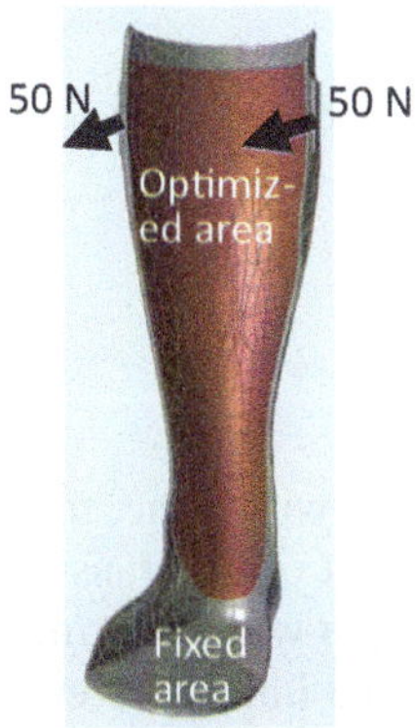

(a) Boundary conditions (front view)

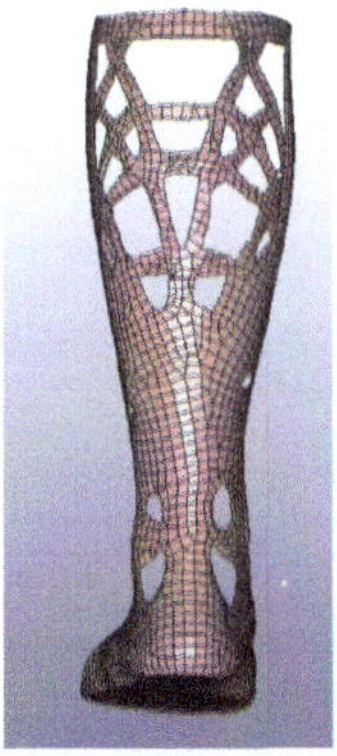

(b) After optimization (rear view)

Fig. 10. Condition of topological optimization and generated shape after the optimization.

was per-formed using the lower thigh and the optimized AFO under the conditions described in the previous section, except for the belt, whose material was strengthened. The results are shown in Fig. 11. The transition of the strain of the posterior part before and after optimization is shown in Fig. 12. The transitions were similar, and the result shows that the stiffness of the posterior part did not change significantly, even after optimization. However, the strain was greater after planter flex-ion, and the strain did not increase during the last part of dorsal flexion.

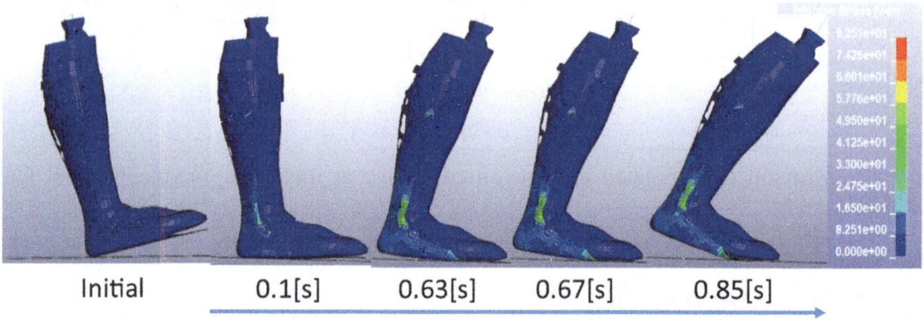

Fig. 11. Result of gait simulation using the AFO after topological optimization.

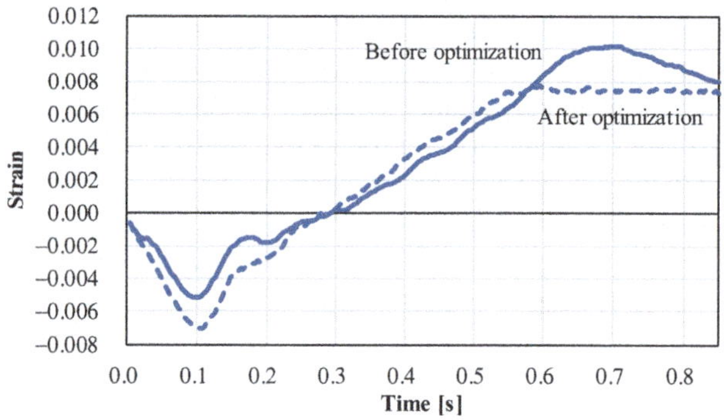

Fig. 12. Transition of strain of posterior part before and after topological optimization.

4 Conclusion

Human models of the residual limb and lower thigh with bone, muscle, and fat were composed. These models were used to evaluate the fit of the transfemoral sock-et and AFO. For the transfemoral socket, insertion simulation was performed, and the validity of the fat part and pressure distribution of the residual limb were evaluated using MRI images and the force measured by sensors built in the socket. For the AFO, gait simulation of the stance phase was performed, and the transition of strain was compared with the measured result by the 3D-printed AFO on which strain gauges were pasted. In addition, by performing topological optimization, the weight of the AFO can be reduced by maintaining the stiffness of the AFO.

The results indicate the qualitative validity of the dynamic FEM for analyzing the prosthesis socket and AFO using a human leg model. However, a difference in the pressure and strain distribution between the simulation and measurement exists, and further improvements of the simulation model and conditions are necessary. For transfemoral socket gait simulation, insertion simulation should be included. For an AFO, topologically optimized AFO needs to be produced and tested. Simulation including the swing phase and that of the patient should be developed.

Acknowledgement. AFO produced by the 3D printer was provided by Konica Minolta Inc. This research was funded by JSPS KAKENHI: JP18K12143.

References

1. Tuan, L.V., Ohnishi, K., Otsuka, H., Agarie, Y., Yamamoto, S., Hanafusa, A.: Finite element analysis for the estimation of the ground reaction force and pressure beneath the foot prosthesis during the gait of trans-femoral patients: J. Biomimetics Biomat. Biomed. Eng. **33**, 1–10 (2017)
2. Jamaludin, M.S., Hanafusa, A., Yamamoto, S., Agarie, Y., Otsuka, H., Ohnishi, K.: Development of an evaluation system for magnetic resonance imaging based three-dimensional modeling of a transfemoral prosthetic socket using finite elements. Appl. Sci. (MDPI) **9**(3662), 1–6 (2019)
3. Jamaludin, M.S., Hanafusa, A., Yamamoto, S., Agarie, Y., Otsuka, H., Ohnishi, K.: Analysis of pressure distribution in transfemoral prosthetic socket for pre-fabrication evaluation via finite element method. Bioengineering **6**(98), 12 (2019)
4. Hanafusa, A., Maruyama, N., Ikeda, T., Nakayama, T.: Dynamic finite element analysis of different width ankle foot orthoses that is incorporated with human lower leg model. In: International Society for Prosthetics and Orthotics (ISPO) World Congress 2015 Abstract Book, pp. 523–524 (2015)
5. Kawamura, T., Hanafusa, A., Yamamoto, S., Mouri, T., Takenouchi, S., Takahashi, A., Ishigai, T.: Development of ankle foot orthosis design support system - comparison of a finite element analysis between the simulation of strength testing machine and gait. Asian Prosthetics Orthot. Sci. Meet. **2018**(APOSM2018), 1 (2016)
6. Bendsoe, M.P., Sigmund, O.: Topology Optimization: Theory, Methods, and Applications, Springer, Heidelberg (2013)

Analyzing the Shear Stress on the Contacting Interface During Ambulation with Transfemoral Prosthesis Socket

Tran Van Thuc[1,2](✉), Shin-ichiroh Yamamoto[1], Akihiko Hanafusa[1], Kengo Ohnishi[3], Hiroshi Otsuka[4], and Yukio Agarie[5]

[1] Shibaura Institute of Technology, 3-7-5 Toyosu, Tokyo, Japan
tran.van.thuc.w2@sic.shibaura-it.ac.jp
[2] Hanoi University of Science and Technology, Hanoi, Vietnam
[3] Tokyo Denki University, Tokyo, Japan
[4] University of Human Arts and Science, Saitama, Japan
[5] Niigata University of Health and Welfare, Niigata, Japan

Abstract. Recently, the development of the technological advances of the prosthesis socket has been considered. However, the successful fitting of the prosthesis socket to the residuum stump is still not satisfied and existing limitations. The investigation of the shear stress contacting interface is important to improve the socket design fitting, and to understand the mechanism that happened inside the socket. In this study, the analysis of the resultant shear stress and the shear angle at the contacting interface of the Manual Compression Casting Technique (MCCT) socket type were presented. The results indicated that the higher the value of shear stress at the anterior proximal and posterior proximal in comparison with the other positions. The outcome would be benefit in clinical and improvement of the socket alignment.

Keywords: Pressure and stress measurement · MCCT socket · Transfemoral prosthesis socket · Gait cycle · Contact interface

1 Introduction

The prosthesis socket is one of the most important parts of the prostheses which are designed to restore the ambulation functionality of the amputee. The socket provides a coupling of the artificial leg and the residual limb. The successful fitting of the prosthetic socket will affect the effectiveness of the forces transferred from the socket to the residual and the walking [1,2]. Despite the development of the technological advances of the prosthesis socket, the proportion of the amputee complain of pain and discomfort was up to 23% [2]. During daily life activities, the pressure and shear stress that happened in the contact interface of the limb residual and socket may cause many problems for deep tissue such as soft tissue adjacent and damage, or ischemia [3]. The contact stress in the

Y. Shiraishi et al. (Eds.): APCMBE 2020, IFMBE Proceedings 82, pp. 102–110, 2021.
https://doi.org/10.1007/978-3-030-66169-4_14

interaction of the residuum and the socket are still the critical consideration of the successful fitting to the socket in clinical.

Recently, many quantify of the stress distribution in the contact interface between the residuum and the socket have been considered and studied. Usually, the methods were either experimental based or computer-aided based such as the finite element analyses [2,4–8]. In these studies, the trans-tibial prosthesis is most considered because the features of the residual limb are more convenient to measure, or finite element analysis causes the less changing of soft tissue volume, the shape of socket nearly the same with residual limb and the simple displacement in human gait. In contrast, few studies investigated the stress distribution of the contacting interface of the trans-femoral or above knee prosthesis socket [9–11]. The limitation, in these studies, is that the application of a few sensors which would not reflect the whole picture of the pressure and shear stress of contacting interface. Furthermore, the transducers that inserted between the socket wall and the liner is also the limitation since the measurement data does not reflect the pressure and shear stress on the skin.

The shear stress in the contact interface is one of the most important factors that affect the comfort walking and cause of the skin problem. Zhang et al. indicated that the shear stresses are no less important than the direct pressures in reducing the skin blood flow [2]. However, only a few studies focused on the investigation of the shear stresses of the trans-tibial prosthesis socket [2, 12]. Furthermore, the shear stress at the contact interface of the trans-femoral prosthesis socket was solely introduced. In this case study, we investigated and analyzed the resultant shear stress at the contact interface of the trans-femoral prosthesis socket during walking. The aim of this measurement experiment is to evaluate the development of a specific type of IRC socket. The clinical results can be used to compare other socket types and improve the current socket design.

2 Method

2.1 Experiment Protocol

In this case study, one amputee (Female, right amputation, 34 years old, 166 cm height) joined our experiment to investigate the resultant shear stresses. The subject walked ten times along 10 m road inside our laboratory while donned a trans-femoral socket. The subject implemented around 10 gait cycles for each walking time. Totally, 100 gait cycles were implemented in this experiment. The trans-femoral prosthesis socket, that is also called as the Manual Compression Casting Technique (MCCT) socket, was used in this experiment. The MCCT socket is an ischial ramus containment (IRC) socket type which has been developed by Agarie [13]. Basically, The MCCT socket is based on the IRC socket developed at the University of California, Los Angeles (so-called UCLA socket) and adjusted the anterior-lateral and sagittal directions. Consequently, the shape of the MCCT socket differs from the UCLA socket, and the stability of the socket is also improved.

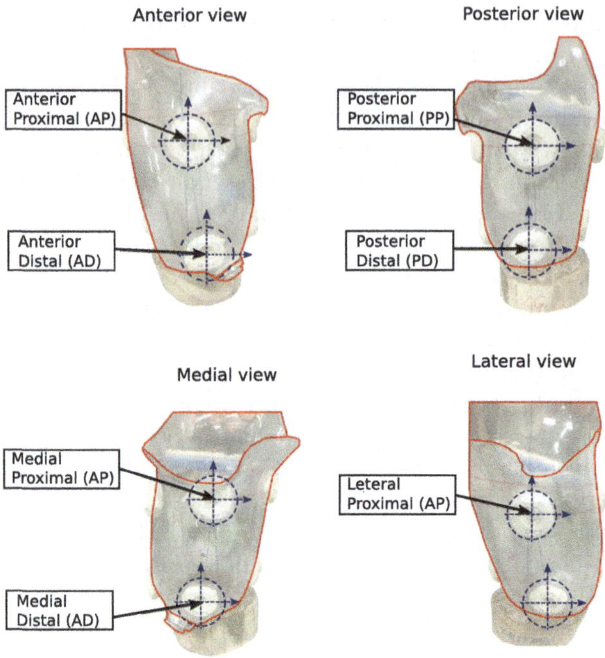

Fig. 1. The placements of the force sensors and abbreviations of the sensor positions.

In this study, in order to investigate the shear stresses, seven triaxial force sensors (MFS20-025 by Liniax) were used in this experiment. The sensors were placed at 7 positions on the socket, particularly, at anterior in both proximal and distal (AP and AD), at posterior in both proximal and distal (PP and PD), at medial both proximal and distal (MP and MD), and at lateral proximal (LP). In Fig. 1, the placements of force sensors on the socket and the positive directions of each axis are showed. The sensor can measure the force in normal and orthogonal directions. The pressure and shear stress were then calculated by dividing the force for the contact area between the tip of the sensor. On the socket, a small hole was drilled for mounting the sensor at the position for the measurement. The sensors were attached such that the measurement tip of the sensor was tightly as closed as possible to the inner surface of the socket. Moreover, two more foot-switch sensors were included in this experiment to detect heel contact and toe-off while subject walking (see Fig. 2). The signals from the sensors were sent to an amplifier with a digital-analog (DA) converter integrated and collected by a measurement software (provided by Kyowa). The sampling frequency was set 1000 Hz. The experiment protocol received the approval from the Shibaura Institute of Technology Ethics Committee before conducting and no change during the experiment.

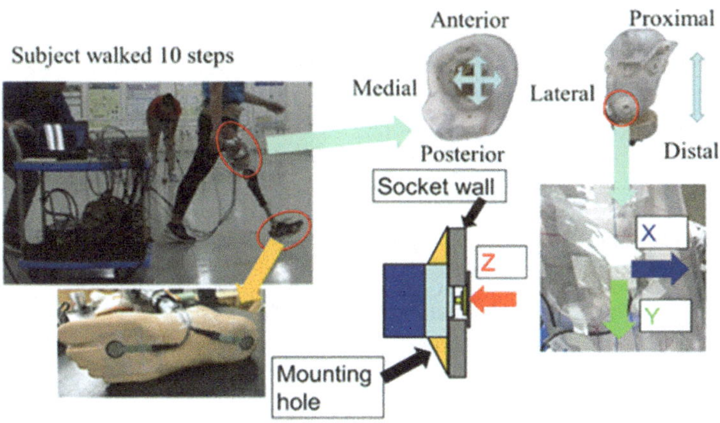

Fig. 2. The placements of the sensors and the experimental procedure.

2.2 Data Analysis

The data collected from the experiment was passed through a low-pass filter before processing. The cut-off frequency of the low pass filter was set 10 Hz. The walking data from sensors for each trial was cut to be several single gait cycles based on the foot-switch signal. Three first gait cycles and three last gait cycles were removed from the data before analyzing, this is mean that only four middle gait cycles were used to be analyzed. The resultant shear stress was calculated based on the Eq. 1, and the resultant shear angle was calculated based on the Eq. 2 as:

$$ResultantShearStress = \sqrt{HorizontalSS^2 + VerticalSS^2} \quad (1)$$

$$ResultantShearAngle = \arctan(\frac{VerticalSS}{HorizontalSS}) \quad (2)$$

in which, HorizontalSS and VerticallSS represent for the shear stress in the horizontal direction and the vertical direction, respectively, measured by the triaxial sensor. In Fig. 3, the vector of the resultant shear stress and the resultant shear angle are shown.

3 Results and Discussion

In this study, the resultant shear stresses at the contact interface of the MCCT socket were considered to analyses. The resultant shear stress profile was normalized to be single-gait cycle which was presented in percentage and showed every phase of the dynamic procedure of walking. Figure 4 shows the average resultant shear stress profile with the standard deviation area while subject walking. In general, the shape of resultant shear stresses at every position is quite consistent

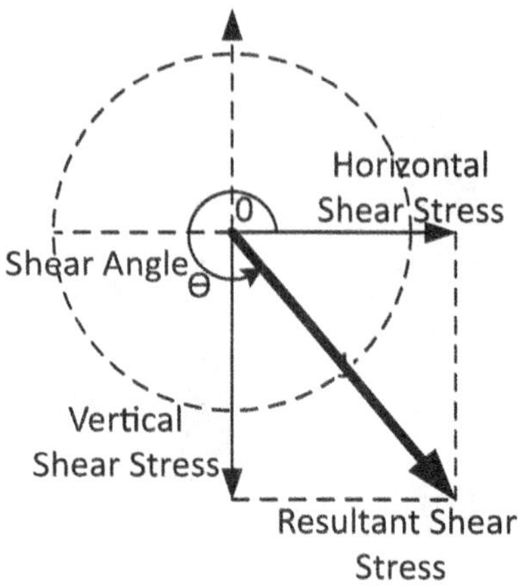

Fig. 3. Vector of the resultant shear stress and the shear angle.

such that they presented two peaks during the stance phase and decreased during the swing phase. The amplitude of the resultant shear stress is quite small in comparison with the direct pressure as shown in Table 1. During the stance phase, the resultant shear stress at AP, PP, and MP are higher in comparison with the rest, and the resultant shear stress amplitude at AP is highest. In contrast, the shear stresses at LP, MD and PD are small, especially at LP, and the variance of the shear stress is small and consistent. From the Table 1, we can see that the largest shear stresses were not necessarily produced at the points of high pressure, since the highest pressure is at MP 49.98 ± 3.16 kPa and AD 50.94 ± 1.99 kPa while the highest value of shear stress is at AP 13.30 ± 0.83 kPa. This result also agrees with the finding from the investigation by Zhang et al. [2]. The shear stress contributes greatly to discomfort while using the prosthesis socket. The increased stress makes him/her even more uncomfortable.

Table 1. The mean and standard deviation values of the peaks of the direct pressures and the resultant shear stresses.

Positions	Direct pressure (kPa)	Shear stress (kPa)
AP	31.76(1.75)	13.30(0.83)
AD	50.94(1.99)	6.12(0.38)
PP	39.23(1.80)	7.90(0.41)
PD	31.53(2.27)	4.58(0.23)
MP	49.98(3.16)	8.56(0.33)
MD	32.83(1.72)	2.97(0.34)
LP	29.17(1.19)	3.58(0.25)

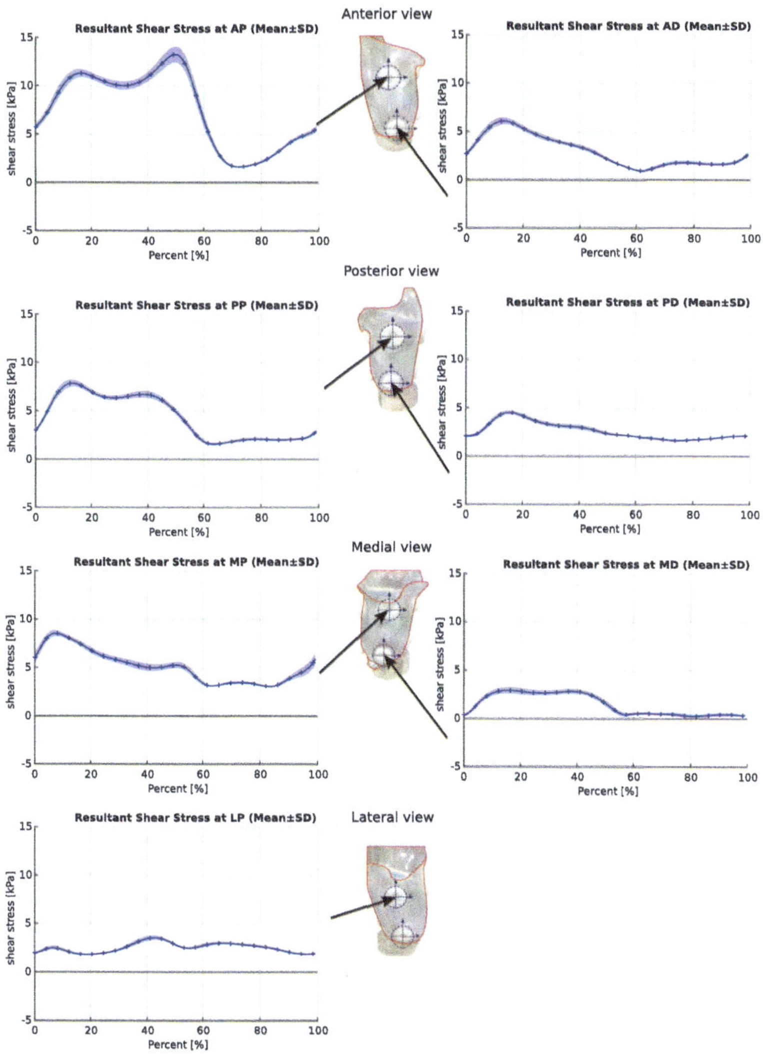

Fig. 4. The average of resultant shear stress and standard deviation presented in percentage of one gait cycle. The bold blue line presents the mean value and the blur area represents for the standard deviation.

In Fig. 5, the shear angles of the resultant shear stresses of a representative gait cycle were presented in polar plots from 0 to 360°. The vector in the polar plot indicated the starting point (heel stride) of the gait cycle while subject walking. From Fig. 5, we can see how the direction of the shear stress changed in one gait cycle. In the polar plot, if we divided the coordination system into 12 arcs, we can see that the variety of the shear angle at AP, AD and PP happened in only one or two arcs. However, the shear angles at PD, MP, MD, and LP

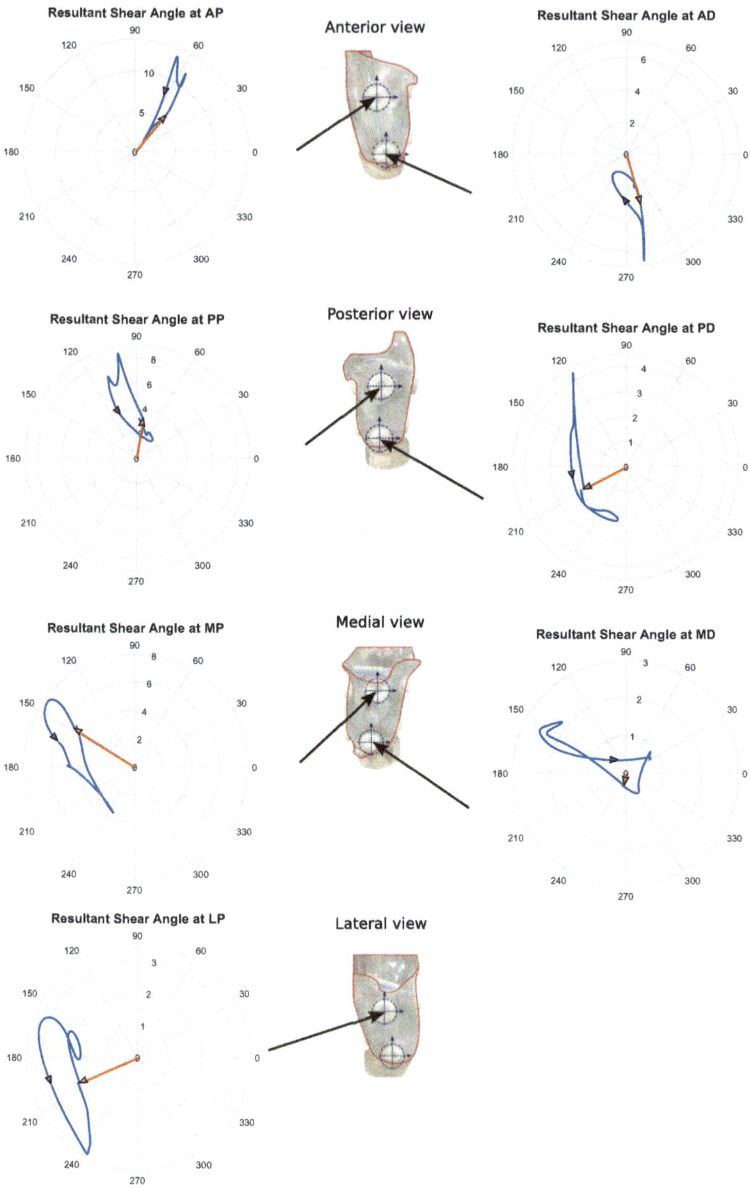

Fig. 5. The resultant shear stress shear angle for the representative gait cycle. The vector presents the starting point of the gait cycle, the curve presents the changes of the shear angle.

changed in more than three arcs, for example, the variety of the shear angle at PD from 120 to 260°. The shear angle at MD changed the largest in almost four quadrants. The more changes of the shear stress may also make the more uncomfortable, however, the amplitude of the shear stress at MD is small and the effect of the direction change could not be significant. From Fig. 5 the variety of the shear angles at AP and PP go upward and are nearly symmetric from the vertical axis. In global coordination, they nearly go the same direction upwards and inwards to the ramus of the ischium. The direction of the shear angle at AP and PP may confirm Radcliffe's principle of lateral stabilization of socket designing [14,15].

4 Conclusion

In this case study, the measurement results of the pressure and shear stresses at the contact interface of the MCCT prosthesis socket were presented. The analysis focused on the dynamic resultant shear stress while subject walking. Seven sensitive triaxial force sensors were attached such that the tip of the sensor as a part of the socket wall and contacted directly to the skin. The data from the sensor was extracted and calculated the resultant shear stress and the shear angle. The results show that the high value of the shear stresses at AP, PP and MP were observed, however, the variety of shear angles at these positions was consistent and confirmed the stabilization characteristic of the socket while walking. The limitation of the study is that the data collected only from one subject with one type of socket then leaked the necessary comparison.

References

1. Beil, T.L., Street, G.M., Covey, S.J.: Interface pressures during ambulation using suction and vacuum-assisted prosthetic sockets. J. Rehabil. Res. Dev. **39**(6), 693–700 (2002)
2. Zhang, M., Turner-Smith, A.R., Tanner, A., Roberts, V.C.: Clinical investigation of the pressure and shear stress on the trans-tibial stump with a prosthesis. Med. Eng. Phys. **20**(3), 188–198 (1998)
3. Oomens, C.W., Loerakker, S., Bader, D.L.: The importance of internal strain as opposed to interface pressure in the prevention of pressure related deep tissue injury. J. Tissue Viability **19**(2), 35–42 (2010)
4. Lee, W.C., Zhang, M.: Design of monolimb using finite element modelling and statistics-based Taguchi method. Clin. Biomech. (Bristol, Avon) **20**(7), 759–766 (2005)
5. Silver-Thorn, M.B., Steege, J.W., Childress, D.S.: A review of prosthetic interface stress investigations. J. Rehabil. Res. Dev. **33**(3), 253–266 (1996)
6. Zachariah, S.G., Sanders, J.E.: Interface mechanics in lower-limb external prosthetics: a review of finite element models. IEEE Trans. Rehabil. Eng. **4**(4), 288–302 (1996)
7. Lee, W.C., Zhang, M., Mak, A.F.: Regional differences in pain threshold and tolerance of the transtibial residual limb: including the effects of age and interface material. Arch. Phys. Med. Rehabil. **86**(4), 641–649 (2005)

8. Zhang, M., Lee, W.C.: Quantifying the regional load-bearing ability of trans-tibial stumps. Prosthet. Orthot. Int. **30**(1), 25–34 (2006)
9. Lacroix, D., Patiño, J.F.: Finite element analysis of donning procedure of a prosthetic transfemoral socket. Ann. Biomed. Eng. **39**(12), 2972–2983 (2011)
10. Zhang, L., Zhu, M., Shen, L., Zheng, F.: Finite element analysis of the contact interface between trans-femoral stump and prosthetic socket. In: Conference on Proceedings of the IEEE Engineering in Medicine and Biology Society, pp. 1270–1273 (2013)
11. Ramirez, J.F., Vélez, J.A.: Incidence of the boundary condition between bone and soft tissue in a finite element model of a transfemoral amputee. Prosthet. Orthot. Int. **36**(4), 405–414 (2012)
12. Sanders, J.E., Daly, C.H.: Normal and shear stresses on a residual limb in a prosthetic socket during ambulation: comparison of finite element results with experimental measurements. J. Rehabil. Res. Dev. **30**(2), 191–204 (1993)
13. Jamaludin, M.S., Hanafusa, A., Yamamoto, S.I., Agarie, Y., Otsuka, H., Onishi, K.: Evaluation of the effect of the geometrical changes in prosthetic socket towards transfemoral residuum via finite element method. In: Proceedings of the IEEE–EMBS Conference on Biomedical Engineering and Science (IECBES), Sarawak, Malaysia, 3–6 December 2018 (2018)
14. Radcliffe, C.W.: Functional considerations in the fitting of above knee prostheses. Artif. Limbs **2**(1), 50–67 (1955)
15. Radcliffe, C.W.: The Knud Jansen Lecture: above-knee prosthetics. Prosthet. Orthot. Int. **1**, 146–160 (1977)

Lower Limb Muscles Activation of Endurance and Power Athletes

Maisarah Sulaiman[1(✉)], Aizreena Azaman[1,2], Noor Aimie Salleh[1], and Azli Yahya[1]

[1] School of Biomedical Engineering and Health Sciences, Faculty of Engineering, Universiti Teknologi Malaysia, Johor Bahru, Malaysia
maisarah47@live.utm.my,
aizreena@biomedical.utm.my, {aimie,azliyahya}@utm.my

[2] Sport Innovation and Technology Centre (SITC), Universiti Teknologi Malaysia, Johor Bahru, Malaysia

Abstract. Power and endurance athletes have different physiological and metabolic properties of their skeletal muscle. The aim of this study was to compare the muscle activation between these two types of athletes by evaluating the muscle performance accessed from various types of lower limb muscles. A total of 27 subjects; 14 power athletes, and 13 endurance athletes voluntarily participated in this study. Electromyographic (EMG) signals were recorded from five lower limb muscles; vastus lateralis, rectus femoris, tibialis anterior, gastrocnemius and soleus muscle of dominant leg using wireless electromyography system. EMG signals were analysed by extracting several time and frequency domain features in order to evaluate the contribution of each muscles in distinguishing between power and endurance athletes. Comparison between groups was quantified and evaluated by Mann-Whitney U Test. Result showed that significant difference ($p < 0.05$) was found only in vastus lateralis, rectus femoris and tibialis anterior muscle for several EMG features. Meanwhile, no group differences were observed for gastrocnemius and soleus muscle in all features. Finding shows that physiological differences between power and endurance athletes was much easier to be accessed from vastus lateralis muscle as indicated by significant test result of most selected EMG features.

Keywords: Muscle activation · Electromyography · Power athlete · Endurance athlete

1 Introduction

Physiological properties of skeletal muscle were noticeably known to be different for athletes from various types of sports. These differences can be accessed especially by studying the muscle performance from power athlete and endurance athletes, as they were thought to have maximal difference in fibre type composition [1, 2]. Relative proportion of Type I muscle fibre was dominant in endurance-trained athlete while on

Y. Shiraishi et al. (Eds.): APCMBE 2020, IFMBE Proceedings 82, pp. 111–117, 2021.
https://doi.org/10.1007/978-3-030-66169-4_15

the other hand, greater proportion of Type II muscle fibre can be observed in power-trained athletes [3–6]. As endurance athletes rely mostly on aerobic mechanism [3], Type 1 fibre are so called slow twitch fibres due to its slow speed of contraction [7]. Contradict to power athletes which require fast speed of muscle contraction, Type II muscle fibre was known as fast twitch muscle fibre [7].

Greater understanding on muscle properties of endurance and power athletes was important as it could help athlete to tailor their training method in order to effectively improve their athletic performance [8]. Muscle activate running ability by developing force which push body forwards while supporting body's weight [9]. Electromyography (EMG) analysis was widely conducted and used in various applications in order to study the muscle activation and muscle performance.

Besides, surface EMG was also act as a global measure of motor unit activity, as it able to provide information on central and peripheral properties of neuromuscular systems [10]. Previous studies which evaluating muscle activation of endurance and power athlete revealed that power athlete has higher initial mean frequency [1], pre-exercise isometric MVC [11], isokinetic MVC [11] and more significant increase in mean amplitude [12] compared to endurance athlete.

However, most of the studies which evaluating the endurance and strength performance among athletes are only focusing on the quadriceps muscles [1, 11–13]. In fact, during running performance, there are a lot of other muscle which also being activated. Study conducted by Hamner et al. revealed that besides quadriceps muscle, plantar flexor muscles was also major contributor during running [9]. Thus, in this study, we would like to evaluate whether the other muscles can also show the distinguishing characteristics between these two types of sportsmen.

2 Method

2.1 Participants

A total of 27 subjects which comprise of two different groups of runners were recruited. 14 power (sprinter; 7 males, 7 females) and 13 endurance (long distance runner; 6 males, 7 females) athletes volunteered for this study. All athletes had competed at either national or international level in which they were recruited among Malaysian junior athletes. The detailed characteristics of participants was shown in Table 1. Participants signed written informed consent form after been thoroughly explained on the purpose, experimental procedures and potential risks or discomfort of this experiment.

Table 1. Participant's characteristics (mean ± SD) (*p > 0.05)

	Sprinter	Long distance runner
Age (years)*	15.31 ± 1.25	15.77 ± 2.2
Weight (kg)	55.85 ± 6.66	45.96 ± 5.49
Height (cm)*	165.64 ± 5.97	162.75 ± 8.0

2.2 Experimental Setup and Procedure

Skin preparation was conducted where excess hair was shaved and dirt was cleaned using 70% isopropyl alcohol from the targeted muscle area prior to the testing. Five lower limbs muscle; vastus lateralis, rectus femoris, tibialis anterior, gastrocnemius, and soleus were chosen in which the selection site for each electrode placement was been conducted based on SENIAM recommendation. Muscle activity was recorded using Delsys Trigno™ Wireless EMG System with sampling rate of 2000 Hz.

Participants were verbally instructed on each exercise prior to the actual testing. Firstly, subject was required to performed three maximum voluntary isometric contraction (MVIC) in which verbal encouragement was provided in order to encourage the subjects to reach their maximum effort. In order to compare the performance of several different muscles, several different static contraction tasks were been conducted for each muscle; knee extension for vastus lateralis and rectus femoris muscle, ankle dorsiflexion for tibialis anterior muscle and ankle plantarflexion for gastrocnemius and soleus muscle. Manual muscle testing technique was being followed for each test.

EMG signals were been processed in order to extract several relevant features which can clearly show the difference characteristics of sprinters and long distance runner. Both time and frequency domain features were extracted from the signals. Basically, time domain features were used to estimate the muscle activation level of individual. For example, root mean square can be used to characterize signal strength [14]. Besides, time domain features commonly been selected in most studies due to its low level of computational complexity[15]. Meanwhile for frequency domain features, as stated by Too et al., frequency domain features usually used for muscle fatigue and muscle force estimation [16]. In this study, frequency domain features were selected as the properties of muscles including the muscle fatigue might be differ between endurance and power athletes.

Butterworth bandpass filter with cut-off frequency of 10-500Hz was used to preprocessed and remove noise and artefact from the raw signals. This signal was being normalized to MVC values of each individual followed by full wave rectified in order to measure time domain features while the original normalized signal was used to calculate frequency domain features. Frequency domain features were estimated from Welch periodogram approach. Time domain features such as root mean square, mean absolute value, and waveform length and frequency domain features such as mean frequency, median frequency, peak frequency, total power and mean power were extracted and their performance were being compared.

2.3 Statistical Analysis

IBM SPSS software version 23 was used to statistically analyze all data. Normality testing based on Saphiro-Wilk test was firstly applied to check the normal distribution of each data. Non-parametric Mann Whitney U-Test was then used to test the significant different of muscle activation between power and endurance athletes. Significant level of 0.05 was selected for all statistical test, thus any differences was considered significant if the probability level was lower than selected significant value ($p < 0.05$).

3 Result

Based on subjects' characteristics shown in Table 1, no significant different were observed between groups for age and height. Due to the problems with some of EMG signals in which it did not meet specified criteria, thus there were several signals that had been excluded.

Based on the assessment of electromyography analysis, both time and frequency domain features were analysed. Sprinter showed lower RMS values for vastus lateralis, rectus femoris and soleus muscle compared to endurance athletes. Similar muscle activation pattern can be observed for MAV feature. However, tibialis anterior and gastrocnemius muscle showed contradict pattern in which higher values can be noticed in sprinters. For frequency domain features, inconsistent pattern can be observed for several features.

Based on statistical test conducted using Mann-Whitney U-Test as represented in Table 2 and Table 3, it can be seen that only muscle activation accessed from vastus lateralis, rectus femoris and tibialis anterior muscle had statistically significant different ($p < 0.05$) when comparing the performance of endurance and power athletes. However, these differences were observed in different EMG features for each muscle. No significant differences were observed for gastrocnemius and soleus muscle in all features.

For vastus lateralis muscle, both time (root mean square, mean absolute value) and frequency (total power, mean power) domain features able to show significant different between the two groups of subjects. However, only one frequency (peak frequency) domain feature was observed for rectus femoris and one time (waveform length) domain features for tibialis anterior muscle to be significant. Meanwhile, the other EMG parameters that had been accessed did not able to show any significant differences between groups.

Table 2. Comparison of time domain features for each muscle. P-value refers to the significant differences between endurance and power athlete (*$p < 0.05$). (VL = vastus lateralis; RF = rectus femoris; TA = tibialis anterior; GM = gastrocnemius; So = Soleus)

Feature	Root mean square					Mean absolute value				
Muscle	VL	RF	TA	GM	So	VL	RF	TA	GM	So
P-value	.020*	.149	.457	.404	.262	.020*	.124	.369	.404	.792

Feature	Waveform Length				
Muscle	VL	RF	TA	GM	So
P-value	.833	1.00	.026*	.920	.262

Table 3. Comparison of frequency domain features for each muscle. P-value refers to the significant differences between endurance and power athlete (*$p < 0.05$). (VL = vastus lateralis; RF = rectus femoris; TA = tibialis anterior; GM = gastrocnemius; So = Soleus)

Feature	Peak Frequency					Total Power				
Muscle	VL	RF	TA	GM	So	VL	RF	TA	GM	So
P-value	.549	.014*	.538	.151	.843	.020*	.149	.427	.404	.262

Feature	Mean Power				
Muscle	VL	RF	TA	GM	So
P-value	.020*	.149	.427	.404	.262

4 Discussion

The present study investigates the contribution of various muscles in discriminating power and endurance athletes by evaluating individual muscle activation measured from surface electromyography analysis. Our aim was to explore and gain better insight on which muscles along with which EMG parameters able to show differences between those two types of athletes.

Our findings indicate that different EMG parameters execute different results for each selected muscle. For example, different muscle activation pattern can be observed for RMS value among different muscles. It is well known that power athletes execute more powerful muscle activation compared to endurance athletes [2]. Thus, it is expected that power athletes will exhibit higher EMG amplitudes compared to endurance athletes. Besides, Farina et al. also stated that EMG amplitude was often used to reflect the intensity of neural drive of muscles along with other muscle fibre membrane's characteristics [17]. In this study, only RMS value extracted from tibialis anterior and gastrocnemius muscle was higher in sprinter, though there is no significant different between them. The results for MAV parameter appear to be slightly contradictory with previous research, in which study conducted by Rainoldi et al. stated that, there was no group differences between sprinters and long distance runner [11]. However, in this study, statistically significant differences can be observed for MAV features, which was recorded from vastus lateralis muscle. As for frequency domain features, similar to study conducted by Gawda et al., analysis of median frequency unable to provide any differences between groups [12].

Among all five tested muscle, only three muscles; vastus lateralis, rectus femoris and tibialis anterior muscle able to show statistically significant different between endurance and power athletes. Even though quadriceps and plantar flexor muscle was said to played important roles during running [9], but in this study, it showed that muscle activation accessed from several selected EMG features unable to show significant different in plantar flexor muscles. This finding suggests that the differences in muscle activation accessed from individual maximal voluntary contraction between endurance and power athletes was mostly and much clearly differ in quadricep muscle especially vastus lateralis muscle. This might be due to the properties of vastus lateralis muscle in which it able to provide excellent signal quality when recorded using surface electromyography compared to other lower limb muscles [18].

5 Conclusion

In conclusion, there are several muscles which can discriminate the muscle activation between endurance and strength athletes. Our data demonstrate that muscle activation of power and endurance athlete can be distinguished mostly by vastus lateralis, followed by rectus femoris and tibialis anterior muscle. Both time and frequency domain features showed statistically significant different when accessed by vastus lateralis muscle. Meanwhile, no group difference can be observed for gastrocnemius and soleus muscles. The limitation of this study can be observed from the feature extraction part in which in this study, only limited time and frequency domain features was being evaluated. Future studies should explore and evaluate numerous other EMG features in finding the best parameters which can differentiating the muscle activation of endurance and power athletes accessed from various types of muscles. Thus, it will be able to provide more understanding on the contribution of different types of muscles especially during running activity.

References

1. Rainoldi, A., Gazzoni, M., Melchiorri, G.: Differences in myoelectric manifestations of fatigue in sprinters and long distance runners. Physiol. Meas. **29**(3), 331 (2008)
2. Loturco, I., et al.: Differences in muscle mechanical properties between elite power and endurance athletes: a comparative study. J. Strength Condit. Res. **29**(6), 1723–1728 (2015)
3. Klaver-Król, E.G., et al.: Distribution of motor unit potential velocities in the biceps brachii muscle of sprinters and endurance athletes during short static contractions at low force levels. J. Electromyogr. Kinesiol. **20**(6), 1107–1114 (2010)
4. Amarantini, D., Bru, B.: Training-related changes in the EMG–moment relationship during isometric contractions: further evidence of improved control of muscle activation in strength-trained men? J. Electromyogr. Kinesiol. **25**(4), 697–702 (2015)
5. Fry, A.C., et al.: Muscle fiber characteristics and performance correlates of male olympic-style weightlifters. J. Strength Cond. Res. **17**(4), 746–754 (2003)
6. de Paula Simola, R.Á., et al.: Muscle mechanical properties of strength and endurance athletes and changes after one week of intensive training. J. Electromyogr. Kinesiol. **30**, 73–80 (2016)
7. Qaisar, R., Bhaskaran, S., Van Remmen, H.: Muscle fiber type diversification during exercise and regeneration. Free Radical Biol. Med. **98**, 56–67 (2016)
8. Lee, E., et al.: Biomarkers in sports and exercise: tracking health, performance, and recovery in Athletes. J. Strength Cond. Res. **31**, 1 (2017)
9. Hamner, S.R., Seth, A., Delp, S.L.: Muscle contributions to propulsion and support during running. J. Biomech. **43**(14), 2709–2716 (2010)
10. Herda, T.J., et al.: A noninvasive, log-transform method for fiber type discrimination using mechanomyography. J. Electromyogr. Kinesiol. **20**(5), 787–794 (2010)
11. Rainoldi, A., et al.: Mechanical and EMG responses of the vastus lateralis and changes in biochemical variables to isokinetic exercise in endurance and power athletes. J. Sports Sci. **26**, 321–331 (2008)

12. Gawda, P., et al.: Differences in myoelectric manifestations of fatigue during isometric muscle actions. Ann. Agric. Environ. Med. AAEM **25**(2), 296–299 (2018)
13. Piasecki, M., et al.: long-term endurance and power training may facilitate motor unit size expansion to compensate for declining motor unit numbers in older age. Front. Physiol. **10**(449), (2019)
14. Yang, C., et al.: SEMG-based multifeatures and predictive model for knee-joint-angle estimation. AIP Adv. **9**(9), 095042 (2019)
15. Phinyomark, A., et al.: Feature extraction and reduction of wavelet transform coefficients for EMG pattern classification. Elektronika ir Elektrotechnika **122**(6), 27–32 (2012)
16. Too, J., et al.: Classification of EMG signal based on time domain and frequency domain features **1**, 25–30 (2017)
17. Farina, D., Merletti, R., Enoka, R.: The extraction of neural strategies from the surface EMG: an update. J. Appl. Physiol. (Bethesda, Md. : 1985) **96**, 1486–95 (2004)
18. Rainoldi, A., Melchiorri, G., Caruso, I.: A method for positioning electrodes during surface EMG recordings in lower limb muscles. J. Neurosci. Methods **134**(1), 37–43 (2004)

Experimental Analysis of Ankle Foot Orthosis Using Pneumatic Artificial Muscle

Nurhanna Zulaikha Ishak[1], Shahrol Mohamaddan[1,2(✉)], Annisa Jamali[1], Shinichiro Yamamoto[2], Helmy Hazmi[3], Low Cheng Yee[4], and Akihiko Hanafusa[2]

[1] Faculty of Engineering, Universiti Malaysia Sarawak (UNIMAS), 94300 Kota Samarahan, Sarawak, Malaysia
mshahrol@unimas.my

[2] College of Systems Engineering and Science, Shibaura Institute of Technology, Saitama City 3308570, Japan

[3] Faculty of Medicine and Health Sciences, Universiti Malaysia Sarawak (UNIMAS), 94300 Kota Samarahan, Sarawak, Malaysia

[4] Faculty of Mechanical and Manufacturing Engineering, Universiti Tun Hussein Onn (UTHM), 86400 Parit Raja, Johor, Malaysia

Abstract. Ankle Foot Orthosis (AFO) is a rehabilitation device which is used to assist patients with weakness over the ankle, foot and part of the leg especially when surgery is not warranted or during the recovery phase of the foot. In this research, the Pneumatic Artificial Muscle was used to develop and actuate the AFO to mimic the movement of the ankle, namely, plantarflexion, dorsiflexion and eversion. Gait pattern analysis was used to gauge the effectiveness of the AFO in human subjects. Two healthy subjects were recruited to simulate gait patterns. The gait analysis showed that the AFO actuated PAM was able to resist plantarflexion while assisting dorsiflexion. During the stance phase, the AFO were able to allow free ankle movement up to approximately 5° dorsiflexion. It also allowed approximately 9° eversion of the ankle from neutral position. Toe drag in the swing phase was avoided as the foot drop was prevented. In conclusion, the AFO performed well in providing support for a foot drop condition and this presents an opportunity for further testing in real patients with foot drop.

Keywords: Gait · Ankle Foot Orthosis · Dorsiflexion · Plantarflexion

1 Introduction

The foot is an important part of the human body that plays role in human mobility and weight bearing. Besides injury, the complexity of the foot causes difficulty in analyzing and understanding the mechanism of motion especially by means of quantification [1]. Neurological disorder such as trauma, stroke, spinal cord injury and cerebral palsy, may disrupt the function of the foot and thus the ability of human to walk properly.

Neurological disorders may affect the foot where it leads to foot drop syndrome, causing a person to walk with an abnormal gait. In foot drop syndrome, the abnormal

Y. Shiraishi et al. (Eds.): APCMBE 2020, IFMBE Proceedings 82, pp. 118–125, 2021.
https://doi.org/10.1007/978-3-030-66169-4_16

gait is caused by the weakness of the distal muscles and extensor hypertonia of the lower limb [2]. There is excessive plantarflexion and weak dorsiflexion during the gait's swing phase [3]. As a result, the patient's toes of the affected foot may "catch" the ground first rather than the normal heel strike [4]. This may lead to injuries to the toes. Therefore, an ankle-foot orthosis (AFO) is useful to assist the toes with dorsiflexion in the swing phase and "push-off" during the late stance phase. The AFO also correct joint instabilities caused by weaknesses of the foot dorsiflexors muscles [5]. The AFO as a rehabilitation device was able to improve the patient's mobility and correct the motor patterns in walking [6]. Besides, the AFO may assist in improving the affected patient's walking pattern by proper control motion of the device [7].

The use of the AFO is not limited to a brace that supports a dropped foot during walking. Through walking gait training, the AFO functions as a physical exercise therapy strengthens the weakened ankle support muscles in addition to improve the walking range of motion [8]. This rehabilitation in turn, serves to passively improve the tones of the affected muscle, thus reducing the severity of muscle contracture and joint stiffness. Both muscle contraction and joint stiffness are complications of prolonged disuse of the muscles. The technology of AFO has improved from being a simple supporting device to one that is dynamic, making them known as an active AFO. Mechanical actuators such as the hydraulic, pneumatic and motor actuators have been fitted into AFOs to improve its functionality [9]. Pneumatic Artificial Muscle (PAM) is one of the actuators that has been used in the AFO mechanism as it has flexibility and elasticity which is similar to skeletal muscle behaviour. The PAM performs in a linear motion and its contractility is controlled by air pressure filled pneumatic bladder [10].

One of the well-known PAM types is McKibben muscle where it is widely used in the active AFO design since it is lightweight, high output power, able to withstand high forces and can mimic the human skeletal muscle's behaviour [11] compared to the other types of artificial actuators. In addition, the overall design is simpler, easily fabricated and cost friendly [12]. The objective of this study is to analyse the effect of an active AFO towards the human gait in terms of plantarflexion, dorsiflexion and eversion of the foot. The three planes were selected considering the complexity of the foot and the difficulty in quantifying motion [1]. A 3-axis accelerometer was used to measure the angle of the ankle in a gait cycle, namely the plantarflexion during the initial contact, the free dorsiflexion during stance and the plantarflexion during the swing phase.

2 Methodology

2.1 Subjects

Two healthy subjects were recruited for the experiment. One of the subjects was a healthy 18-years-old adult female who was recruited as a pilot sample for the experiment. The pilot experiment aimed to observe any differences in movement in two scenarios i.e. with the pilot sample wearing versus without wearing the AFO. After the pilot experimentation, the experiment was carried out on a healthy 12-years-old male, the main subject for this research. The child subject was selected since the main focus of this research is to develop an AFO for the children patient.

2.2 Orthosis Hardware

In designing the AFO for the 12-year-old child subject, anthropometric data of the ankle from [13] were applied. A total of 18 dimensions were used and inputted in the Computer Aided Software (CAD) in the design of the AFO. The completed design of the AFO contraption is then constructed using a 3D printer. This ensures that the AFO contraption fits the child and that the child is comfortable and can retain a normal walking gait as much as possible during the gait simulation.

The AFO was constructed using 3D printer. The shank and foot section were made from thermoplastic elastomer (TPE) rubber. A hinge between the shank and foot section was made from polylactic acid (PLA) material. The hinge is needed to support the walking process and allowed a free ankle movement. PAM fabrication was constructed in house and the performance was discussed previously in [13]. The PAM is attached with metal hook at the foot section in order to create the dorsiflexion and plantarflexion movement. The metal hook can be easily attached and detached from foot section to the 5th metatarsal head. In effect, the AFO can provide eversion movement on top of dorsiflexion and plantarflexion by attaching the metal hook to the metatarsal head.

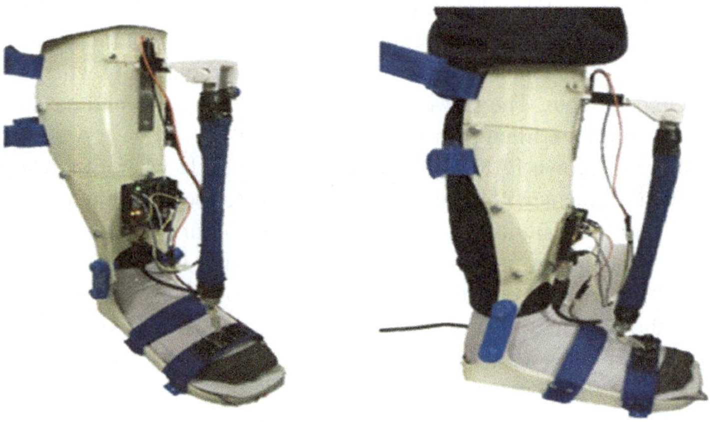

Fig. 1. The final prototype of fabricated AFO

Figure 1 show the AFO prototype in use by the subject. The 250 mm PAM is placed parallelly and anteriorly of tibialis anterior (shank section). It is controlled by FRS (Force Resistive Sensor) which is placed at the bottom of foot pad. The decision to place the sensor was based on the prominent pressure distribution provided by Tekscan F-scan system [14]. The sensor acts by changing its resistance based on the amount of force given. Low resistance is given to the electric current as more force is detected. The signals will be sent to the micro controller (Arduino), which is connected to the motor driver. Those control signals will be sent to the solenoid valve to supply air to activate the PAM.

The 3-axis accelerometer (ADXL335) is placed on the foot and shank section where it measured the angle of plantarflexion and eversion during walking. The power source to the PAM is the mini air compressor JD-2540B (F), which is connected via the silicone rubber tube. The ½ inch both side male switch of the solenoid valve 220V was used to control the flow of air from the air compressor. Closed solenoid valve was used because it can easily open when an electronic current is passed through and returned to the closed position when there is no current applied. The relay is used to control high voltage since the high voltage of solenoid valve is used in circuits. Overall, the circuit relay was purposely connected to activate (switch on) and deactivate (switch off) the solenoid valve. The relay used was a single channel BB relay 5V breakout.

2.3 Biomechanical Assessment

The subject was requested to walk with the AFO in two different conditions; passive and active AFO to observe the adaptation process. In the passive AFO condition, the subject was requested to walk with AFO without activating the PAM. The subject was asked to walk continuously for five to 10 min to get accustomed to the AFO before the measurements were taken. This process ensures a steady state dorsiflexion and plantarflexion assistance is achieved by the wearer. In the active AFO condition, the subject was asked to walk with the AFO while the PAM is activated. As in the passive AFO condition, the subject was asked to walk continuously for 5 to 10 min for familiarization. The data collected was used as the baseline of ankle angle measurement to observe any improvements or differences between passive and active AFO.

2.4 Data Acquisition and Analysis

The 3-axis accelerometer was used to collect the ankle angle data during gait. The accelerometer, which was placed at the shank and the foot, uses a high-pass second order Butterworth filter (20-Hz cut off frequency) to rectify the full wave, thus removing movement artefact. A low-pass Butterworth filter with a second-order (10-Hz cut off frequency) was applied to ensure a smooth data signal. In order to better examine the gait mechanics, the background noise was eliminated by adjusting the gain scale of the control signal and the threshold cut off point. For both active and passive gait simulation using the AFO, the data for the ankle angle was taken for every five full strides. One stride is defined from the beginning to the end of the gait cycle, i.e. from the heel strike to toe off at the stance phase and to the heel strike again. The range of 0–100% is used to define completion of a gait cycle.

3 Results and Discussion

3.1 Pilot Experiment with Adult Subject

Figure 2 shows the ankle angle of the adult subject under two conditions; passive (orange dashed line) and active AFO (baseline, blue line). The graph are the average values of all the attempts in one cycle gait. The vertical line in the middle of graph represents the toe off point. The 0% to 60% range of the gait cycle area represents the initial contact of the foot to the ground (heel on ground) to the stance phase (the whole foot pad in contact with the ground); while the 60% to 100% range of gait cycle area represents the toe off and swing phases of walking gait. In the initial contact during passive AFO, some degree of plantarflexion up to 6° were noted. However, this was reduced by around 2° after active AFO was applied.

During the midstance, AFO allowed the natural dorsiflexion where the subject ankle is in neutral position (0°) which means it allowed free ankle movement to the subject. Excessive plantarflexion of the passive AFO during the toe off and pre-swing phase were corrected during active AFO, where there is an increase in the ankle dorsiflexion by ~ 6°. Even though there was an unexpected excessive reduction in dorsiflexion observed during the swing phase, the results still showed pre-empted AFO in toe clearance. The unexpected excessive reduction in dorsiflexion was occurred due to the AFO design that based on children's anthropometric data. The size of the AFO may interrupted the adult subject's movement (e.g. imbalanced during walking gait with AFO).

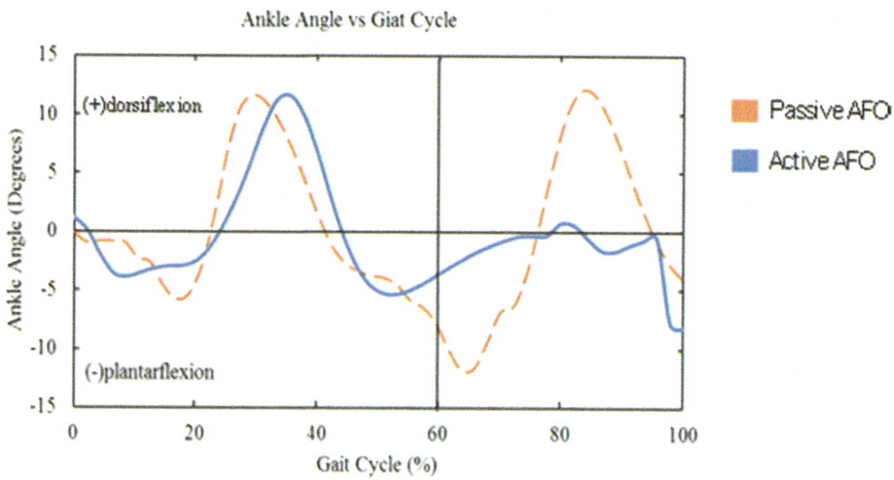

Fig. 2. Ankle joint range of motion. The vertical lines in the middle of graph represents the toe off (pre-swing) by the subject during walking. Before the vertical line area represents stance phase and after the vertical line area represents swing phases of walking gait

3.2 Experiment with Child Subject

Figure 3 shows the experiment result with healthy child subject. The result shows that the AFO allowed plantarflexion after the initial contact, even after it reduced plantarflexion in the passive AFO phase by about 2°. In mid-stance, the AFO allowed natural dorsiflexion which indicate free ankle movement allowance to the subject. In the swing phase, plantarflexion was corrected by ~5° dorsiflexion. The result shows that AFO is able to control excessive plantarflexion, which is predominant in foot drop conditions, thus preventing toe drag on patients. Based on Fig. 2 and 3, the newly developed AFO is able to perform in assisting patients with foot drop syndrome by providing support and providing proper gait pattern. In addition, the gait cycle data demonstrated that the AFO allowed free ankle movement during the stance phase, which is crucial to provide space enough to push the toe off before the swing phase. In the stance phase, the AFO provides proper plantarflexion. During the swing phase, the AFO can avoid toe drag as it provides proper dorsiflexion.

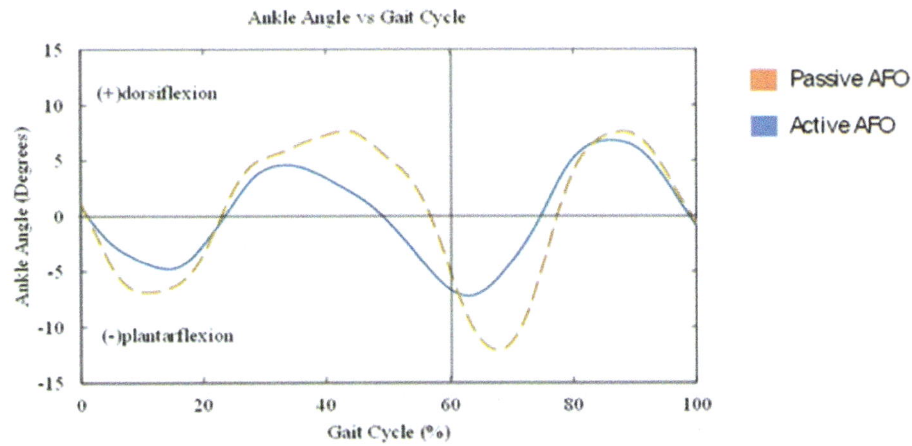

Fig. 3. Ankle joint range of motion of the healthy child subject

3.3 Joint Kinematics – Eversion

Figure 4 shows the eversion angle during AFO assistance. The eversion angle ranged from 0° to 9° from the neutral position. Since there was no previous study analyzing the inversion and eversion of the ankle by using the AFO, the findings in this study can be used as a benchmark for future research. Based on Fig. 3 and 4, the newly developed AFO is able to perform in assisting patients with foot drop syndrome by providing support and providing proper gait pattern. In addition, the gait cycle data demonstrated that the AFO allowed free ankle movement during the stance phase, which is crucial to provide space enough to push the toe off before the swing phase. In the stance phase, the AFO provides proper plantarflexion.

During the swing phase, the AFO can avoid toe drag as it provides proper dorsiflexion. However, the subjects are sometimes challenged with keeping a balance during walking with the AFO on. This occurred due to the inconsistent fitting of the AFO prototype on the legs. The walking distance of the subjects is also limited due to the heavy air compressor and power source.The activation of PAM reduced the plantarflexion where the AFO assist subject by smoothly lowering the fore foot strike to the ground. In general, positive works from AFO might benefit the user's walking pattern.

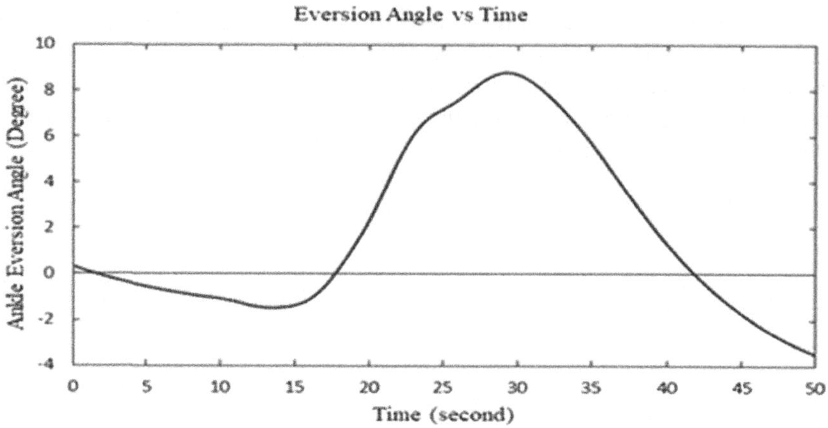

Fig. 4. Eversion angle during AFO assistance

4 Conclusions

The performance of the AFO allowed free ankle movement during stance phase and proper dorsiflexion, which is important to avoid toe drag during swing phase. This is crucial in avoiding the drop foot characteristics. The AFO showed the capability to angulate up to ~12° from the resting (neutral) foot position for both dorsiflexion and plantarflexion. In addition, the AFO mechanism provided a proper gait pattern framework for patients who might have lost the ability to do so. The findings of this study provide an impetus for possibilities of future works with real subjects with foot drop syndrome.

Acknowledgement. This research is funded by the Ministry of Higher Education Malaysia (MOHE) under the Research Acculturation Collaboration Effort (RACE) grant [Grant No: RACE/c(4)/1333/2016(6)]. The authors would like to thank Universiti Malaysia Sarawak (UNIMAS) and Shibaura Institute of Technology (SIT), Japan for providing the facilities for this research.

References

1. Boehler, A.W., Hollander, K.W., Sugar, T.G., Shin, D.: Design, implementation and test results of a robust control method for a powered ankle foot orthosis (AFO). In: Proceedings of IEEE International Conference on Robotics and Automation, Pasadena, CA, USA, pp. 2025–2030 (2008)
2. Gait abnormalities. https://stanfordmedicine25.stanford.edu/the25/gait.html. Accessed 08 Jan 2020
3. AFO for foot drop. https://walkwellstaywell.wordpress.com/2014/03/25/afos-for-foot-drop/. Accessed 02 Jan 2020
4. Romkes, J., Hell, A.K., Brunner, R.: Changes in muscle activity in children with hemiplegic cerebral palsy while walking with and without ankle foot orthoses. Gait Posture **24**(4), 467–474 (2006)

5. Chen, W.P., Tang, F.T., Ju, C.W.: Stress distribution of the foot during mid-stance to push-off in barefoot gait: A 3-D finite element analysis. Clin. Biomech. **16**(7), 614–620 (2001)
6. Kao, P.C., Ferris, D.P.: Motor adaptation during dorsiflexion-assisted walking with a powered orthosis. Gait Posture **29**(2), 230–236 (2009)
7. Park, Y.L., Bor, R.C., Young, D., Stirling, L., Robert J.W., Eugene G., Radhika N.: Bio-inspired active soft orthotic device for ankle foot pathologies. In: IEEE/RSJ International Conference on Intelligent Robots and Systems, San Francisco, CA, USA, pp. 4488–4495 (2011)
8. Chou, C.P., Hannaford, B.: Measurement and modeling of McKibben pneumatic artificial muscles. IEEE Trans. Robot. Autom. **12**(1), 90–102 (1996)
9. Manzoor, A., Hesham, E., Ekwaneen, Z.: Adaptive control of foot orthosis. Master Thesis in Electrical Engineering, Halmstad University (2007)
10. Frank, D., Dirk, L.: Pneumatic artificial muscles: Actuators for robotics and automation. Eur. J. Mech. Environ. Eng. **47**(1), 11–21 (2002)
11. Shorter, K.A: The design and control of active ankle-foot orthoses. Ph.D. Dissertation, University of Illinois (2011)
12. Hamill, J., Knutzen, K.M., Derrick, T.R.: Biomechanical Basis of Human Movement, 4th edn. Lippincott Williams and Wilkins, New York (2015)
13. Ishak, N.Z., Mohamaddan, S., Kamaruddin, A.M.N.A., Khamis, H., Yamamoto, S., Dawal, S.Z.M.: Development of Ankle Foot Orthosis (AFO) Using Pneumatic Artificial Muscle for Disabled Children. In: MATEC Web of Conferences, p. 02031. EDP Sciences (2017)
14. Tekscan. https://www.tekscan.com/resources/ebook/foot-pressure-measurement-in-clinicalsetting. Accessed 02 Jan 2020

Conceptual Design and FEM Analysis of an Exoskeleton Suit for Post-stroke Patient: A Lower Limbs Exo Suit

Muhamad Fikhri bin Abdul Hamid[1], Mohd Hanif bin Mohd Ramli[1](✉), Noor Ayuni Che Zakaria[2], and Zulkifli Mohamed[1]

[1] Sport Engineering and Artificial Intelligence Centre, Faculty of Mechanical Engineeing, UiTM, Shah Alam, Malaysia
haniframli@uitm.edu.my

[2] Neoteric Rehabilitation Sciences Research Group, Faculty of Mechanical Engineeing, UiTM, Shah Alam, Malaysia

Abstract. This paper presents a study on the conceptual design of an exoskeleton suit for post-stroke rehabilitation. A lower-limb exoskeleton suit is proposed to facilitate the post-stroke patients in the restoration of their gait motion. The aim of this study is to design and analyze the mechanical properties and strength via the finite element method (FEM). The exoskeleton suit is capable to assist the victims of post-stroke patients from sitting to stand-up positions and vice versa. The design aspect takes the basis of the human anatomy of Asian people with an average mass of 80 kg. The most effective and optimum design of the exo-suit model is evaluated through FEM static stress analysis using CATIA and ANSYS. Results proved that the proposed design is able to fully support the aforementioned average mass with the maximum Von-Mises stress of 9.5887e7 N/m^2 (Standing Position) and 160.16e6 N/m^2 (Bent Knee Position) of which did not exceed the yield strength of 180e6 N/m^2

Keywords: Post-stroke · Exoskeleton · Lower limb · Finite element method

1 Introduction

Generally, post stroke is an emotional disturbance and sometimes give rise to personality change due to physical effects, which caused by brain damage. In addition, the stroke may cause paralysis of one part of the body and the patient might find it difficult to moves. Stroke is known to be a significant global health problem and stroke was ranked as the second commonest cause of death and the third most common cause of disability-adjusted life-years (DALYs) [1]. Most stroke survivors tend to have emotional disorder and affect their self-confidence. A mixed feeling of fear, apprehension, and uncertainty are the norm amongst the stroke patients. One of the main problems that post stroke survivors face is they are trying to be independent. Often survivors feel like a burden to those around them and become depressed. These feelings may lead to more difficult of

Y. Shiraishi et al. (Eds.): APCMBE 2020, IFMBE Proceedings 82, pp. 126–134, 2021.
https://doi.org/10.1007/978-3-030-66169-4_17

a recovery where post stroke patient having limited movement issue that make such as to stand up and walk.

In the recent Stroke Registry Report (2009–2016), Malaysia's National Stroke Registry (NSR) captured over 11000 patients from sixteen participating source data providers (SDP). Stroke predominantly affected men (56%). Slightly over than 40% of the total cohort sustained stroke below the age of 60 years. Hypertension remained the major risk factor (70%), followed by diabetes (41%) and hyperlipidemia (24%) [2]. There are several alternatives to regain the confidence and back on normal life after a stroke. The most common approach is through a rehabilitation program. However, not all rehabilitation (rehab) centers are close to patients. Commuting back and forth to the rehab is sometimes very challenging for patients living far from the centers. According to statistics, only 68.9% of the patients had been referred for neurorehabilitation [3]. Post stroke care at public primary care health centers showed benefits in stroke risk factors control (i.e. hypertension and dyslipidaemia). However, the lack of knowledge and awareness on post stroke rehabilitation services are identified as causes which prevent delivery of comprehensive post stroke care for patients residing at home in the community. A more structured coordination is needed to optimize post stroke care beyond acute phase management for patients who reside at home in the community [3]. In this regard, one of the effective ways is to adopt a robotic technology in the loop. For this purpose, an exoskeleton based robotic system is proposed to assist post stroke patients in gaining their ability to stand up and walk again.

Exoskeleton suit (exo-suit) is a wearable device that is powered by a system of electric motors, pneumatics, levers, hydraulics, or a combination of technologies to assist limb movements especially lower limb with increased strength and endurance [4]. Typically, its design aims to provide back support, sense the user's motion, and send a signal to motors which manage the gears. There are variants of exo-suits that support the shoulder, waist and thigh, and assists movement for lifting and holding heavy items, while lowering back stress. Recently, there is a number of of exo-suits produced to facilitate stroke patients, for example; The Robotic Orthosis Lokomat developed by Hocoma (Zurich, Switzerland), Active Leg Exoskeleton (ALEX) from University of Delaware (Newark, DE, USA), and Hybrid Assistive Limb (HAL) by Japan's Tsukuba University [5–7].

In this paper, the fundamental challenge in realizing exo-suit is addressed in terms of conceptual design and analysis. The design aspect takes the basis of human anatomy of Asian people with an average mass of eighty (80) kilograms. The most effective and optimum design of the exo-suit model is then evaluated through FEM of static stress analysis using dedicated software packages (CATIA and ANSYS). The analysis considers three major joints, i.e., hips, knee and ankle joints.

2 Modelling

2.1 Pre-processing Stage: Geometry Modeling

Figure 1 shows a reference mannequin used in getting the size of geometrical design of the exo-suit. The total length from hip to the ground is approximately 1000 mm and width 320 mm. The average body mass is set to 80kg, and due to gravitational effect, the body weight acting on the exo-suit will be around 784.8N. The expected outcome for structural design is to ensure that the load of patient body weight is fully supported without structure failure.

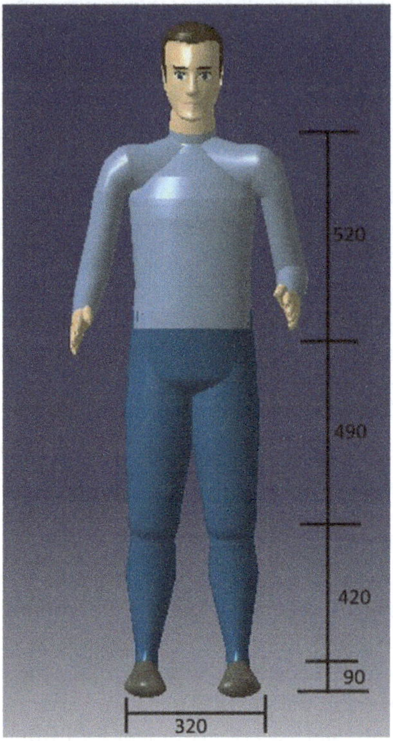

Fig. 1. Reference body dimension in unit mm

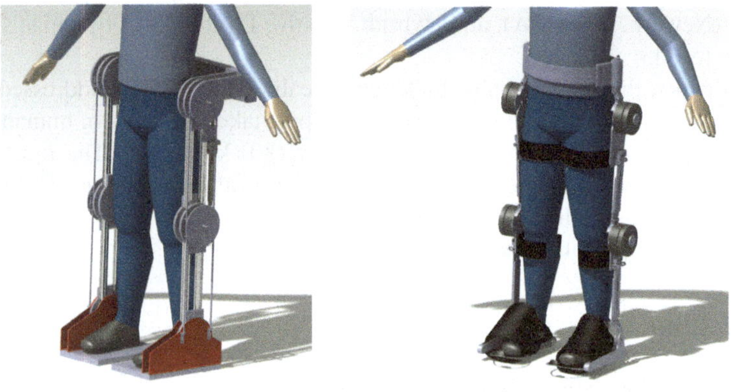

Fig. 2. (right) Preliminary structure design and (left) Finalized structure design.

Figure 2 shows the initial and finalized design in 3D environment. The overall mass of the final design is 10 Kg including the four motors integrated to the structure. The structural design is also incorporated with a height adjustable mechanism as shown in Fig. 3. In addition, a planetary gear system is properly designed and fitted to model of brushless motor GBM110-150T of which capable of producing 49.03 Nm to 78.45 Nm of torque (see Fig. 3). A strapping mechanism is integrated to ergonomically increase user's comfortability.

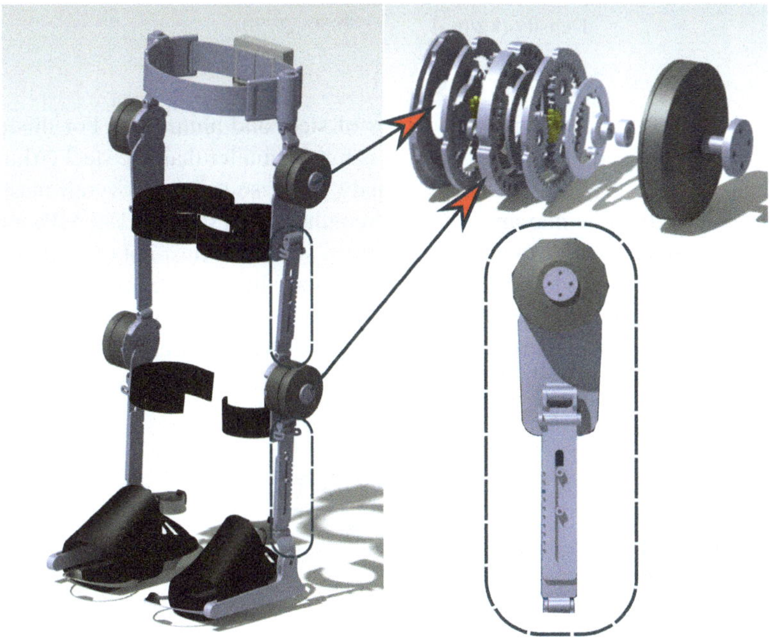

Fig. 3. (left) Isometric view of the finished exo-suit (right) Exploded view of planetary gears and Dc Motor.

2.2 Post-processing Stage

Once the final design is completely modelled in CAD, next is to carry out static or dynamic analysis on the structure. First is to consider the material of the structure, and then export the 3D model into static/dynamic analysis software package, ANSYS and establish the meshing, loads and boundary conditions.

Table 1. Material properties of steel and aluminum.

Properties	Steel	Aluminum
Young Modulus (GPa)	200	71
Yield Strength (MPa)	280	180
Poisson's ratio	0.3	0.33
Density (kg/m^3)	7850	2770

Table 1 shows the mechanical properties of steel and aluminum. For this project, aluminum is chosen due to its density which is much smaller than the steel (Aluminum: - 2770kg/m^3; steel: - 7850kg/m^3). Low density give rise to lower overall mass of the structure. Meanwhile, the maximum yield strength of aluminum is 180 MPa and Steel 280 Mpa. The simple rule of thumb, the designed structure must not exceed the tensile yield strength while supporting the load of the patient. In this case, only two conditions are analyzed, the stress distributions during knee bend and standing position as illustrated in Fig. 4. While the patient is in sitting position, the stress concentrations/distributions is not significant at all as all the load is distributed to the sitting object.

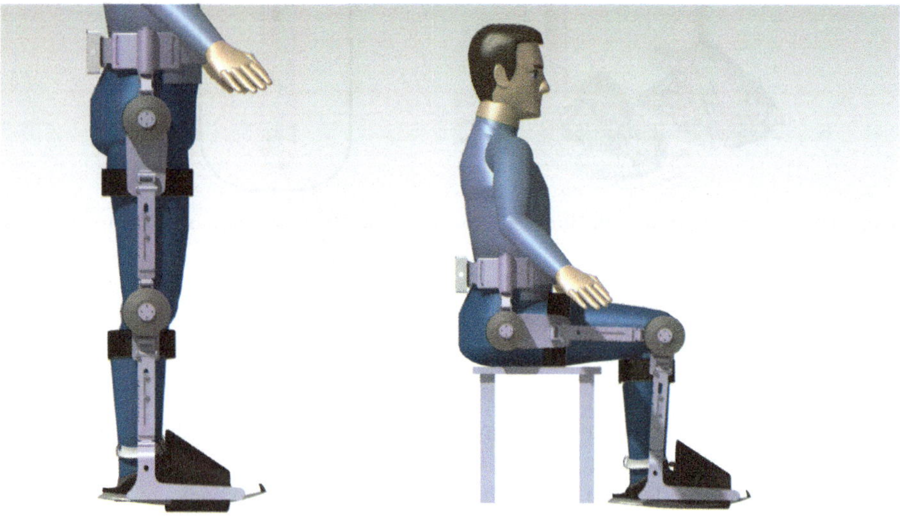

Fig. 4. The movements that exo-suit to assist/actuate: - sitting to standing positions.

The complexity of the structure design reduces the number of nodes to establish on the structure. The highest number of nodes found to be optimal is 91,968 nodes with 54,445 elements networked on the structure. Mesh networks of exo-suit is illustrated in Fig. 5.

Fig. 5. Mesh networks on the exo-suit structure.

Meanwhile, Fig. 6 shows boundary conditions and applied load on the exo-suit structure. Force of 784.8N exerted in vertical axis (negative Z-axis) on top hip joint and fixed support boundary condition is applied on the bottom of the foot surface.

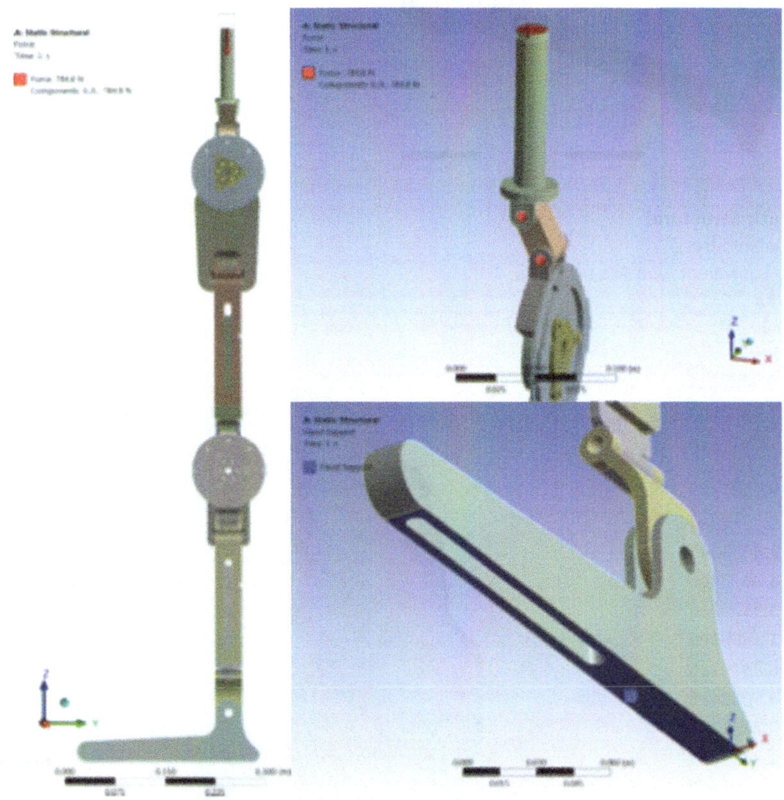

Fig. 6. Force exert in vertical axis (negative Z-axis) on top hip joint, fixed support boundary condition is applied on the bottom of the foot surface.

3 Results and Discussions

This section discusses the findings of the simulated stress distributions on exo structure by comparing their maximum and minimum stress (Von-Mises) to validate the yield strength of the aluminum characteristics. For this model, static analysis is produced using the structural ANSYS. The geometry is imported with IGS file from CATIA CAD to ANSYS to execute static analysis (focus only on static analysis).

The forces to be analyzed are at the hip, knee and ankle main joint. Besides that, the analysis is also simulated at the joints where the motor case is located and focused only on one side of the structure.

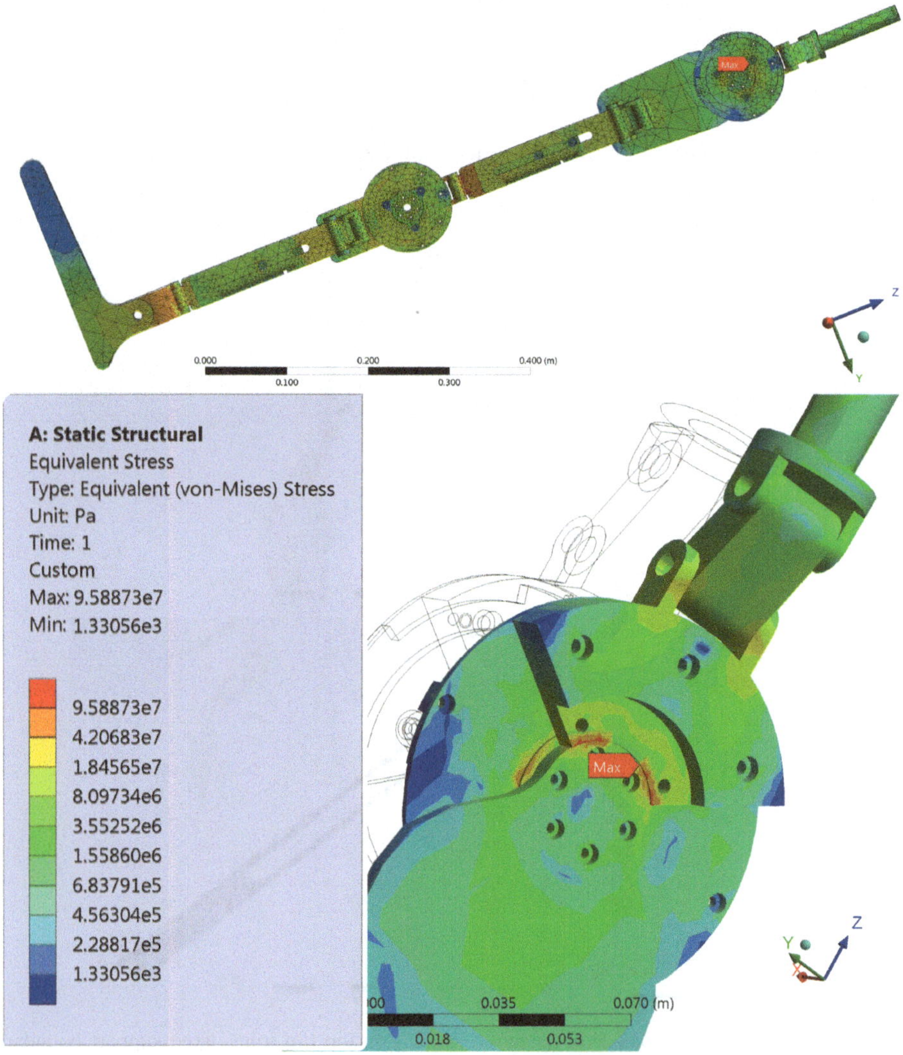

Fig. 7. Stress contour and maximum static stress (von-Mises) on the structure and motor hub during standing position.

3.1 Standing Position

Static stress analysis is conducted by applying fixed support boundary condition on bottom part of the structure i.e., foot surface and exert the force of 784.8N in Z-direction downward to determine the maximum stress. The results stress contour is depicted in Fig. 7 with maximum of 95.88 MPa and the minimum stress on the structure is 1.33 kPa. The maximum stress concentration is located around the motor hub.

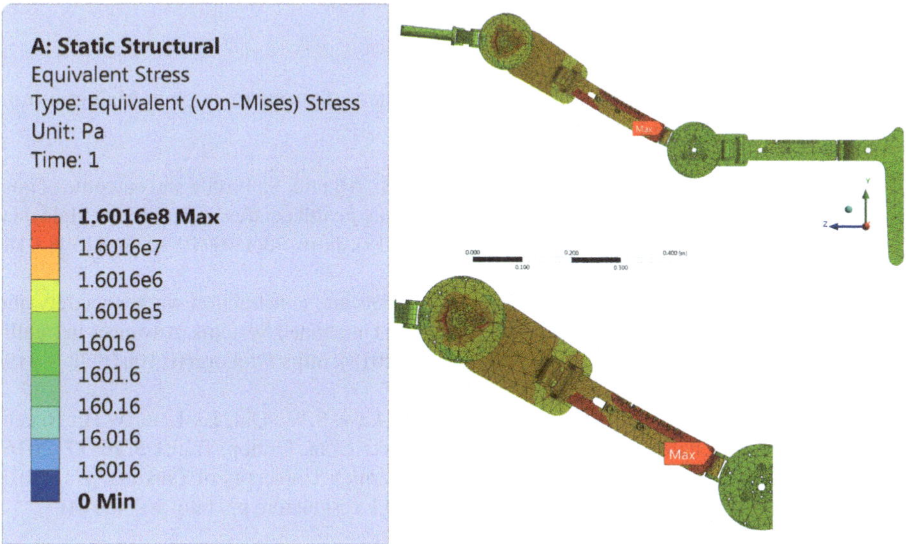

Fig. 8. Stress contour and maximum static stress (von-Mises) on the structure during knee-bent position.

3.2 Knee-Bent Position

The static stress continues by applying force to the structure while the knee part of the structure is bent and force is also applied downwards in component z-direction as the standing position, but the knee joint is set as rigid where the knee position set to bent and the result of maximum stress is 160.16 MPa as shown in Fig. 8. It can be seen that the maximum stress concentration is around the thigh section. However, the maximum stress is less than the aluminum yield strength of 180 MPa.

4 Conclusions

The aim of this project is to model a conceptual mechanical exoskeleton for the lower limb that can assist the recovery ability of the post-stroke patient. The choice of material i.e., aluminum is sufficient for this project due to the lower density compared to steel. The area of analysis is very critical because the area is exposed to higher loads such as each bolt and nut joint will easily deform and collapse if the material properties beyond the elastic region. Result of the analysis are obtained by conducting finite element analysis by comparing the maximum Von-Mises stress for standing position and bent-knee positions. The maximum stress for both positions 9.5887e7 N/m^2 (standing position) and 160.16e6 N/m^2 (knee-bent position) does not exceed the yield strength of the material (180 MPa). For future research, the dynamic analysis could be done to the next level due to this conceptual design such that optimized geometry design could be achieved to increase ergonomics and comfortability to the post stroke patient.

References

1. Kooi, C.W., Peng, H.C., Aziz, Z.A., Looi, I.: A review of stroke research in Malaysia from 2000–2014. Med. J. Malaysia **71**, 58–59 (2016)
2. National Stroke Report, 2009–2016
3. Aziz, A.A., Ali, F., Yusof, M., Che'Man, Z., Sulung, S., Aljunid, S.: Profile and outcome of post stroke patients managed at selected public primary care health centres in Peninsular Malaysia: a retrospective observational study. Sci. Rep. **8** (2018). https://doi.org/10.1038/s41598-018-36154-0
4. Li, R.M., Ng, P.L.: Wearable robotics, industrial robots and construction worker's safety and health. In: Advances in Human Factors in Robots and Unmanned Systems. Advances in Intelligent Systems and Computing, vol. 595, pp. 31–36 (2018). https://doi.org/10.1007/978-3-319-60384-1_4. ISBN 9783319603834
5. Chen, B., Ma, H., Qin, L.Y., Gao, F., Chan, K.M., Law, S.W., Qin, L., Liao, W.H.: Recent developments and challenges of lower extremity exoskeletons. Orthop. Transl. **5**, 26–37 (2016)
6. Sankai, Y.: HAL: hybrid assistive limb based on cybernics. University of Tsukuba, p. 10 2019
7. Aguilar-Sierra, H.: Hybrid actuation. A Lower Limb Exoskeleton, pp. 696–967 (2014)

Preliminary Study on the Use of Kinect Camera for Observational Gait Analysis System

Aizreena Azaman[1,2](✉), Husnir Nasyuha Abdul Halim[2], Muhammad Fariz Shafiq Abd. Aziz[2], and Sagida M. A. Bilal[1,2]

[1] Sport Innovation and Technology Centre, Institute of Human Centered Engineering, Universiti Teknologi Malaysia, 81310 Skudai, Johor, Malaysia
aizreena@biomedical.utm.my

[2] School of Biomedical Engineering and Health Sciences, Faculty of Engineering, Universiti Teknologi Malaysia, 81310 Skudai, Johor, Malaysia

Abstract. This study introduced Kinect as motion capturing device and its potential as analysis platform by developing graphical user interface (GUI) for observational gait analysis system. Spatiotemporal parameters and kinematic gait parameters of three normal individuals and two individuals with knee osteoarthritis were recorded using the proposed system and VICON system simultaneously. Correlation study were conducted between the proposed system and VICON system. The proposed system successfully generates and display the observational gait analysis results. The generated results show a strong correlation with VICON system and have no significant difference. As a conclusion, the developed system for gait analysis has high potential to be utilize in clinical settings. Some improvements should be done to increase the practicability of the system.

Keywords: Gait analysis · Kinect · Spatiotemporal · Gait kinematic

1 Introduction

Gait analysis is the analysis of locomotion and movement pattern especially walking. Quantitative gait analysis is based on concepts of gait biomechanics including spatiotemporal parameters, kinematics, kinetics and electromyography [1]. Gait abnormalities can indicate various condition due to injuries, arthritis and neurological problems such as Parkinson, hemiplegic gait and more [2].

There is various parameter to determine gait abnormities like types of devices and equipment for motion capture in laboratories to acquire data for gait analysis such as inertial measurement unit (IMU), force plates platforms, and multi-camera system [1]. However, these tools are expensive, difficult to access, complex and requires high level of skill in its use [3]. Although these devices are already established being accurate and reliable, they are fairly expensive, have a specific setup in laboratory, require technical expert and require placement of markers on subjects' body [4].

Alternatively, observational evaluations assessment such as Visual Gait Assessment Scale (VGAS) is viable in most clinical setting since it has low cost and does not require

Y. Shiraishi et al. (Eds.): APCMBE 2020, IFMBE Proceedings 82, pp. 135–140, 2021.
https://doi.org/10.1007/978-3-030-66169-4_18

complex set up and high skills [3]. Nevertheless, according to William et al. (2009), low accuracy reported on visual observational gait analysis related to kinematic variable among experience observer. Individual experiences also do not improve accuracy of assessment.

Recently, many studies proposed Kinect camera developed by Microsoft initially invented for gaming purpose, to be utilized as an alternative for motion capture for quantitative gait analysis [5]. It consists of RGB camera, infrared light source, depth sensor and microphone array and could detects human motion without the need of any attachment of sensors or markers [6].

The Kinect characteristics such as its economical price, portable, and easily operated will be useful for motion capture specifically as an alternative to devices that have been used in laboratory [7].

Despites the growing interests of researchers to use Kinect as gait data acquisition device, the development of system of user interface for gait analysis is still lacking. By employing the software development kit (SDK) for Kinect that has been provided by Microsoft, various types of application actually can be developed including for gait analysis system. This especially benefits clinicians for example physiotherapist in terms of acquiring quantitative and fast results as alternative of laboratory-based systems that usually inaccessible in clinical settings [7]. This study aims to develop an observation gait analysis system using Microsoft Kinect camera and to measure correlation of proposed device with a "gold standard device", VICON. Clinician acceptance of the proposed device were measured to determine potential use in Malaysia.

2 Methodology

2.1 System Development

Skeleton data were captured using a unit Kinect for Windows V2 camera. To record, Kinect Studio from software development kit (SDK) provided by Microsoft was used and Skeleton Recorder apps developed by [8] was used for extracting data. The skeletal data are 3D coordinates of the 25 joints specified by Kinect. By using the 3D coordinates, coding for pre-processing raw data and computation of kinematic and spatiotemporal parameters were built based on Eq. (1) and (2) below.

For knee angle;

$$\theta = 180 - (\cos^{-1}(\frac{((Yh - Yk) \times (Ya - Yk)) + ((Zh - Zk) \times (Za - Zk))}{(\sqrt{(Yh - Yk)^2 + (Zh - Zk)^2} \times \sqrt{(Ya - Yk)^2 + (Za - Zk)^2})}))$$

For hip angle:

$$\theta = \tan^{-1} \frac{(Zk - Zh)}{(Yk - Yh)}$$

where X, Y, and Z are 3D coordinates of respective joint where h = hip, k = knee, a = ankle. Kinematics joint angle formula adapted from [9]. For spatial temporal parameter, equation developed by [10] were adapted in this study. Kinect for Windows V2 with

Table 1. System requirement

Component	Specification
Processor	64-bit (x64) processor
CPU	Physical dual-core 3.1 GHz
USB port	USB 3.0
Operating System	Windows 8 and above version
Installed memory (RAM)	10.0 GB and above

field of view of 70 × 60° and 30 frames per second of frame rate is use in this system. It is connected with a computer via USB 3.0 port that provided by Microsoft. The system requirement for Kinect V2 is list in Table 1.

A graphical user interface was created by using GUIDE function from MATLAB software. The interface has divided into 3 main panel. The first panel for data acquisition which is used to preview and extract the data from Kinect sensor. The second panel is used to display kinematic measurement of knee and hip extension and flexion, meanwhile the third panes is for spatiotemporal measurement. Spatial temporal panel used display value the spatial temporal parameter based on VGAS.

2.2 System Validation

A correlation study is conducted to examine the reliability of Kinect for gait analysis with a VICON system. 5 volunteer subjects were involved in this experiment. Details of subject information as listed in Table 2 below. Each subject was asked to walk in their comfortable speed. Before the recording start, each subject was given a time to familiarize with the runaway. All kinematic of lower limb and spatial temporal data were recorded. Correlation degree between parameter measured using proposed device and VICON were determined using Pearson's correlation coefficient.

Table 2. Subject details.

Subject	Age	Gender	Condition	BMI
S1	20	Female	Normal	20.6 (normal)
S2	20	Female	Normal	24.3 (normal)
S3	24	Female	Normal	23.1 (normal)
S4	55	Male	Knee OA	21.0 (normal)
S5	52	Female	Knee OA	29.1 (overweight)

Knee OA = knee osteoarthritis.

Besides that, response from seven physiotherapist aged 34 ± 8.50 years old from Universiti Teknologi Mara, Puncak Alam were collected using IsoMetrics questionnaire to identified the suitability of the system for the task in clinical setting. The focus was on the first sub-part of the questionnaire which is 'suitability for the task'. There were 15 statements that the participants need to indicate the extent which they agree or disagree by marking from column 1 for "predominantly disagree" to column 5 for "predominantly agree".

3 Results and Discussion

Basically, this portable gait analysis system is developed to ease and make sure the observational gait analysis is compatible for early detection. Figure 1 shows the GUI for this portable gait analysis system. GUIDE function from MATLAB used to design the interface of this system. For this system, it is using Kinect Studio to monitor and record the data. Then, Kinect SDK were used in order to extract the skeleton data. The data is then saved in.txt file format. The.txt file is then were automatically loaded into MATLAB. The results were displayed the result in a GUI.

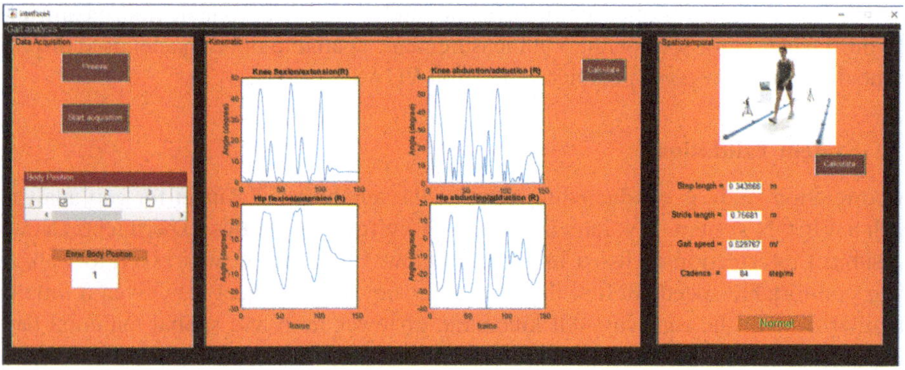

Fig. 1. Graphical User Interface for the proposed system.

The normalized kinematic and spatiotemporal parameter computed by the proposed system were compared with results generated by VICON system. Based on correlation analysis of normalized kinematic data, in average, the proposed results have a strong correlation with VICON system as shown in Table 3. However, low correlation recorded for left knee for subject S5.

For spatiotemporal parameters as shown in Table 4, in average, the percentage different between the proposed and VICON system was 9.5% with no significant difference were measured ($p > 0.05$). Minimal detectable change of gait parameters is crucial as it could be clinically important [11].

This project has some issues that can be improved in the near future. Currently, the GUI comprise the recording and extracting data process too so that make it more user-friendly. Moreover, the extracted data used for gait parameters computation were from a trial which include several strides or gait cycles instead of normalization into single gait cycle.

Table 3. Pearson's correlation of normalized kinematic data

Subject	Right knee (flex/ext)	Left knee (flex/ext)	Right hip (flex/ext)	Left hip (flex/ext)
S1	0.875	0.986	0.922	0.981
S2	0.895	0.900	0.850	0.899
S3	0.983	0.929	0.999	0.963
S4	0.965	0.963	0.982	0.986
S5	0.774	0.477	0.965	0.924

Table 4. Average Spatiotemporal parameter

Device	Step length ± std (cm)	Stride length ± stdx (cm)	Stride velocity ± std (m/s)	Cadence ± std (step/min)
Left Leg				
Kinect	42.89 ± 6.78	96.60 ± 15.53	0.79 ± 0.21	96.88 ± 20.32
VICON	50.93 ± 4.35	98.20 ± 13.84	0.74 ± 0.18	89.60 ± 14.89
Right Leg				
Kinect	41.39 ± 5.56	98.73 ± 11.24	0.80 ± 0.19	96.57 ± 20.99
VICON	47.13 ± 11.06	92.40 ± 20.16	0.71 ± 0.22	90.49 ± 16.60

From the questionnaire results, the system acceptance and usability for gait analysis task were rated satisfactory with average score 3.83 (SD ± 0.70). All participant agreed that the functions implemented in the system support them in performing their work. However, all participant agreed that the system should be more user friendly with less steps to be performed to deal with the system.

Other than that, the GUI can be improved by providing selections of templates for different physical performance-based tests such as timed up and go test, walk test, sit-to-stand test, high jump test and others; therefore, the analysis can be done according to the test specific needs.

As a conclusion, the developed system for gait analysis has high potential to be utilize in clinical settings. Some improvements should be done to increase the practicability of the system so that it can be used widely and consequently physiotherapist or clinicians could made quantitative and fast analysis with simpler device, the Kinect camera.

Acknowledgement. We would like to thank Assoc. Prof Dr Mohd Haidzir Abd Manaf from Universiti Teknologi Mara, Puncak Alam for his guidance and support. This research was supported by UTM GUP Tier 2 (.18J11) grant.

References

1. Whittle, M.W.: Chapter 4 - Methods of gait analysis, in Gait Analysis (Fourth Edition). Butterworth-Heinemann, Edinburgh (2007)
2. Pirker, W., Katzenschlager, R.: Gait disorders in adults and the elderly: a clinical guide. Wien. Klin. Wochenschr. **129**(3–4), 81–95 (2017)
3. Ridao-Fernández, C., Pinero-Pinto, E., Chamorro-Moriana, G.: Observational gait assessment scales in patients with walking disorders: systematic review. BioMed Res. Int. (2019)
4. Latorre, J., Llorens, R., Colomer, C., Alcañiz, M.: Reliability and comparison of Kinect-based methods for estimating spatiotemporal gait parameters of healthy and post-stroke individuals. J. Biomech. **72**, 268–273 (2018)
5. Springer, S., Yogev Seligmann, G.: Validity of the kinect for gait assessment: a focused review. Sensors (Basel, Switzerland) **16**(2), 194–194 (2016)
6. Wiedemann, L.G., Planinc, R., Nemec, I., Kampel, M.: Performance evaluation of joint angles obtained by the kinect V2. Paper presented at the IET International Conference on Technologies for Active and Assisted Living (TechAAL) (2015)
7. Mentiplay, B.F., Perraton, L.G., Bower, K.J., Pua, Y.-H., McGaw, R., Heywood, S., Clark, R.A.: Gait assessment using the Microsoft Xbox One Kinect: concurrent validity and inter-day reliability of spatiotemporal and kinematic variables. J. Biomech. **48**(10), 2166–2170 (2015)
8. Zhu, Y., Zhao, Y., Zhu, S.: Understanding tools: task-oriented object modeling, learning and recognition. Paper presented at the 2015 IEEE Conference on Computer Vision and Pattern Recognition (CVPR) (2015)
9. Tanaka, R., Takimoto, H., Yamasaki, T., Higashi, A.: Validity of time series kinematical data as measured by a markerless motion capture system on a flatland for gait assessment. J. Biomech. **71**, 281–285 (2018)
10. Kale Moyano, R.: Human Gait Characterization Using Kinect. Universitat Pompeu Fabra, Barcelona (2015)
11. Latorre, J., Colomer, C., Alcañiz, M., et al.: Gait analysis with the Kinect v2: normative study with healthy individuals and comprehensive study of its sensitivity, validity, and reliability in individuals with stroke. J. NeuroEng. Rehabil. **16**, 97 (2019)

Finger Exoskeleton for Early Acute Post Stroke Rehabilitation: Control Design and Performance Analysis

Mohd Nor Azmi Ab Patar[1,4(✉)], Muhammad Naim Abdu Salam[1,4], Muhammad Amir Hazim Jamaluddin[2,4], Hokyoo Lee[2,4], and Akihiko Hanafusa[3,4]

[1] Faculty of Mechanical Engineering, Universiti Teknologi MARA, Shah Alam, Malaysia
azmipatar@uitm.edu.my

[2] College of Engineering, Universiti Tenaga Nasional, 43000 Kajang, Selangor, Malaysia

[3] Department of Mechanical and Control Engineering, Niigata Institute of Technology, Niigata, Japan

[4] Department of Bio-science and Engineering, Shibaura Institute of Technology, Tokyo, Japan

Abstract. According to World Health Organization, the number of stroke patients related to upper extremity hemiparesis increases yearly. Due to this, many types of rehabilitation devices have been invented to help these patients, especially to recover their hand functions. These include exoskeleton devices which aims to deliver high intensity therapy for a longer period. Every exoskeleton rehabilitation device is made up of three components; exoskeleton mechanism, actuator and control system. Control system is central to rehabilitation devices. The accurate control algorithm implementation is very crucial for effective rehabilitation. Therefore, the performance and of the control algorithm for any rehabilitation device must be investigated to assess its efficiency. This study aims to investigate the performance of position control based on a new control strategy system of an index finger exoskeleton namely FiCoPRS, by comparing the Heuristic and Ziegler Nicol method based on various controller configurations. The parameter controlled was position of index finger measured through MCP joint angle. PID parameters were tuned using heuristic method and Ziegler Nichols' closed loop method. In order to assess the performance of the control system, actual data from position control was compared to reference values. This study suggested that the PID controller managed to control the parameter according to the set point better than P, PI and PD controller. The device could be manipulated to move to any position within the range of exoskeleton motion accurately. The results of the study also suggested that the performance of the control system for rehabilitation was very good as the error was less than 10%. The current findings prove that the study is useful and has contributed significant knowledge in understanding better the control system of an index finger exoskeleton.

Keywords: PID tuning · Heuristic method · Ziegler nichols · Finger exoskeleton · Acute post stroke · Rehabilitation · Performance characteristic

Y. Shiraishi et al. (Eds.): APCMBE 2020, IFMBE Proceedings 82, pp. 141–150, 2021.
https://doi.org/10.1007/978-3-030-66169-4_19

1 Introduction

According to World Health Organization, cerebrovascular accidents (stroke) are the second leading cause of death and the third leading cause of disability globally [1]. While the incidence rate of stroke is decreasing in developed countries, the number of patients diagnosed with acute stroke in Asia pacific regions are increasing. The number of stroke patients in Malaysia continues to rise 29.5% annually ac-cording to a study conducted from 2009 until 2014 [2]. Majority of stroke survivors suffer weakness of one side of the body known as hemiplegia post stroke [3]. As hands are very important for basic activities of daily living (ADL), stroke rehabilitation is very important to ensure motor recovery of stroke patient.

In order to achieve maximum rehabilitation results, patient must finish high-intensity training for the first three months after stroke incident [4]. Rehabilitation of stroke patients suffering from loss of motor functions involve intense repetitions of coordinated motor activities [5]. This study focuses on robot-assisted therapy using finger rehabilitation exoskeleton. Rehabilitation exoskeletons are defined as any exoskeleton specifically designed to treat disabilities of patients in a clinical setting [6].

After the introduction of ARMIN in 2005, the employment of rehabilitation robots is becoming more popular [7]. Since then, a variety of robotic exoskeleton as Mano [8], Maestro [9] and GraspyGlove [10] have been developed for rehabilitations purposes. Robotic-assisted training offers a variety of benefits and advantages. The main advantage is it can provide consistent and precise therapy for a long period of time with reduced fatigue to both patients and therapist [11]. Improvement of motor function can be effectively achieved with the repetitive exercise of the affect-ed limb [12]. Besides, robotic exoskeleton can be programmed in different modes such as passive or active assistance according to the therapy program.

Recently, an index finger exoskeleton has been developed and tested [13, 14], nevertheless, the performance and of the control algorithm for this rehabilitation device has not been be investigated thoroughly to assess its efficiency. Therefore, this study aims to investigate the performance of position control of the index finger exoskeleton based on various controller configurations.

2 Methods and Materials

2.1 Position Control

The design of a robotic finger exoskeleton regardless of its complexity can be divided into three subsystems, which are exoskeleton mechanism, actuator mechanism and control mechanism. For this study, the position control of an index finger exoskeleton is studied based on MCP joint angle as illustrated in Fig. 1.

At present, the studies in developing and improving hand exoskeletons are creating more attention and interest, especially for the applications in biomedical engineering and rehabilitation. The hand exoskeleton is an integral part of rehabilitation robotics that provides physical exercises and support in activities of daily living (ADL), such as gripping and grasping. Many studies have demonstrated that training with different robotic control strategies can significantly reduce motor impairment as assessed with standard clinical outcome measures. There are many different types of control system currently deployed for position control. Table 1 summarizes few hand and finger exoskeletons utilized for position control. This study is novel in such a way that it employs a new control strategy system applied onto an index finger exoskeleton namely FiCoPRS based on the input and optimized parameter from Heuristic and Ziegler Nicol methods.

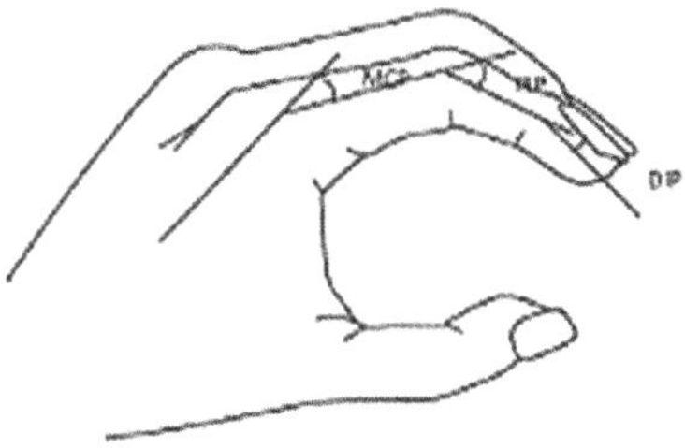

Fig. 1. MCP, PIP and DIP joint angles.

Table 1. Finger and hand exoskeletons for position control

Exoskeleton	Actuator	Control Strategy	Sensor
1. SMA-actuated Exoskeleton [15]	Shape memory alloy actuator	Variable Structure Controller (Joint Motion Control)	Angle sensor
2. Wearable Hand Exoskeleton with Compact Actuator Module [16]	Linear motor with spring and linkage structure	PID controller (Joint Angle control)	Potentiometer embedded in the linear motor
3. CAFÉ [17]	Servo motor and cable actuated finger exoskeleton	PID Controller (Joint Angle Control)	Optical Encoder
4. N-Shaped linkage exoskeleton [18]	Stepper motor with rigid serial links	Voice control	Encoder
5. FiCoPRS	Servo motor with slotted lead screw mechanism	Heuristic and Ziegler Nicol Method tuning PID Control	Rotary Encoder

The models of hand and finger exoskeleton presented show variations of control algorithms, parameters con-trolled and electronics hardware used for position control. In this paper, a new control strategy system was implemented by comparing the Heuristic and Ziegler Nicol method in tuning PID Control and obtained the optimized parameter to implement into FiCoPRS for the position control of and index finger. MCP joint angle is the parameter to be controlled via Arduino UNO board utilized as the micro-controller integrated with motor driver circuit to rotate servo DC motor clockwise and counterclockwise.

2.2 Control Design

The whole experimental setup for index finger exoskeleton rehabilitation device [20, 21], is illustrated in Fig. 2. The pulse counts measured from the encoder were converted to MCP joint angle by; a) driving the motor through a certain number of counts, b) manually measuring the joint angle using goniometer at the respective count number, and c) steps a and b are repeated until enough data are recorded to visualise the performance characteristic.

In this study, the control algorithm to be applied is Proportional-Integral-Derivative (PID) control. The mathematical function of the algorithm is described as in Eq. 1. Kp, Ki and Kd which represent proportional, integral and derivative terms are tuned to obtain a stable process.

$$y(t) = K_p \cdot x(t) + K_i \int_0^t x(t)dt + K_d \frac{dx(t)}{dt} \tag{1}$$

The control strategy employed to manipulate the exoskeleton for rehabilitations are summarized in Fig. 3. The process started with initialization and declaration of input and output parameters such as encoder and dc motor parameter. Then the current position of the finger exoskeleton was measured. Error was calculated. If the error is not equal to zero, then PID control signal was calculated. The calculated PID control signal was determined the pulse with modulation (PWM) duty cycle. The motor driver drove the motor according to the calculated PWM duty cycle until error reaches zero. Then, the process stopped.

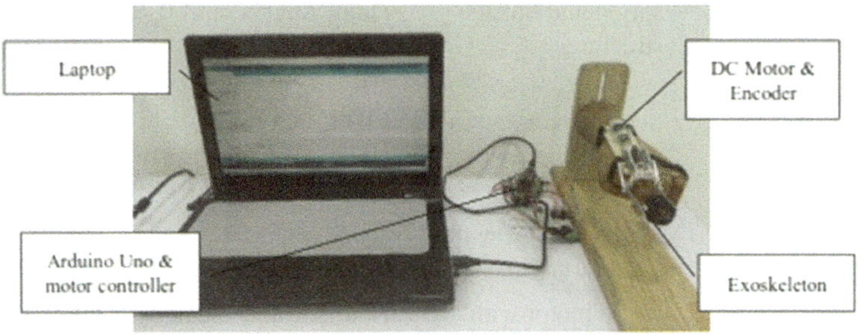

Fig. 2. Experiment setup for position control

2.3 Tuning of PID Parameters

Heuristic Method

PID parameters were tuned manually. The effect of increasing each of the controller parameter Kp, Ki and Kd are summarized in the Table 2.

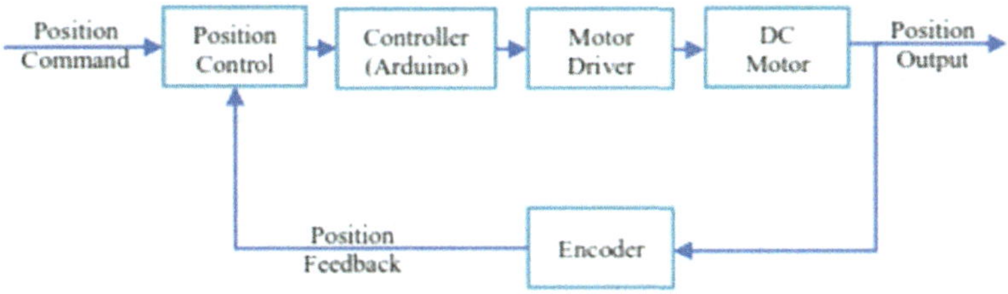

Fig. 3. Control strategy block diagram

Table 2. Effect of increase in PID parameters.

Response	Rise Time	Overshoot	Settling Time	Steady State Error
Kp	Decrease	Increase	NT	Decrease
Ki	Decrease	Increase	Increase	Eliminate
Kd	NT	Decrease	Decrease	NT

*NT: No definite trend. Minor changes.

Ziegler-Nichols Closed Loop Method
Firstly, PID controller was turned into Proportional, Kp gain by setting integral and derivative gain to zero; Ki = 0 and Kd = 0. Then, proportional gain, Kp was increased until there were sustained oscillations of the signals in the control system. Finally, the controller parameter values were calculated using the following equation where the critical value of Kcr is the value of Kp when sustained oscillation occurred while Pcr is the period of the sustained oscillation.

$$\left.\begin{aligned} K_p &= 0.6K_{cr} \\ K_i &= \frac{1.2K_{cr}}{P_{cr}} \\ K_d &= 0.075K_{cr} \cdot P_{cr} \end{aligned}\right\} \quad (2)$$

3 Results and Discussion

3.1 Performance Characteristic

In position control performance analysis, the reference graph is constructed based on Fig. 4. The joint angle increases linearly from 0% to 50% of movement cycle, then the angle decreases linearly from 50% to 100% of movement cycle. An extra step is added between the two movements. The exoskeleton has moved the patient's finger to a certain position, hold it for a certain period before returning the finger to its initial position.

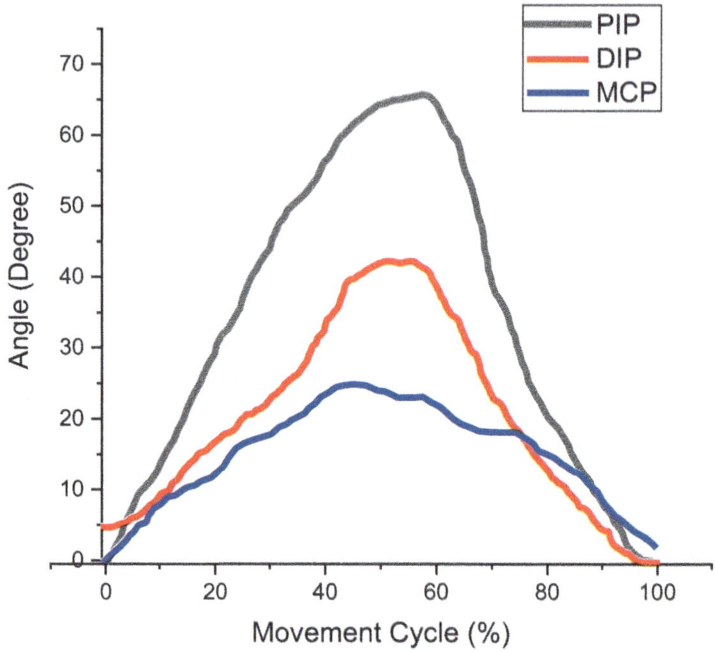

Fig. 4. Flexion and extension movement of the human finger

3.2 Position Control: Different PID Controller Configuration

Experiments of position control using different PID configuration were conducted to assess the most suitable controller for index finger exoskeleton. The results are summarized in Table 3. Resulting rise time of each controller is similar; 1.56s. PID controller leads to lower overshoot, settling time and steady state error compared to P, PI, and PD controller. Hence, it is concluded that PID controller is best suited for the system and has been applied for conducting performance analysis.

Table 3. Comparison of position control results using different PID configurations.

Controller	Rise time	Overshoot	Settling time	Steady state error
P	1.56	1.44	-	-
PI	1.56	1.72	3.84	0.0098
PD	1.56	1.70	3.57	0.0154
PID	1.56	0.28	1.98	0.0014

3.3 PID Tuning

The results of PID position control demonstrated in Table 3 is obtained by comparing results of two tunings methods; heuristic and Ziegler-Nichols Closed-loop method. The results of PID tuning using the two methods are illustrated in Fig. 5 and Table 4.

Table 4. Comparison between heuristics and Ziegler Nichols (Z-N) in PID tuning methods.

Controller	Rise time (s)	Overshoot (%)	Settling time (s)	Steady state error (°)
Heuristics	1.56	0.28	1.98	0.0014
Z-N	1.56	2.86	2.97	0.24

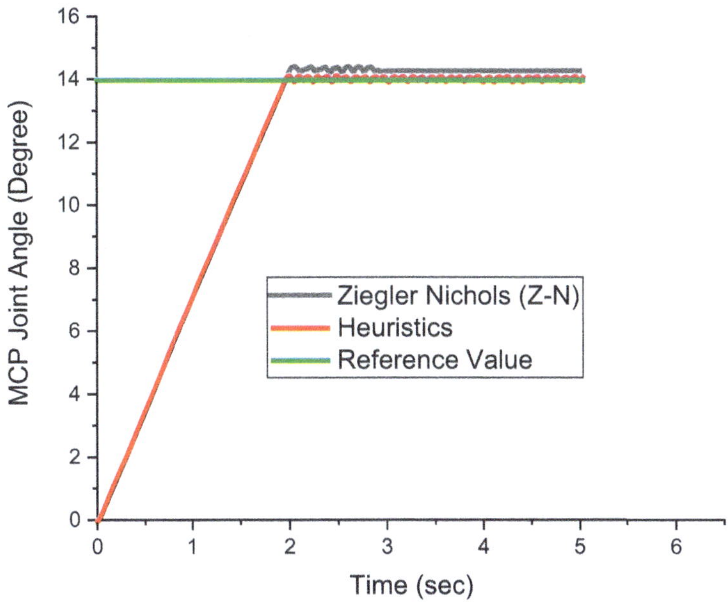

Fig. 5. Comparison of PID tuning methods between heuristic and Z-N method

PID tuning using heuristic methods lead to lower over-shoot, settling time and steady state error compared to Ziegler-Nichols methods. Hence, it is concluded that PID tuning using heuristic methods lead to better performance compared to Ziegler-Nichols method. PID parameters tuned using Heuristic method is selected to be used for performance analysis. The results obtained using heuristic methods are used in for comparison between different PID controller configurations.

3.4 Performance Analysis

In order to demonstrate the function of index finger exoskeleton for rehabilitation of stroke patients with loss of hand motor function, performance analysis was executed based on repetitive dynamic stretching therapy. Index finger exoskeleton in extension position (MCP Joint Angle = 0°) was driven to flexion at 14°. The exoskeleton remained at the desired MCP angle for a period of 2.0 s before returning to original extension position. All the stages mentioned before mimic dynamic stretching, part of hand therapy performed by therapist to stroke patients during middle stage of stroke rehabilitation (Fig. 6).

The actual value of index finger exoskeleton is com-pared to the reference value as shown in Position Performance Analysis Results. The performance of the position control of index finger for rehabilitation is measured based on the error. Errors are calculated by finding the differences between the actual values and reference values. According to a study [19], the error between the desired and measured values must lie between the limit of ±10%. Based on Fig. 7, which is showing the values of error against time, the highest error is found to be 0.854°, which is equivalent to 6.10%. Hence, the performance of the position control system is acceptable as the maximum error is less than 10%. This proves that the current index finger exoskeleton with the implemented PID position control system can be used as rehabilitation device to provide a complete cycle of hand motion, from hand opening to hand closing. In addition, the target position of the index finger exoskeleton can be set depending on the users' need during therapy session.

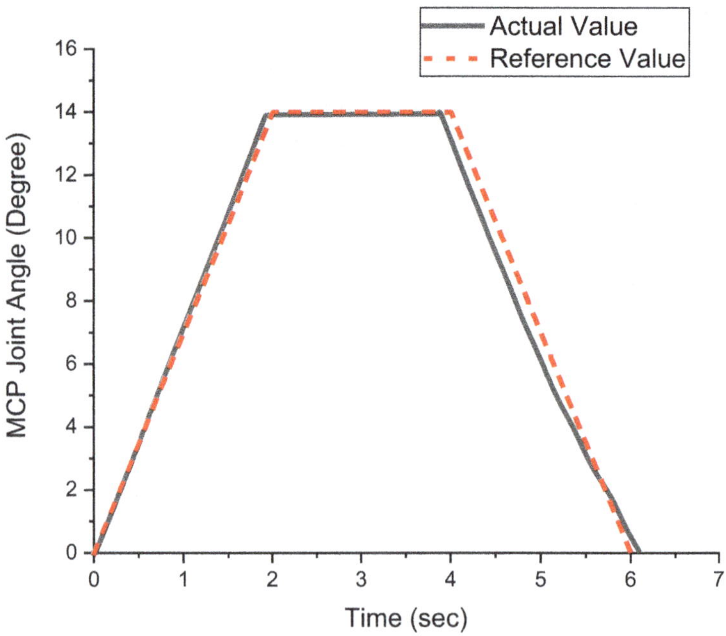

Fig. 6. Position performance analysis of MCP joint angles

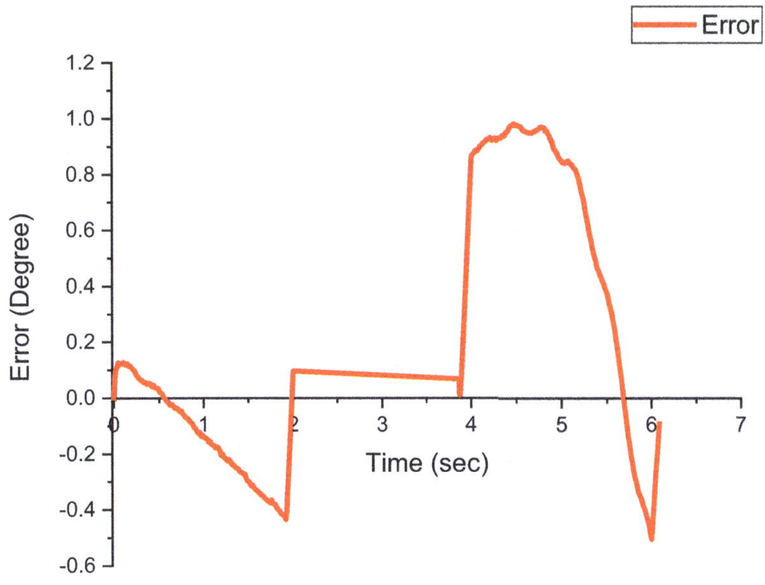

Fig. 7. Error analysis in position control

4 Conclusion

Based on the results of the experiments, it can be concluded that the main objective of this study has been achieved successfully. The measured error of less than 10% proves that the current shows that the current index finger exoskeleton with the implemented PID position control system has the potential to be used as a rehabilitation device in providing a complete cycle of hand motion, from hand opening to hand closing. By applying the PID controller, it leads to lower percentage overshoot, settling time and steady state error compared to P, PI and PD controller. It is also found that both Heuristic and Ziegler Nicol methods employed in tuning PID Control are very useful in providing optimum parameters to be implemented into FiCoPRS. The results proved that heuristic methods lead to better performance compared to Ziegler-Nichols method. The new data provided in this study could be useful for other researchers. Therefore, it can be concluded the current study is useful and has contributed new knowledge in understanding better the control system of an index finger exoskeleton.

References

1. Walter, J., Oyere, O., Mayowa, O., Sonal, S.: Stroke: a global response is needed (2016). https://www.who.int/bulletin/volumes/94/9/16-181636/en/
2. Aziz, Z.A., Lee, Y.Y.L., Ngah, B.A., Sidek, N.N., Looi, I., Hanip, M.R., et al.: Acute stroke registry Malaysia, 2010–2014: results from the national neurology registry. J. Stroke Cerebrovasc. Dis. **24**, 2701–2709 (2015)
3. Lang, C.E., Bland, M.D., Bailey, R.R., Schaefer, S.Y., Birkenmeier, R.L.: Assessment of upper extremity impairment, function, and activity after stroke: foundations for clinical decision making. J. Hand Ther. **26**, 104–115 (2013)

4. Bianchi, M., Cempini, M., Conti, R., Meli, E., Ridolfi, A., Vitiello, N., et al.: Design of a series elastic transmission for hand exoskeletons. Mechatronics **51**, 8–18 (2018)
5. Maciejasz, J.E.P., Eschweiler, Jörg., Gerlach-Hahn, K., Jansen-Troy, A., Leonhardt, S.: A survey on robotic devices for upper limb rehabilitation. J. Neuroeng. Rehabil. **11**(1), 3 (2014)
6. Sarac, M., Solazzi, M., Frisoli, A.: Design requirements of generic hand exoskeletons and survey of hand exoskeletons for rehabilitation, assistive, or haptic use. IEEE Trans. Haptics **12**(4), 400–413 (2019)
7. Bataller, A., Cabrera, J.A., Clavijo, M., Castillo, J.J.: Evolutionary synthesis of mechanisms applied to the design of an exoskeleton for finger rehabilitation. Mech. Mach. Theory **105**, 31–43 (2016)
8. Randazzo, L., Iturrate, I., Perdikis, S., Millan, J.D.R.: mano: a wearable hand exoskeleton for activities of daily living and neurorehabilitation. IEEE Robot. Autom. Lett. **3**(1), 500–507 (2018)
9. Agarwal, P., Deshpande, A.D.: Subject-specific assist-asneeded controllers for a hand exoskeleton for rehabilitation. IEEE Robot. Autom. Lett. **3**, 508–515 (2018)
10. Popov, D., Gaponov, I., Ryu, J.H.: Portable exoskeleton glove with soft structure for hand assistance in activities of daily living. IEEE/ASME Trans. Mechatron. **22**, 865–875 (2017)
11. Adnan, N.M.N., Ab Patar, M.N.A., Lee, H., Yamamoto, S.-I., Jong-Young, L., Mahmud, J.: Biomechanical analysis using Kinovea for sports application. In: IOP Conference Series: Materials Science and Engineering, vol. 342 (2018)
12. Brokaw, E.B., Black, I., Holley, R.J., Lum, P.S.: Hand spring operated movement enhancer (handsome): a portable, passive hand exoskeleton for stroke rehabilitation. IEEE Trans. Neural Syst. Rehabil. Eng. **19**, 391–399 (2011)
13. Nazri, A.F.A., Ab Patar, M.N.A., Majeed, A.P.P.A., Lee, H., Mahmud, J.: System integration of finger contracture prevention system device for early post stroke rehabilitation. Int. J. Eng. Technol. **7**(4.36), 398–403 (2018)
14. Ab Patar, M.N.A., Komeda, T., Yee, L.C., Mahmud, J.: Model-Based Systems Engineering Of A Hand Rehabilitation Device. Jurnal Teknologi **76**(4), 101–106 (2015)
15. Kazeminasab, S., Hadi, A., Alipour, K., Elahinia, M.: Force and motion control of a tendon-driven hand exoskeleton actuated by shape memory alloys. Ind. Robot: Int. J. **45**, 623–633 (2018)
16. Jones, C.L., Wang, F., Morrison, R., Sarkar, N., Kamper, D.G.: Design and development of the cable actuated finger exoskeleton for hand rehabilitation following stroke. IEEE/ASME Trans. Mechatron. **19**, 131–140 (2014)
17. Hsu, T.H., Chiang, Y.C., Chan, W.T., Chen, S.J.: A finger exoskeleton robot for finger movement rehabilitation. MPDI J. (2017)
18. Arata, J., Ohmoto, K., Gassert, R., Lambercy, O., Fujimoto, H., Wada, I.: A new hand exoskeleton device for rehabilitation using a three-layered sliding spring mechanism. In: 2013 IEEE International Conference on Robotics and Automation, pp. 3902–3907 (2013)
19. Jo, I., Bae, J.: Design and control of a wearable and force controllable hand exoskeleton system. Mechatronics **41**, 90–101 (2017)
20. Patar, M.N.A.B.A., Komeda, T., Mahmud, J., Low, C.Y.: Model based design of finger exoskeleton for post stroke rehabilitation using a slotted link cam with lead screw mechanism. In: Gen, M., Kim, K., Huang, X., Hiroshi, Y. (eds.) Industrial Engineering, Management Science and Applications 2015. Lecture Notes in Electrical Engineering, vol. 349. Springer, Heidelberg (2015)
21. Ab Patar, M.N.A.B., Komeda, T., Mahmud, J.: Force assisted hand and finger device for rehabilitation. In: International Symposium on Technology Management and Emerging Technologies, Bandung, 2014, pp. 133–138 (2014)

Biomedical Database

Analysis of Occupational Injury Reports Related to Patient Care Activities Using Text Mining Technique

Kodai Kitagawa[1(✉)], Takayuki Nagasaki[2], Sota Nakano[3], Mitsumasa Hida[1,4], Shogo Okamatsu[1,5], and Chikamune Wada[1]

[1] Graduate School of Life Science and Systems Engineering, Kyushu Institute of Technology, 2-4 Hibikino, Wakamatsu-ku, Kitakyushu 808-0196, Japan
kitagawakitagawa156@gmail.com, shogo.okamatsu182505@gmail.com, wada@brain.kyutech.ac.jp

[2] Department of Rehabilitation, Faculty of Medical Science and Welfare, Tohoku Bunka Gakuen University, 6-45-1 Kunimi, Aoba-ku, Sendai 981-8551, Japan
nagasaki@rehab.tbgu.ac.jp

[3] Department of Rehabilitation, Kyushu University of Nursing and Social Welfare, 888 Tomio, Tamana 865-0062, Japan
nakano@kyushu-ns.ac.jp

[4] Department of Physical Therapy, Faculty of Rehabilitation, Osaka Kawasaki Rehabilitation University, 158 Mizuma, Kaizuka 597-0104, Japan
hidam@kawasakigakuen.ac.jp

[5] Department of Physical Therapy, Kitakyushu Rehabilitation College, 1575 Kamikatashima Kanda-machi, Miyako-gun, Fukuoka-ken 800-0343, Japan

Abstract. Many caregivers experience occupational injuries due to patient care activities. An analysis of actual occupational injury reports will be effective to prevent such occupational injuries. However, analyzing these reports is difficult because they comprise text data that have not been quantified or standardized. In this study, we employ a text mining technique and analyze 422 occupational injury reports that include the word "patient." We obtained Severe Injury Reports published by the Occupational Safety and Health Administration. These reports were investigated using cluster analysis and a co-occurrence network based on Jaccard coefficients. The cluster analysis results indicate that are fall accidents could be caused by cords on the floor, finger injuries were associated with doors, and lower back, or neck pain was related to lifting in patient care activities. The generated co-occurrence network showed that fall accidents and lower back injury are common occupational injuries among caregivers. These results indicate the need to prevent fall accidents and lower back pain among caregivers by improving the work environment and workspace strategies, such as cord placement and caregiver postures.

Keywords: Occupational injury · Patient care activities · Text mining

Y. Shiraishi et al. (Eds.): APCMBE 2020, IFMBE Proceedings 82, pp. 153–158, 2021.
https://doi.org/10.1007/978-3-030-66169-4_20

1 Introduction

Many caregivers experience serious occupational injury [1, 2]. Ways to prevent occupational injury among caregivers have been investigated [2, 3]. Iwakiri et al. administered questionnaires to investigate the relationship between lower back pain and care strategies [2]. This investigation identified the importance of training and using proper care equipment [2]. Bell et al. reported that most falls among nurses could likely be prevented by improving floor conditions, such as removing water or loose cords [3]. These investigations provided effective knowledge for preventing occupational injuries; however, the studies may have been biased due to questionnaires and participants. Therefore, to avoid bias, a retrospective study based on actual case reports is necessary. However, most case reports comprise text data that are not quantified and standardized. The objective of this study is to investigate actual occupational injury reports using a text mining technique [4] in order to prevent occupational injuries among caregivers.

2 Method

2.1 Dataset

We analyzed Severe Injury Reports published by the Occupational Safety and Health Administration (OSHA) [5]. The OSHA requires that Severe Injury Reports be submitted for occupational injuries that resulted in hospitalization, an amputation, or loss of an eye. We collected reports files between 1/1/2015 and 6/30/2019 (Accessed 12/24/2019) [5]. This study focused on patient care activities; therefore, descriptions of the incident in 422 occupational injury reports that included the word "patient" were extracted and analyzed. Examples of incident descriptions are provided in Table 1. These reports were prepared by different people; consequently, different words were used to describe the same type of injury. For example, in Table 1 the words "fractured," "fracturing," and "breaking the bone" are used.

2.2 Text Mining Technique

The KH Coder [6], which is open source software, was used for this study. The KH Coder provides text mining techniques that are implemented using the Stanford POS Tagger for word extraction, the R programming language for statistical analysis, and MySQL, a relational database management system, for organizing, and retrieving data [7, 8]. In addition, the KH Coder is open for verification by third parties and has been used for more than a thousand academic studies related to text mining [7, 8]. Thus, the KH Coder was considered suitable for text mining in this study.

This study investigated occupational injury reports using cluster analysis and a co-occurrence network based on the Jaccard coefficient [9, 10] provided by the KH Coder. The Jaccard coefficient is the rate, at which any two words occur in the same description [9, 10]. The Jaccard coefficient was calculated as follows:

$$J_{XY} = \frac{A}{A + B + C} \tag{1}$$

where JXY is the Jaccard coefficient for words X and Y, A is the number of descriptions that contain both words X and Y, B is number of descriptions that contain word X only, and C is number of descriptions that contain word Y only. In the cluster analysis, words that occurred at least 15 times in all data were classified into eight classes based on a high Jaccard coefficient. The co-occurrence network was generated by words that occurred at least 15 times in all data and 50 relationships of these words with the highest Jaccard coefficient. Each word node was colored based on centrality in the co-occurrence network, and nodes with more centrality were painted a darker color.

Table 1. Examples of severe injury report incident descriptions. [5].

Case	Description of the Incident
1	While lifting a patient in the bathroom, an employee fractured her back.
2	An employee was transporting a patient on a stretcher out of a room. In the process, the door handle trapped her pinky finger between the door handle and the stretcher resulting in a fingertip amputation.
3	Employer reported that at about 10:15 a.m. on Friday, January 23rd, an employee was leaving a patient's room and tripped and fell, breaking the bones at the shoulder joint.
4	An employee tripped over a power cord in a patient's room and fell, fracturing her right humerus.
5	While adjusting a patient in bed, an employee experienced abdominal pain due to a hernia, requiring surgery and hospitalization.

3 Results and Discussions

Table 2 shows clusters and included words obtained from the cluster analysis for the occupational injury reports. Words in the same cluster are in a co-occurrence relationship in each report. Cluster 3 indicates that a fall, such as a trip or a slip, was likely caused by cord on the floor. Cluster 4 indicates an injury to a finger or a hand was due to a door. In addition, cluster 6 shows that a painful injury to the lower back or neck could be due to lifting during patient care activities. Furthermore, cluster 8 shows that care activities around beds and chairs could cause occupational injury. From these results, it is considered that injury-prevention among caregivers should focus on falling accidents, finger and hand injuries due to doors, neck and back pain caused by lifting, and activities around beds and chairs. Previous investigations also reported the risks for caregivers associated with falling and lifting [2, 3]; however, severe occupational injuries associated with doors, bed, and chairs were not reported.

Figure 1 shows the co-occurrence network that was generated from the occupational injury reports. In this network, "back," "injury," "suffer," and "fall" nodes are colored in dark gray since these words are located in near the center, of the network, and therefore have a significant number of relationships with other words. This network also indicates the need to prevent lower back pain and falling accidents among caregivers by improving the work environment and developing a workspace strategy, such as adjusting cord placement and teaching low load postures based on ergonomics and using assistive devices.

In this study, based on an analysis of 422 occupational injury reports using a text mining technique, we found information that will be applied to prevent severe injuries among caregivers. A potential limitation of this study was that different words that have the same meaning, such as "fracture," "fracturing," and "breaking bone," were not considered as the same word. In future, we will realize a more detailed analysis by improving preprocessing for text mining, and the extraction method of the target dataset. In addition, we will analyze Japanese occupational injury reports because Japan also has serious problems related to caregivers' occupational environment [2]. Finally, based on the knowledge obtained in this study, we will develop a strategy, and an assistive device to prevent occupational injuries.

Table 2. Clusters and words included in occupational injury reports.

	Cluster							
	1	2	3	4	5	6	7	8
Word	be	surgery	room	hand	head	side	hospital	assist
	employee	hospitalization	cord	left	strike	pain	admit	back
	patient	require	trip	door	lose	lower	care	they
	she	in-patient	arm	finger	patient	neck	provide	chair
	when	result	right		cause	lift		use
		inpatient	broken		hit	suffer		help
		that	hip		knee	injury		move
			break		shoulder	back		bed
			leg		injure			attempt
			ground		he			have
			slip		right			work
			floor		fracture			try
			fall		ankle			become
			home		left			who
			walk		sustain			foot
					fractured			approximately
					hospitalize			get
					face			catch
					run			

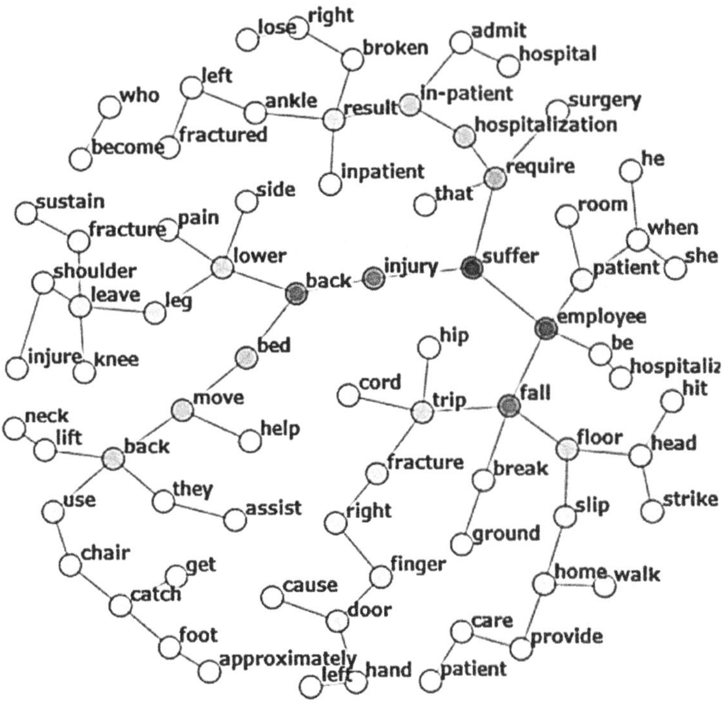

Fig. 1. Co-occurrence network obtained from occupational injury reports

4 Conclusion

In this study, to identify ways to prevent occupational injuries among caregivers, we analyzed 422 occupational injury reports using a text mining technique. The results showed that falls were probably caused by cords on the floor, finger or hand injuries were associated with doors, and lower back or neck pains was due to lifting patients. These occupational injuries should be prevented by improving the work environment and developing a workspace strategy, such as improved cord placement and improved caregiver postures.

References

1. Sinha, A., Chaitanya, A., Wasim, A., Runu, R.: Caregivers need care too–work related musculoskeletal disorders in nursing staff. J.f medical science and clinical research, 4(1), 8993e8998 (2016)
2. Iwakiri, K., Takahashi, M., Sotoyama, M., Liu, X., Koda, S.: Priority approaches of occupational safety and health activities for preventing low back pain among caregivers. J. Occup. Health **61**(5), 339–348 (2019)
3. Bell, J.L., Collins, J.W., Tiesman, H.M., Ridenour, M., Konda, S., Wolf, L., Evanoff, B.: Slip, trip, and fall injuries among nursing care facility workers. Workplace Health Safety **61**(4), 147–152 (2013)

4. Fleuren, W.W., Alkema, W.: Application of text mining in the biomedical domain. Methods **74**, 97–106 (2015)
5. Occupational Safety and Health Administration: Data & Statistics: Severe Injury Reports. https://www.osha.gov/severeinjury/. Accessed 24 Dec 2019
6. Higuch, K.: KH Coder 3. https://khcoder.net/en/. Accessed 24 Dec 2019
7. Higuchi, K.: A two-step approach to quantitative content analysis: KH Coder tutorial using Anne of Green Gables (Part I). Ritsumeikan Soc. Sci. Rev. **52**(3), 77–91 (2016)
8. Higuchi, K.: A two-step approach to quantitative content analysis: KH Coder tutorial using Anne of Green Gables (Part II). Ritsumeikan Soc. Sci. Rev. **53**(1), 137–146 (2017)
9. Higuchi, K.: KH Coder 3 reference manual. Ritsumeikan University, Japan (2016)
10. Niwattanakul, S., Singthongchai, J., Naenudorn, E., Wanapu, S.: Using of Jaccard coefficient for keywords similarity. In: Proceedings of the International Multiconference of Engineers and Computer Scientists, , Hong Kong, pp. 380–384 (2013).

Proposed Module for Establishment of National Cardiovascular Database in Low Middle Income Countries

Md. Ashrafuzzaman(✉), Md. Sakib Abrar Hossain, Monjurul Ahsan, and Nabil Islam

Military Institute of Science and Technology, Dhaka, Bangladesh
ashezaman@gmail.com, blksakib@gmail.com,
monjurulahsan1441@gmail.com, nabilislam22@gmail.com

Abstract. Cardiovascular Disease has a severe mortality rate in Bangladesh. Due to the unique pattern of cardiovascular diseases in Bangladesh, local health professionals are facing immense difficulties in both the cases of diagnosis and treatment. Machine learning and Deep Learning based approaches to diagnose CVD in early stages will surely provide an edge to tackle this alarming situation. But proper database of cardiovascular patients is still not established due to cost effectiveness (as per local perspective) and other local standpoints. The current situation of CVD has become a barrier to achieve Health Related Sustainable Development Goals set by the government of Bangladesh. Our research group is focused to solve this problem by proposing a new model to build a centralized CVD database after minutely analyzing the local health infrastructure. This cost effective and viable approach will surely lead the way for future machine learning and deep learning based researches in order to detect CVD in early stages for the benefit of local population. The benefit of such model goes beyond the boundaries of Bangladesh, as low income localities with infirm health infrastructure can also adapt such model to build centralized CVD database.

Keywords: CVD · Central database · Medical instrumentation · Android system · Biomedical engineering in developing countries · Technologies for preventive healthcare

Classification: Biomedical engineering in developing countries · Technologies for preventive healthcare

1 Introduction

1.1 Local Context

Bangladesh has a total population of 162.9 million of which 62% lives in rural areas [2]. It have been estimated that, by 2030 urban population will be 40% but currently 35% of the urban population in the capital Dhaka is under the poverty line [3]. Where, Bangladesh has a total expenditure on health per capita of $31 [4].

Y. Shiraishi et al. (Eds.): APCMBE 2020, IFMBE Proceedings 82, pp. 159–172, 2021.
https://doi.org/10.1007/978-3-030-66169-4_21

The Ministry of Health and Family welfare along with local administration plays the main role for planning the health care system of Bangladesh. With that the total health system is illustrated by "shortage, inappropriate skill mix and inequitable distribution" of its health care workers. The numbers of current registered doctors are 64,434, the current doctor to nurse to technologist ratio is 1:0.4:0.24 whereas the recommended ratio of WHO is 1:3:5 [5].

Bangladesh has a unique urban health governance system where it's organized into three different levels such as: city corporations, municipal cities and towns (paurashavas), and upazila headquarters. In short the entire health administration can be described by the following block diagram (Fig. 1):

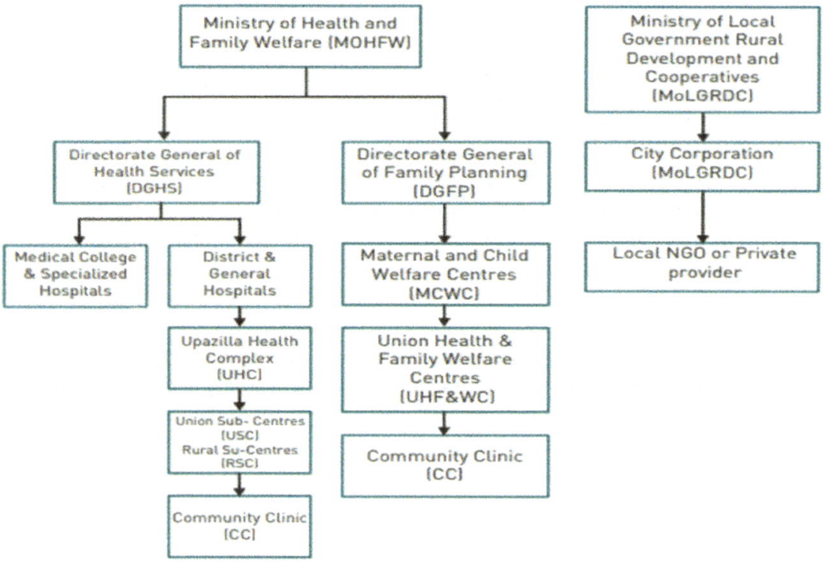

Fig. 1. Health system of Bangladesh

There are currently 490 medical facilities in upazila level, 1,450 medical facilities in Union level and 13,442 medical facilities in Ward level, the detail distribution of health facilities in upazila and union level is provided [6] (Fig. 2):

Now a days the health and nutrition policies and projects are mainly focused on the rural area in Bangladesh. Due to this the poor urban people could not able to enjoy sufficient access to quality health and nutrition services. We came to know that the expenditure for medicine (per population) in the capital Dhaka is 2.2% of the total earnings, in other urban area it is 4.5% and in rural areas it is 4.8% [7].

Level	Type of facility	Type of service	Total no. of facilities	Total no. of beds
Upazila	Upazila health complex (50-bed)	Hospital	297	14,850
	Upazila health complex (31-bed)	Hospital	112	3,472
	Upazila health complex (10-bed)	Hospital	11	110
	Upazila health complex (0-bed)*	OPD	4	-
	Subtotal of upazila health complexes		424	18,432
	Upazila health office	OPD	60	-
	31-bed hospital	Hospital	4	124
	30-bed hospital	Hospital	1	30
	25-bed hospital	Hospital	1	25
	Subtotal of hospitals outside health complexes		66	179
	Total of upazila-level facilities		490	18,611
Union	20-bed hospital	Hospital	32	640
	10-bed hospital	Hospital	19	190
	Subtotal of union-level hospitals		51	830
	Union subcenter and Union health & family welfare center	OPD	1,399	-
	Total of union-level facilities		1,450	
Ward	Community clinic (functional at present)	OPD	13,442	-
Grand total of primary-care hospitals (Upazila and below)			477	19,441
Grand total of primary health facilities (Upazila and below)			15,382	19,441
*Hospital not yet started				

Fig. 2. Accommodation distribution in upazila and union levels

To overcome the inability of public services and government administration to reach the entire population, particularly to the poor, a large NGO sector play a crucial role as they emerged as "third sector" of health providers in Bangladesh. The role of NGOs are as a bridge between the donors and the less privileged person where they carry the funds towards the under privileged persons directly. Not only funds but also they are involved in helping the local government about financing, planning, service delivery, capacity building, and monitoring and evaluation have produced some health gains [8].

Bangladesh has been focusing on harmonizing its community health system, at upazila level the planning is overseen by the upazila health management committees (UHMCs). Still the health system suffers from shortage of doctors, insufficient time given by doctors to patients, ill behavior of the doctors towards the poor patient, long waiting time, exploitation by the middlemen and many other factors [9]. In rural areas there are two major problems, which are: Retention and absenteeism.

Above all these analyses we came to a result that we have to introduce the modern technologies such as telemedicine, e-health and m-health systems which can play a significant role to improve the current health situation of Bangladesh.

The Health Ministry and Welfare have recently created a Medical Information System but the collected data are not so good quality. Thus the managers and policy planners in Bangladesh rely mainly on survey data, not routine MIS data [10].

Current data suggest that there are 13 million mobile phones with over 32,000 village phones operating in 52 districts. According to the report of WHO in 2011, 98% of Bangladeshi adults are using mobile phone. There were reported to have 105 million active mobile phone subscribers in June 2013. Now NGO and various non-profitable organizations is being involved in the m-health sector. There are many e-health and m-health services in Bangladesh. But most services include health advice lines [11].

1.2 Current Situation of Cardio Vascular Disease (CVD) in Bangladesh

In 2010, All Age Mortality Rates/100 000 for cardiovascular and circulatory diseases was 94.2. The ischemic heart and cerebrovascular disease (stroke), were among the top ten causes of death in 2011. Recent survey suggests that cardiovascular disease is one of the significant reasons of death in Bangladesh. Mortality rate of CVD in medical college hospital level was 54.24% in 2016, in district level hospitals it was 47.53% and in upazila level hospitals it was 32.42%. Mortality rate due to CVD was 39.63%. The survey population was over five years old (Figs. 3 and 4).

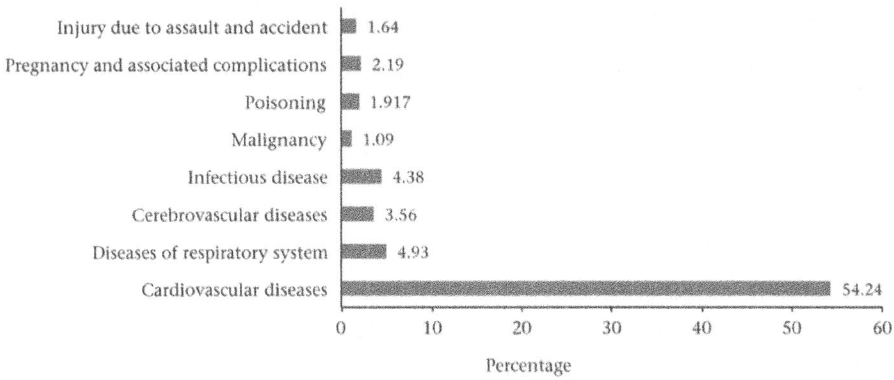

Fig. 3. Major mortality causes in Bangladesh (2016)

According to current (latest) survey of 2017 (DGHS Health Bulletin 2018), mortality rates due to CVD outnumbers other major causes of death like Cerebrovascular diseases, diabetes, pregnancy associated, hypertension by a huge margin.

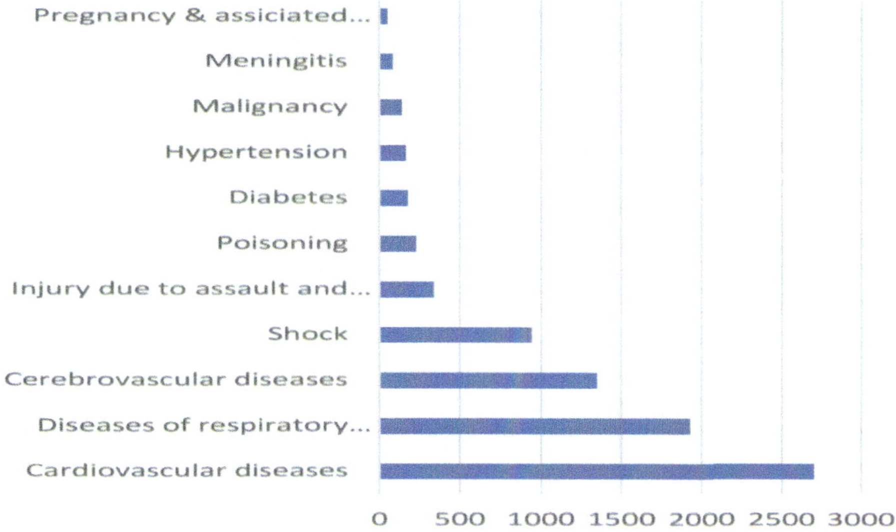

Fig. 4. Major mortality causes in Bangladesh (2017)

1.3 Aim of the Research

Our survey suggests that, there is no existence of any decent digital database for cardiovascular diseases in Bangladesh.

Bangladesh Government has taken 17 health related sustainable development goals for ensuring healthy lives and promotes wellbeing for all ages is governments 3rd health related sustainable development goals. A cardio vascular disease database could certainly help achieving such goals in current local context. It should provide decent machine learning approaches to solve diagnosis and analysis of critical problems in the near future.

We aim to implement the low cost ECG module that was developed to all upazila, union and ward level medical accommodations for collecting CVD data.

2 Methods

We plan to collect CVD data using low cost three lead ECG module and Google's Firebase server.

Firebase is outcome of Google's cloud research. The Firebase server can be integrated with play store, data studio, BigQuery, Slag, Google's Marketing Platform. In the app industry, Firebase is trusted by large apps such as The New York Times, Shazam, Duolingo, trivago, The Economist and many others [12].

We have used Raspberry pi 3 B+ board along with Arduino Mega board. The features of Raspberry Pi 3 are 1.2 GHz 64-bit quad-core ARMv8 CPU with built in 802.11n wireless lan, 1 GB RAM with 4 USB ports. It also has 40 GPIO pins, full HDMI ports, Ethernet port, combined 3.5 mm audio jack and composite video, CSI, DSI, micro SD card slot and video IV 3D graphics core. It uses a Linux base OS, we have used

Raspbian. The Arduino Mega 2560 is a microcontroller board. It significantly use 54 digital input/output pins (of which 15 can be used as PWM outputs), 16 analog inputs, 4 UARTs (hardware serial ports), a 16 MHz crystal oscillator. It also has a USB connection, a power jack, an ICSP header, and a reset button. It contains everything needed to support the microcontroller; simply connect it to a computer with a USB cable or power it with an AC-to-DC adapter or battery to get started [13].

A three lead ECG module is made of various cascaded filters, isolation circuitry and an arduino board. The arduino board basically works as an ADC (analog to digital converter). The module uses AD620 IC as its instrumentation amplifier. The AD620 is known for high efficiency amplifier which is very low in price [14].

The main target is to send the ECG module data to firebase server. Where it works step by step to complete its task. First the ECG module is connected with the arduino board through I2C connection where the arduino board sends the ECG data to the raspberry pi board using USB connection. Finally the raspberry pi board collects the data and pass it to the firebase server.

The data of the firebase server is accessed and visualized through website and android app (Figs. 5 and 6).

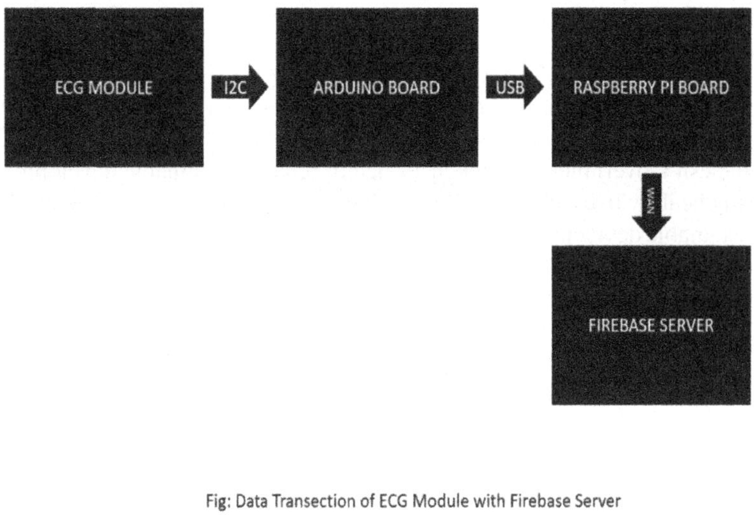

Fig. 5. Data transection process of the ECG module

Multiple ECG modules placed at various locations can send data to the firebase server by using the raspberry pi board. The ECG modules are planned to be placed in the upazila and union level medical accommodation and collect CVD electrocardiogram.

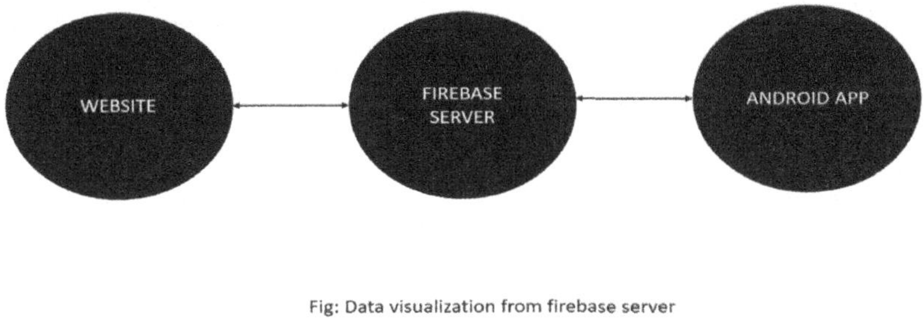

Fig. 6. Data transection process of the ECG module

3 Results

The design of the ECG module has been completed along with fabrication. The ECG module distributes its gain in two stages for attenuation of noise (Fig. 7).

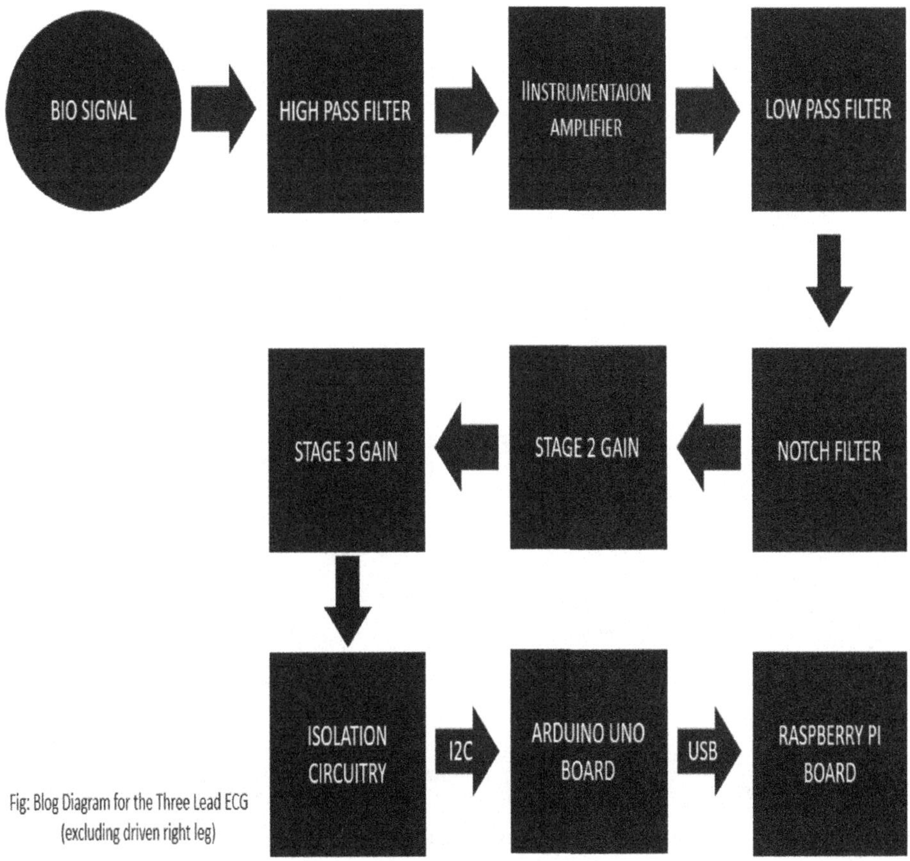

Fig. 7. ECG module block diagram

The schematic design was conducted in Proteus. TL series opamps are being used due to their good noise attenuation capability.

The figure bellow shows the detailed schematic of the three lead ECG [15–18] (Fig. 8).

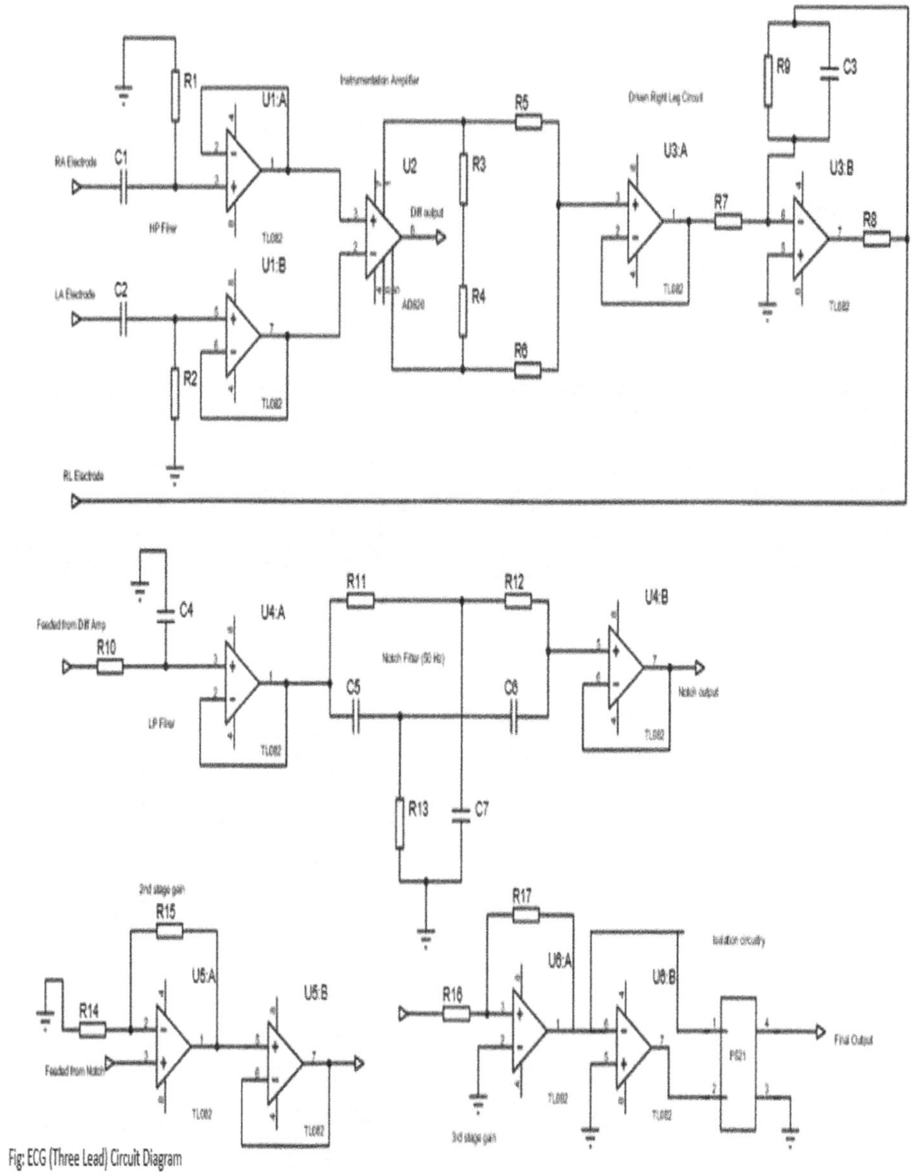

Fig. 8. Full proteus schematic of the ECG module

The bio-signal at first is feeded to a high pass filter from then to instrumentation amplifier (AD620). The DRL circuitry enables to reduce noise along with it also deals with safety issues (Figs. 9 and 10).

The AD620 is being used as a diff amplifier.

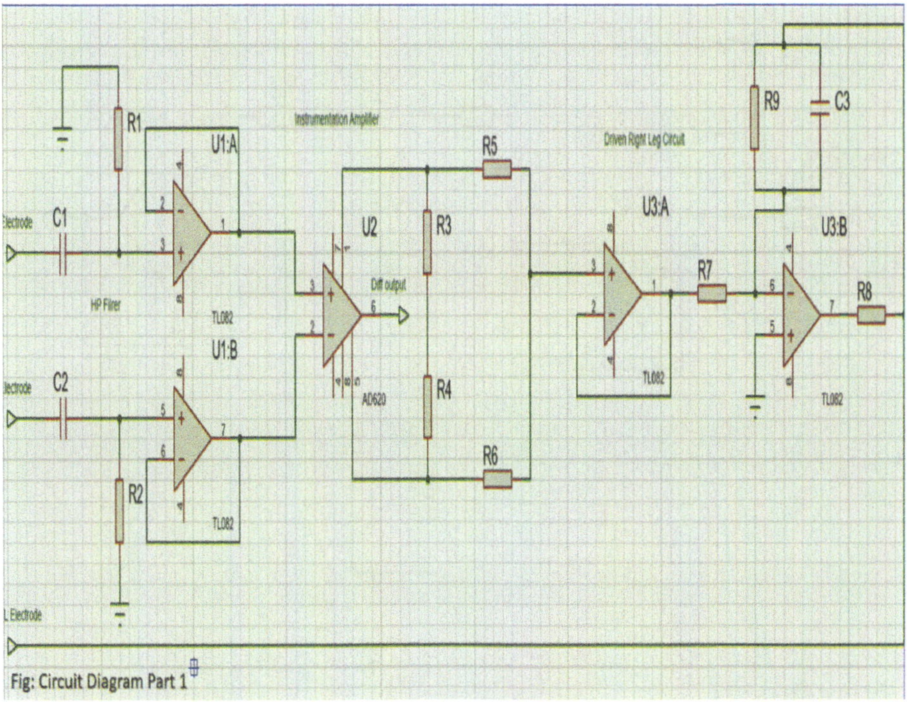

Fig. 9. Proteus schematic of the instrumentation amplifier and Driven right leg circuit

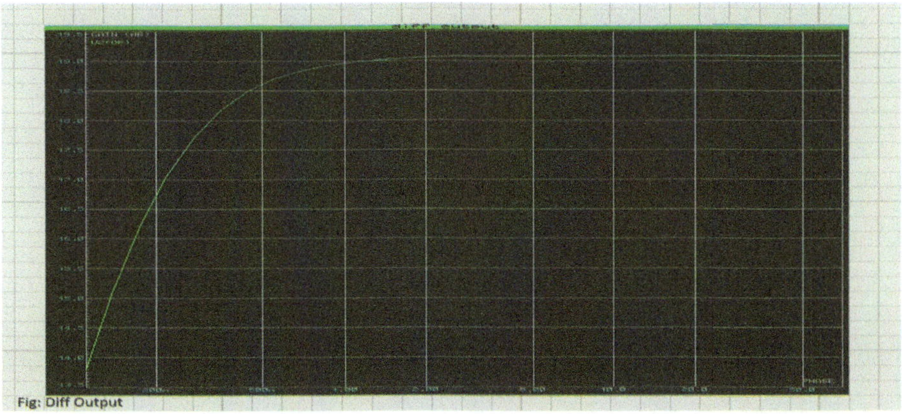

Fig. 10. Proteus simulation output from diff amplifier

The diff output is then feeded to a low pass filter followed by a notch filter (Figs. 11 and 12).

The notch filter trunks off the 50 Hz power line noise very sharply.

As the 50 Hz signal is attenuated, two gain stages are introduced (Fig. 13).

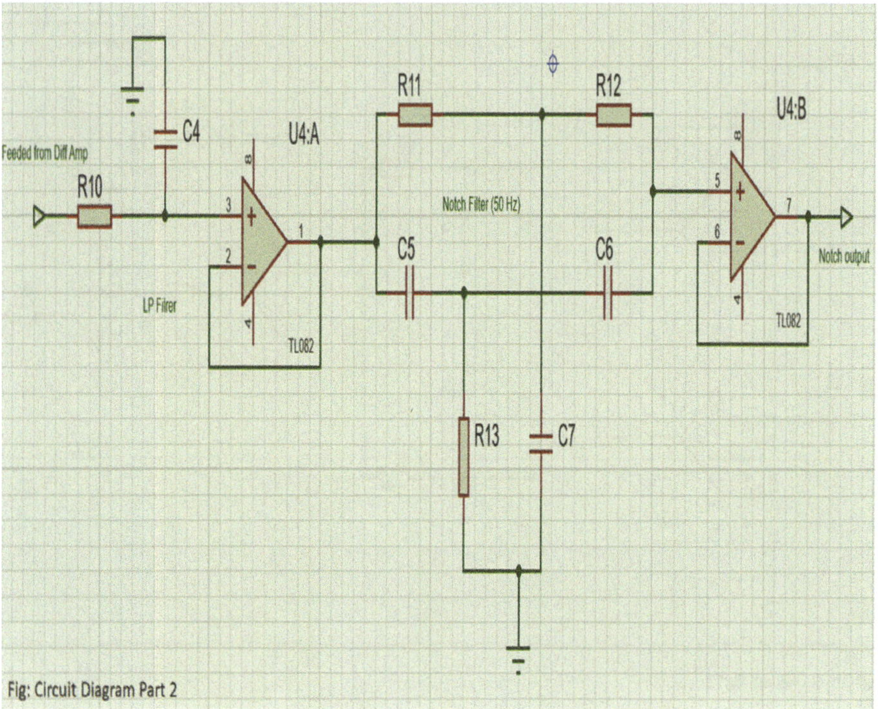

Fig. 11. Proteus schematic of cascading filters

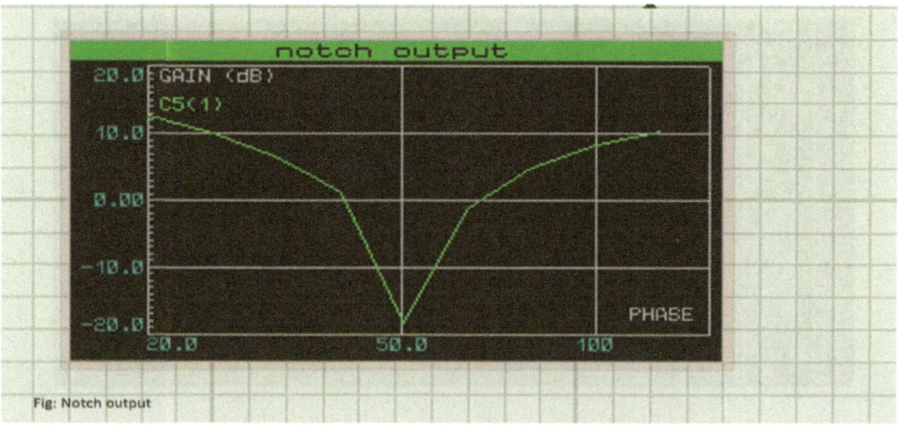

Fig. 12. Proteus simulation output of the notch filter

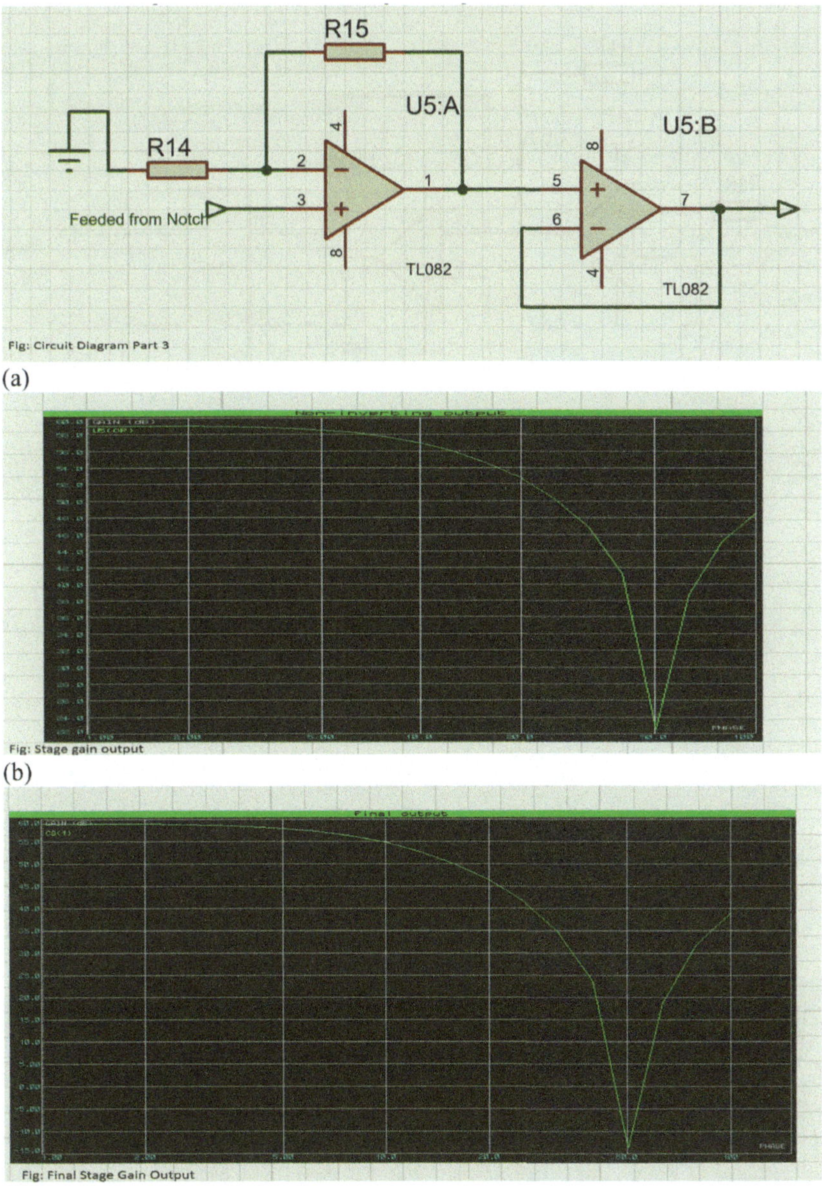

(a)

(b)

(C)

Fig. 13. (a) Proteus schematic for two gain stages. (b) Proteus simulation output from first gain stage. (c) Proteus simulation output from the final stage gain

The final output is feeded to a P521 IC for power isolation (Fig. 14).

The output of the fabricated device was visualized by a java software provided by local vendor. The output of the fabricated device is provided bellow (Fig. 15):

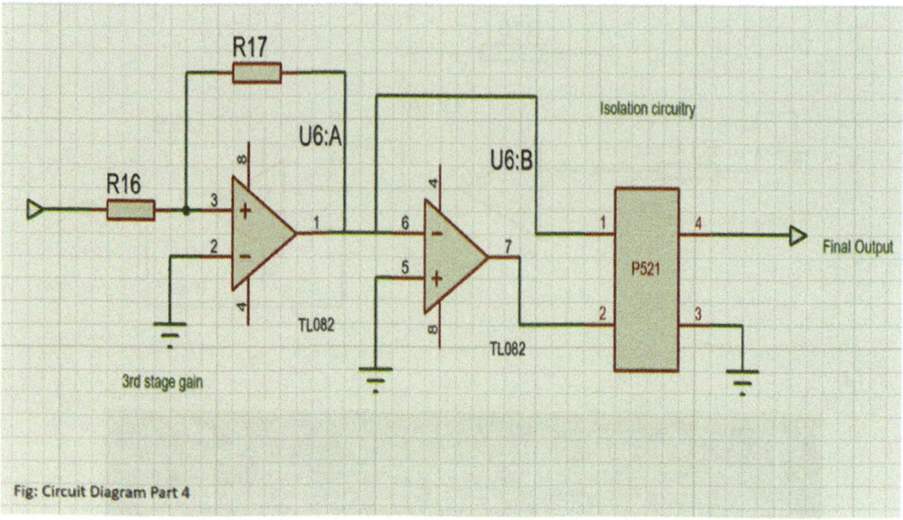

Fig. 14. Proteus schematic for isolation circuitry

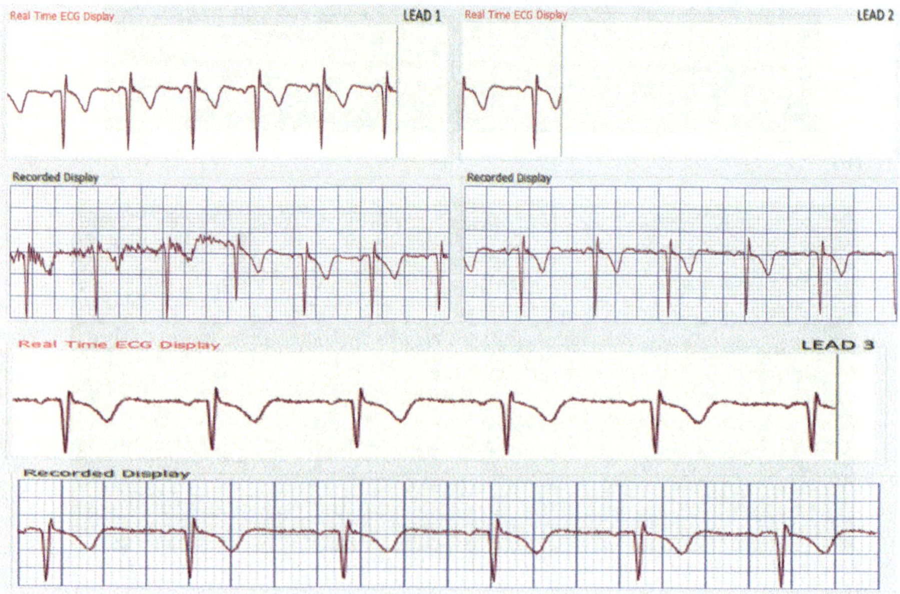

Fig. 15. Output from the fabricated three lead ECG device

A snap shot of the firebase server data and android app interface is provided bellow (Fig. 16):

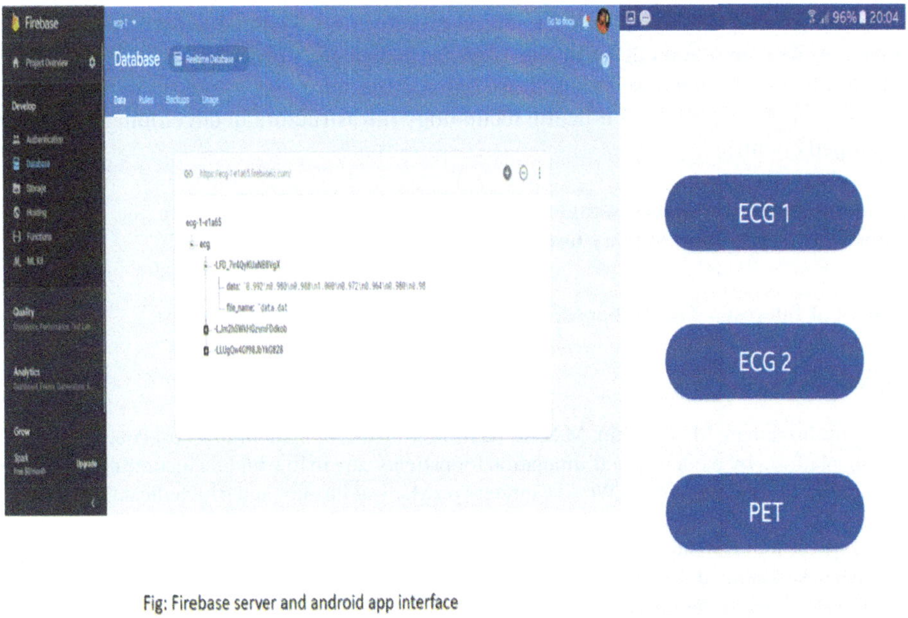

Fig. 16. Snapshot of firebase server and android app GUI.

As the fabrication phases have been accomplished, the current research phase involves data acquisition. With the developed ECG modules data acquisition of various patients is currently undergoing.

4 Conclusions

From the analysis of local context it is eminent that establishment of a CVD database is indispensable due to high mortality rate of CVD and such establishment will require a completely different approach due to unique local health infrastructure. In such distinctive local perspective this low cost and viable model of CVD database establishment should be prolific. Beyond the context of Bangladesh, the severity of CVD can be seen in other developing and lower income localities. This low cost and viable model can play a vital role in confronting CVD in such areas. Initiatives of such scale will surely pave the way for researches in the field of Machine Learning and Deep Learning on the local context for detecting Cardiovascular Complications in local population. The scarcity of proper technical manpower, resources and lack of interest has been a great barrier for researches which solely focuses on local population. However, despite of many disadvantageous factors dept. of Biomedical Engineering, MIST has been conducting researches solely focusing on the welfare of the local population. This research which aimed to seek a viable model for establishing central CVD database was inspired by previous researches of the department, such as low cost central monitoring system development and many others. Active involvement form the likes of IFBME/IUPESM/JACE/ACCE/CEASA will surely excel such researches. Last but not the least; such initiatives will surely pave the way for an efficient e-health technology infrastructure in developing and under developed countries.

Acknowledgments. Authors would like to acknowledge the support from the Biomedical Engineering Department, Military Institute of Science and Technology (MIST).

Conflict of Interest. The Authors declare that they have no conflict of interest.

References

1. Ashrafuzzaman, M., Hossain, M.S.A., Aziz, M.T.: Development of low cost central monitoring platform by modeling and simulation for patients care in low middle income countries. In: IFMBE Proceedings, 2018 World Congress on Medical Physics and Biomedical Engineering, pp. 737–744 (2019)
2. Pay Research Bureau Report (2016)
3. Ashiq-Ur-Rahman, M.: Housing the Urban Poor in Bangladesh: A Study of Housing Conditions, Policies and Organisations (Ph.D. thesis) (2012). https://www.ros.hw.ac.uk/handle/10399/2600?show=full
4. World bank Annual Report (2016). https://documents.worldbank.org/curated/en/782691475489708512/World-Bank-annual-report-2016-organizational-information-and-lending-data-appendixes
5. Bangladesh Health System Review, WHO (2015). https://apps.who.int/iris/handle/10665/208214
6. Govindaraj, R., Raju, D., Secci, F., Frere, J.-J.: Health and Nutrition in Urban Bangladesh: Social Determinants and Health Sector Governance (2018). https://doi.org/10.1596/978-1-4648-1199-9
7. HEALTH BULLETIN 2017 and 2018, Management Information System Directorate General of Health Services
8. Bangladesh 2016 Politics, Governance and Middle Income Aspirations Realities and Challenges An Empirical Study
9. Community Health Systems Catalog Country Profile: Bangladesh, June 2017
10. Health Information Systems and eHealth (HIS & e-Health) Director (MIS), DGHS: Inter-country Conference on Data for Decision (D4D)
11. Ahmed, T., Bloom, G., Iqbal, M., Lucas, H., Rasheed, S., Waldman, L., Khan, A., Islam, R., Bhuiya, A.: E and M Health in Bangladesh: Opportunities and Challenges... IDS Evidence Report (2014)
12. https://firebase.google.com
13. https://www.raspberry-projects.com/pi/pi-hardware/raspberry-pi-3-model-b/rpi3-model-b-hardware-general-specifications
14. https://store.arduino.cc/usa/arduino-mega-2560-rev3
15. AD620 Datasheet
16. Gakare, P., Patel, A., Vaghela, J., Awale, R.: Real time feature extraction of ECG signal on android platform, pp. 1–5 (2012). https://doi.org/10.1109/ICCICT.2012.6398230
17. Rabbani, K.S.E.: Neuro-Physiological Measurements for Diagnosis using Evoked Responses. Bangladesh J. Med. Phys. **5**(1), 1–24 (2012)
18. Payne, E.K., Harmah, D.J.: Multi-parametric telemetry patient monitoring system. Int. J. Sci. Res. Publ. **5**(11), 272–276 (2015)

Diagnostic Evaluation

Time-Frequency Analysis of the First-Order Coefficient of Auto-Regressive Approach to Recognize the Normal Breath or Wheeze

Bing-Yuh Lu[1(✉)] and Chin-Yuan Lin[2]

[1] Guangdong University of Petrochemical Technology, Guangdong, People's Republic of China
[2] Tungnan University, Taiwan, Republic of China
franklinlu888@outlook.com

Abstract. This study presents a method of computing the coefficients of auto-regressive (AR) model at first and then showing the spectrogram of the 1st order coefficient of the AR model. The spectrogram is a popular time-frequency analysis for studying the lung sound. Researchers usually employed the method to find the components of the specific frequencies to enhance the recognition of the lung sounds. The AR model is a powerful mean for time series analysis. The proposed method was tested by segments of normal breaths sound and those of wheezing. The results showed that the normal breaths seldom display the continuous stripes, but the breaths of wheeze (asthma, bronchiolitis, and COPD) usually present the continuous stripes in a manner of the vertical trend in the spectrograms of the 1st order coefficient of AR model. In conclusion, we find a faster computing process (80% off time-consuming) to recognize the wheeze as a manner of better performance than spectrogram only.

Keywords: Time-frequency Analysis · Auto-Regressive · Wheeze · Spectrogram

1 Introduction

This study introduced an innovative method to analyze the lung sound. At first, the lung sound was transferred to a 4th-order auto-regressive (AR) model, and then showed the spectrogram of the first order coefficient of the AR model. Compared with the method of directly computing the spectrogram, the method we proposed presented a higher speed in computing time, and clearer recognition in the visual presentation.

2 Methods

The functional block diagram of the processing of this study was presented in Fig. 1. In Fig. 1(a), the lung sound signal was preprocessed by normalized (Nz) and moving averaged (MA) and then was analyzed by time-frequency method to show the spectrogram. The proposed method was presented in Fig. 1(b). The acoustic signal was normalized and moving averaged to transfer the AR approach, then the 1st order coefficient (1OC)

Y. Shiraishi et al. (Eds.): APCMBE 2020, IFMBE Proceedings 82, pp. 175–180, 2021.
https://doi.org/10.1007/978-3-030-66169-4_22

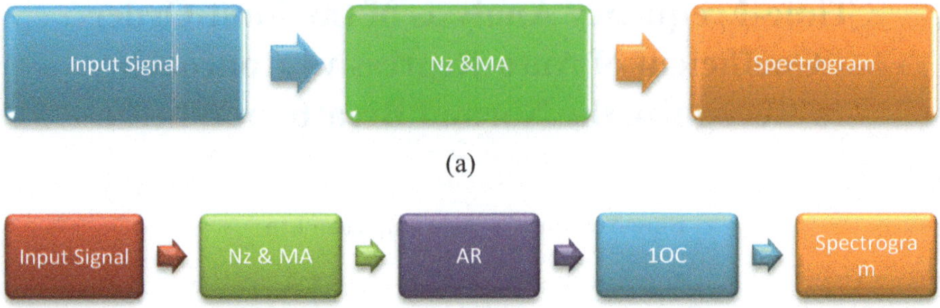

(a)

(b)

Fig. 1. The functional block diagram: (a) spectrogram only; (b) Spectrogram of the 1OC of AR

of the AR model was analyzed by time-frequency method to show its spectrogram. The computation time of (a) and (b) were recorded in the computer promgram.

The spectrogram $S_\tau(j\omega)$ of a signal s(t) at time t can be written as [1],

$$|S_\tau(j\omega)|^2 = \left| \int_{-\infty}^{\infty} s(t)w^*(t-\tau)e^{-j\omega t}dt \right|^2 \tag{1}$$

where $w^*(t-\tau)$ is a window function. Therefore, $w^*(t-\tau)e^{-j\omega t}$ represents an envelope function which operates a time-shift of τ with a pair of sine waves whose difference of phase is 90°. The operator $e^{-j\omega t}$ is the same to Fourier transform which is also known as the short-time Fourier transform (STFT), has numerous applications in speech, sonar, and radar processing. The spectrogram of a sequence is the magnitude of the time-dependent Fourier transform versus time.

The AR model is a powerful mean for time series analysis. The polynomial was written as,

$$\Phi(z) = 1 - \sum_{k=1}^{p} \varphi_k z^k \tag{2}$$

where $k = 0,1,2\ldots$, φ_k denotes for the samples of the modeled signal, $\Phi(z)$ for the AR coefficients, and p for the order of the AR model. The coefficients of AR model are approached by taking the minimum of the variance [2].

The simulation was executed in a MATLAB 2016b (Mathwork, USA) developing environment. A laptop PC was used (ACER Aspire V Nitro, Taiwan, ROC) with Intel CPU Core i7-4720HQ, NVIDIA GPU GeForce GTX 960M (4GB GDDR5) with a 16 GB DDR4 memory to run the computer simulation. The operating system was Microsoft Windows 8.1.

3 Methods

Table 1 compares the computing times of functional block diagrams of (a) and (b) in Fig. 1. It is disclosed that the proposed method has less computation time.

Table 1. Comparison of the computing time

	Spectrogram only (s)	Spectrogram of 1OC (s)
Wheeze (10.5 s)	14.6094	2.8438
Normal breath(15 s)	20.1563	4.6250

The results presented the computation of a wheeze and a normal breath, respectively. In Fig. 2, (a) is the signal of a wheeze; (b) is the spectrogram of (a) after moving averaged; (c) is the 1OC of AR model of (a) after moving averaged; and (d) is the spectrogram of (c). In Fig. 3, (a) is the signal of a normal breath; (b) is the spectrogram of (a) after moving averaged; (c) is the 1OC of AR model of (a) after moving averaged; and (d) is the spectrogram of (c). The other breaths of wheezes, i.e. asthma, bronchiolitis, and chronic obstructive pulmonary disease (COPD) are presented in Fig. 4 as the spectrograms of the 1OC of AR model.

4 Discussion

The AR model whose computations are all in time domain is a simple real number calculation. Therefore, the computing time has less computation time than that of spectrogram. Furthermore, the segmentation size of AR model was 256. The number of points for the computation of spectrogram is 1/256 times of the spectrogram only. That is the dominant operation for the computation speedup.

The wheeze is usually presented some horizontal stripes in the spectrogram, because of the disclosing of the higher harmonic components of the wheeze. The stripes usually maintain 20 to 100 ms. Therefore, the spectrogram should be enlarged to find them. However, in Fig. 2(d), the stripes present in a manner of vertical lines, and easy to find. Furthermore, the results of the normal breath are presented in Fig. 3 for comparing with the results of wheeze. The normal breath does not display the stripes in the both spectrograms. The method we proposed presents the all of the wheezes own more continuous stripes and higher intensity than those of the normal breath in the spectrogram of the 1OC in AR model (Fig. 4).

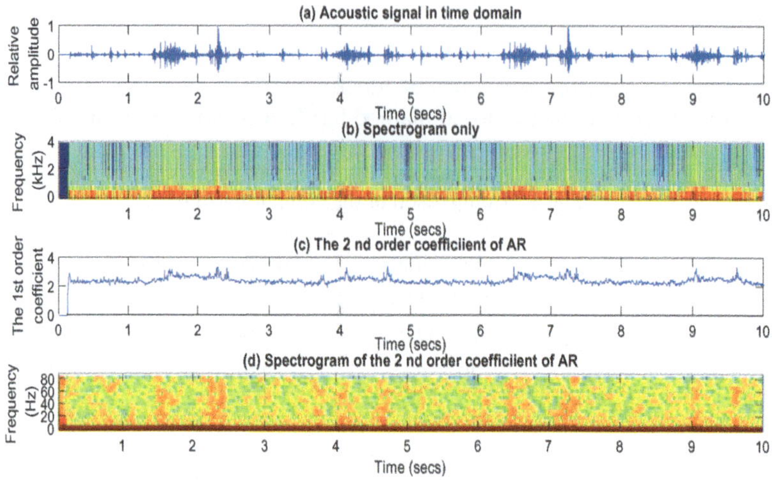

Fig. 2. Wheeze

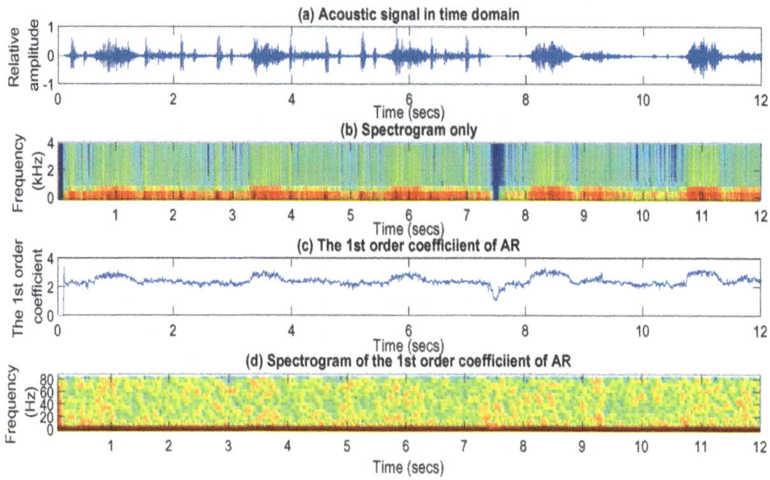

Fig. 3. Normal breath

Lu et al. [3] studied on the real-time noise cancelation method. The adaptive filter has been employed to reduce the sound of siren on the ambulance. The results disclosed the performance of the adaptive filter. Some of the time delay of the computation exists, but do not influence the total performance because of the unstable time delays of the network communication. The method that this study proposed can aid such study to reduce the computing time and the volume of network communication for recognizing the wheezing sounds [4, 5].

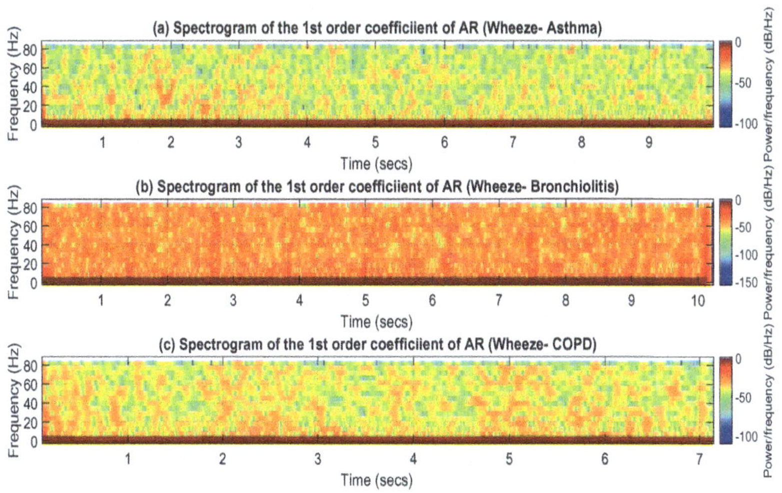

Fig. 4. Wheeze: asthma, bronchiolitis, and COPD

5 Conclusion

We find a faster computing process to recognize the wheeze as a manner of better performance than spectrogram only. The computing time of the method we proposed is reduced to 20% of spectrogram only. A novel method to identify the wheeze is successfully presented by the spectrogram of the 1OC of the AR approach. The method was tested by segments of normal breaths sound and those of wheezing. The results showed that the normal breaths seldom display the continuous stripes, but the breaths of wheeze (asthma, Bronchiolitis and COPD) usually present the continuous stripes in a manner of the vertical trend in the spectrograms of the 1OC of AR model. The studies of the elementary components of the coefficients of the AR model will be found in the further researches.

Acknowledgments. The authors thank for the support from www.Thinklabs.com. In this study, the lung sounds which were recorded by a Thinklabs digital stethoscope were downloaded from the website of Thinklabs (https://www.thinklabs.com/copy-of-lung-sounds) under the agreement of the copyright notice for www.Thinklabs.com. We also thank Dr. Gwo-Ching Chang, Associate Professor at Department of Information Engineering, I-Shou University, Taiwan, ROC for his sincere personal communications.

References

1. Haykin, S., Van Veen, B.: Signals and Systems. Wiley, New York (1998)
2. Chang, G.C., Lai, Y.F.: Performance evaluation and enhancement of lung sound recognition system in two real noisy environments. Comp. Meth. Prog. Biomed. **97**, 141–150 (2010)
3. Lu, B.Y., Hsueh, M.L., Wu, H.D.: Reducing the ambulance siren noise for distant auscultation of the lung sound. Acoust. Austr. **45**(2), 381–387 (2017)

4. Khan, T.A., Vijayakumar, P.: Separating heart sound from lung sound using LabVIEW. Int. J. Comp. Electr. Eng. **2**(3), 524–533 (2010)
5. Gnitecki, J., Moussavi, Z.M.: Separating heart sounds from lung sounds. IEEE Eng. Medi. Biol. Mag. **26**(1), 20 (2007)

Analysis of Representations of 3-Dimensional Objects in the Cell Populations in the Temporal Association Area Using Machine Learning

Jun-ya Okamura(✉), Yusuke Yamamoto, Lulin Dai, Yoshihiro Uto, Yousuke Yamada, and Gang Wang

Kagoshima University, 1-21-40 Korimoto, Kagoshima 890-0065, Japan
jokamura@ibe.kagoshima-u.ac.jp

Abstract. Three-dimensional objects can be recognized regardless of the viewing angles. It was reported that the cells in the temporal association area responded tolerantly in a certain viewing angle range to the objects experienced in discrimination at the same viewing angles. In the present study, machine learning was applied to the population response vectors in the temporal association area. A classifier was trained to create a hyperplane that separated an object from similar object at the same viewing angles and then tested by the population response vectors to the object images at a different viewing angle. The discrimination performance evaluated from true positives, false positives, and d' values was higher in the objects experienced in discrimination at the same viewing angles than those without discrimination experience in the whole 100–600 ms time window. For the objects experienced in discrimination at the same viewing angles, the d' values were low in the early phase of the responses of 100–220 ms time window, but became higher in the late phase of 220–600 ms time window. The d' values for the objects experienced by learning association of different views of the objects were high in the early and late phases of the responses. The results demonstrate computational models for object representation created by using cellular responses in the temporal association area, and differences of the representation of the viewing angle tolerance in different time windows.

Keywords: Temporal association area · Population responses · Machine learning

1 Introduction

Two-dimensional images of an object changes depending on the viewing angles, but we can recognize the object regardless of the viewing angles. It was assumed that the view-invariant object recognition capability develops through successively experiencing different views of an object when the object rotates or when the observer moves around the object. Different views of an object are associated with each other [1]. Recently, it was found that learning association of different views was not required to develop the view-invariant object recognition capability [2, 3]. Experiencing discrimination of similar objects at the same viewing angles can develop view-invariant object recognition

Y. Shiraishi et al. (Eds.): APCMBE 2020, IFMBE Proceedings 82, pp. 181–187, 2021.
https://doi.org/10.1007/978-3-030-66169-4_23

capability in a certain range of the viewing angle. Following electrophysiological studies found that cells in the anterior part of the temporal association area responded tolerantly in certain viewing angle range to the objects experienced in discrimination at the same viewing angles [4, 5]. In the present study, to understand the neural representations of the three-dimensional objects, machine learning was applied to the responses of populations of the cells in the temporal association area. A classifier was trained to discriminate the objects at the same viewing angles, and then tested by the population responses to the object images at different viewing angles. Discrimination performance was compared between the objects experienced in discrimination at the same viewing angles and those without discrimination experience. The discrimination performance to the objects experienced in discrimination at the same viewing angles was also compared with those experienced learning association of different views.

2 Methods

2.1 Animals

Three male macaque monkeys, weighting 10, 10, and 8 kg, were used. Experimental procedures were performed in accordance with the guidelines of the Japan Neuroscience Society and approved by the Animal Experiment Committee of Kagoshima University. Rearing of the animal, anesthesia, surgery, and electrophysiological experimental procedures were described in the previous reports [4, 5]. Responses of the cells, recorded in our previous reports [4, 5] in the anterior part of the temporal association area, were used for the machine learning analysis in the present study.

2.2 Stimulus Object Images

Objects were created using 3D graphics software (Shade 9, e-frontier, Tokyo, Japan). Details of the object creation were described in our previous reports [2–5]. In brief, four daughter objects were created by deforming a prototype object in four different directions in three-dimensional space. The four objects were rotated from 0° to 90° in 30° intervals around an axis that was perpendicular to the axis connecting the viewer's eyes and the object. A total of 16 object images comprised an object set (Fig. 1A). Similarities of the objects within a set was adjusted by a human psychophysics so that about 80% of the discrimination performance was obtained in any pair of the objects. Six object sets were created from distinct protypes.

2.3 Training Task

A task started when a monkey, sitting in a monkey chair, held a lever placed in front of the animal. A fixation point appeared for 500 ms before the image of the first object appeared for 500 ms. The first object appeared 1–4 times, each separated by the appearance of a fixation point, before the image of the second object appeared. The monkey had to keep the lever pressed during the appearances of the first object, and release the lever when the second object appeared to get a juice reward. There were three versions of the task (Fig. 1B). One version was Object task, in which the first object appeared with the viewing angle changed before the appearance of the second object selected from the same object set as the first object. The monkey learned association of different views of an object in Object task. The second version was Within-set Image task, in which the first object appeared at the same viewing angles before the appearance of the second object selected from the same object set at the same viewing angles as the first object. No association of different views of an object was required in Within-set Image task. The third version was Across-set Image task, in which the first object appeared at the same viewing angles before the appearance of the second object selected from different object sets. Across-set image task was designed just to expose the object images to the monkey. The monkeys experienced two object sets in each of the three versions of the task.

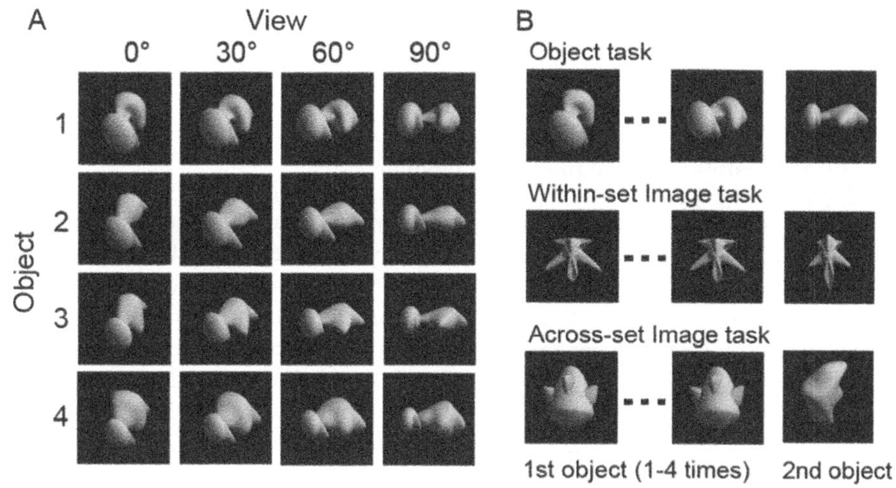

Fig. 1. An example of an object set (A), and presentation sequences of the object images in Object, Within-set Image, and Across-set Image tasks.

2.4 Machine Learning Using Population Response Vectors

In the present study, data of the spike rates of the cells recorded in our previous studies in the responses to the object images that the monkeys had experienced in the three versions of the tasks were used. Spontaneous spike rates during 500 ms preceding the stimulus onset were subtracted from the spike rates during 100–600 ms period after the stimulus onset. Because the responses of the cells in the temporal association area contain two phases [5] in the early (100–220 ms) and late (220–600 ms) phases, spike rates in the early and late phases were also used for analysis after the spontaneous spike rates were subtracted. The spike rates were normalized by the following formula,

$$R_{normalization} = \frac{R_i - R_{min}}{R_{max} - R_{min}},$$

where R_{max} and R_{min} are, respectively, the largest and the smallest spike rates among the responses to the 16 images in an object set, and R_i is the spike rates in the responses to the ith stimuli in a total of 16 images in an object set. Population response vectors were created by aligning the normalized responses of each cell to the image into vectors. Six population response vectors, corresponding to six repetitions of the presentation of each image, were created for each image. Response vectors were created for each object image that the monkeys had experienced in the three versions of the tasks. The dimension of the vectors corresponded to the numbers of the cells showing statistically significant responses to at least one of the images in an object set. Total numbers of the cells were 78, 108, and 75 for the objects experienced in Object, Within-set Image, and Across-set Image tasks, respectively. Support vector machine (SVM) was applied by using python. A classifier was trained to create a hyperplane that separated the object from the other object in an object set at the same viewing angles, and then tested by the response vectors to the object images at a viewing angle difference of 30°.

3 Results

3.1 Discrimination Performance of a Classifier

Discrimination performances of the classifier trained to create a hyperplane separating an object from the other object at the same viewing angles were tested by the population response vectors created by aligning the responses to the object images at a viewing angle difference of 30°. One of the four objects were labeled as 1, and the other object was labeled as −1. Labels of the object images at a viewing angle difference of 30° was predicted by the test. Among the object images, in which the predicted labels in the test was 1, true positive was a proportion of the object images correctly labeled as 1, and false positive was a proportion of the object that was wrongly labeled as 1 at a different viewing angle. True positive and false positive rates of the objects experienced in Object, Within-set Image, and Across-set Image tasks were shown in Fig. 2. Proportions of the true positives were significantly larger than those of the false positives in the objects experienced in Object, Within-set Image, and Across-set Image tasks ($p = 1.3 \times 10^{-47}$, 1.3×10^{-46}, and 2.5×10^{-14}, respectively, chi-squared test) in the whole 100–600 ms period. In the whole 100–600 ms period, the differences between the true and false positives were smaller in the objects experienced in Across-set Image task than those in the Object and Within-set Image tasks. In the early phase of 100–220 period, although the difference between the true and false positives were larger in the objects experienced in Object task than those in Within-set and Across-set Image tasks, the proportions of the true positives were significantly larger than the false positives in the objects experienced in Object, Within-set Image, and Across-set Image tasks ($p = 2.1 \times 10^{-21}$, 1.6×10^{-4}, and 3.4×10^{-7}, respectively, chi-squared test). In the late phase of 220–600 ms period, true positives were significantly larger than the false positives in the Object, Within-set Image, and Across-set Image tasks ($p = 1.5 \times 10^{-42}$, 4.4×10^{-51}, and 1.2×10^{-25}, respectively, chi-squared test).

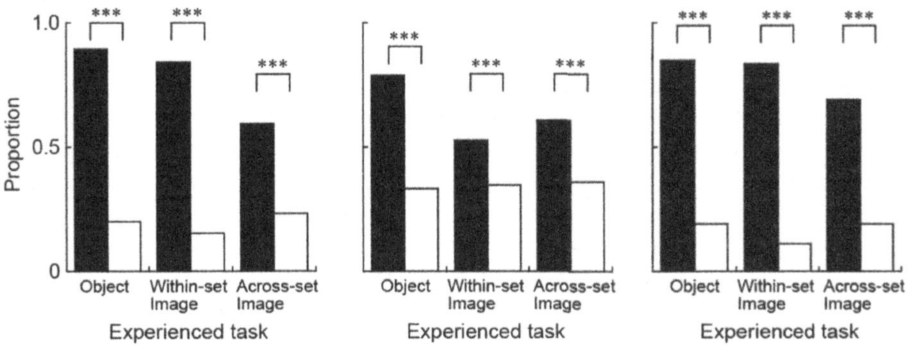

Fig. 2. Proportions of the true positive (black) and false positive (white) in the test of the classifier with the population response vectors for the images at a viewing angle difference of 30° from the trained view in the objects experienced in Object, Within-set Image, and Across-set Image tasks at time windows of 100-600 ms (left), 100-220 ms (middle), and 220-600 ms (right). *** $p < 0.001$.

The discrimination performance was evaluated by d' calculated from the proportions of the true and false positives (Fig. 3). The d' values in the objects experienced in Object and Within-set Image tasks were significantly larger than that in the objects experienced in Across-set Image task ($p = 0.005$ and 0.02, respectively, Kruskal-Wallis test followed by Bonferroni correction) in the whole 100–600 ms period. In the early phase of 100–220 period, The d' value for the objects experienced in Object task was significantly larger than those for the objects experienced in Within-set and Across-set Image tasks ($p = 0.007$ and 0.039, respectively). In the late phase of 220–600 ms period, the d' value for the objects experienced in Within-set Image task was significantly larger than that for the objects experienced in Across-set Image task ($p = 0.045$), and comparable to that for the objects experienced in Object task ($p = 1.0$). The d' value for the objects experienced in Object task was significantly larger than that for the objects experienced in Across-set Image task ($p = 0.045$) in the late phase.

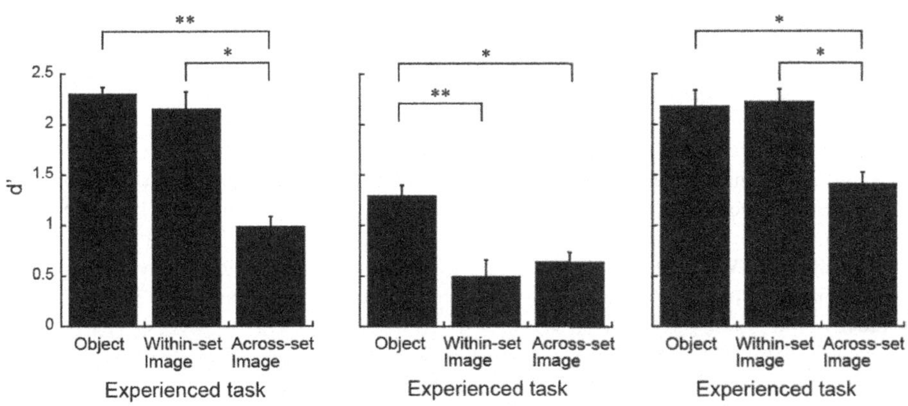

Fig. 3. The d' values calculated from the proportions of the true positive and false positive for the objects experienced in Object, Within-set Image, and Across-set Image tasks at time windows of 100-600 ms (left), 100-220 ms (middle), and 220-600 ms (right). * $p < 0.05$, ** $p < 0.01$.

4 Discussion

In the present study, machine learning was applied to the responses of the cells in the temporal association area to examine the generation of the viewing angle tolerance of the cells. Population response vectors were created and machine learning was applied in order to train a classifier to create a hyperplane that separated the objects within an object set at the same viewing angles, and the classifier was tested by the population response vectors in the responses to the object images at a viewing angle difference of 30° from the trained view. The proportion of the true positive was significantly larger than the false positive in the objects experienced in Object, Within-set Image, and Across-set Image tasks, but the differences between the true and false positives tended to be larger in the objects experienced in Object and Within-set Image tasks than in the objects experienced in Across-set Image task in the whole 100–600 ms time window. The differences between the true and false positives tended to be small in the objects experienced in Within-set Image task at the early phase of 100–220 ms time window, but the difference was large at the late phase of 220–600 ms time window. To evaluate the differences between the true and false positives, d' values were calculated. The d' values for the objects experienced in Object and Within-set Image tasks were significantly larger than that for the objects experienced in Across-set Image task in the whole time window. The d' value for the objects experienced in Object task was significantly larger than those for the objects experienced in Within-set Image and Across-set Image tasks in the early phase, while the d' value for the objects experienced in Within-set Image task became larger than that for the objects experienced in Across-set Image task, and comparable to that for the objects experienced in Object task. These results suggest after the discrimination experience of the objects in Within-set Image task, the viewing angle tolerance of the populations of the cells in the temporal association area may be represented in the late phase of the responses, while after learning association of different views of the objects in Object task, the viewing angle tolerance may be represented from the early phase of the population responses. Further study such as following the discrimination performances in the response time courses may be conducted in order to reveal the temporal dynamics of the representation of the viewing angle tolerance for the objects experienced in the three versions of the tasks.

References

1. Stryker, M.P.: Temporal associations. Nature **354**, 108–109 (1991)
2. Wang, G., Obama, S., Yamashita, W., Sugihara, T., Tanaka, K.: Prior experience of rotation is not required for recognizing objects seen from different angles. Nat. Neurosci. **8**(12), 1768–1775 (2005)
3. Yamashita, W., Wang, G., Tanaka, K.: View-invariant object recognition ability develops after discrimination, not mere exposure, at several viewing angles. Eur. J. Neurosci. **31**(2), 327–335 (2010)

4. Okamura, J.Y., Yamaguchi, R., Honda, K., Wang, G., Tanaka, K.: Neural substrates of view-invariant object recognition developed without experiencing rotations of the objects. J. Neurosci. **34**(45), 15047–15059 (2014)
5. Okamura, J.Y., Uemura, K., Saruwatari, S., Wang, G.: Difference in the generalization of response tolerance across views between the anterior and posterior part of the inferotemporal cortex. Eur. J. Neurosci. **48**(12), 3552–3566 (2018)

X-ray Projection and Fluid Analysis of Contrast Agent Dynamics Through Stenosis

Ryotaro Kaneko(✉), Asuka Hatano, and Satoshi Izumi

Department of Mechanical Engineering, Graduate School of Engineering, The University of Tokyo, 7-3-1, Hongo, Bunkyo, Tokyo, Japan
kaneko.ryotaro@fml.t.u-tokyo.ac.jp

Abstract. Diagnosis using the contrast CT images is a simple and minimally invasive method for estimating the functional severity of coronary stenosis. Since the contrast CT images get the distribution of the contrast agent, it is important to clarify the advection and diffusion of contrast agent in the diagnosis of stenosis severity using a CT value gradient such as Transluminal Attenuation Gradient (TAG). In this study, contrast agent flowing through stenosed tubes was imaged by X-ray simple projection. Contrast agent flowed in water or blood under steady flow conditions. The tubes were diameter of the tube was 3 and 2.1 mm, artificially stenosed with area reduction of 62% and 69% simulating a stenosed coronary artery for the case of water and blood respectively. In the case of water, the stenosis spread the contrast agent uniformly over the entire cross section. On the other hand, in the case of blood, the region with high concentration of the contrast agent existed along the tube wall. In the analysis simulating the case of blood, the jet flow was off the center of the axis and along the tube wall. However the analysis of contrast agent dynamics did not agree with the experiment. This may be affected by the fact that the analysis is sorely in consideration of advection.

Keywords: Contrast agent · Stenosed tube · X-ray simple projection

1 Introduction

Myocardial infarction is a major cause of death in developed and developing countries. The disease is necrosis of the heart wall and necrosis is caused by stenosis in the coronary arteries. In order to select an appropriate treatment method for myocardial infarction, it is necessary to accurately evaluate the functional severity of the stenosis. Currently, the basic evaluation method is cardiac catheterization. This method can be expected to be highly accurate evaluation [1], but has a problem in that it involves invasiveness. Therefore, TAG has attracted attention as a simple and minimally invasive method. According to Wong, TAG calculated with 320-row CT is 77% sensitivity and 74% specificity. It is inferior to FFR-CT [2]. However, the explanation of these values from the viewpoint of fluid dynamics is insufficient, and the relationship between stenosis shape and TAG value is unknown.

In our laboratory, fluid structure interaction analysis and contrast agent dynamics analysis have been performed to consider the relationship between stenosis shape and

Y. Shiraishi et al. (Eds.): APCMBE 2020, IFMBE Proceedings 82, pp. 188–194, 2021.
https://doi.org/10.1007/978-3-030-66169-4_24

TAG value. For the fluid structure interaction analysis, the validity of the calculation results of flow rate and pressure was confirmed by comparison with experiments in a bifurcation system model that assumed a coronary artery [3]. On the other hand, it is necessary to confirm the validity of the analysis of the contrast agent dynamics, but no experiments have been performed on the projecting of the contrast agent dynamics in tubes with stenosis. In this study, in order to clarify the effect of stenosis of the tube on the dynamics of the contrast agent, we performed an experiment of projecting the dynamics of the contrast agent flowing a stenosed tube simulating a stenosed blood vessel. In order to clarify the phenomena, fluid analyses simulating the experiment was performed.

2 Method

2.1 Experiment

In this study, assumed a coronary artery with stenosis, an experiment equipment shown in Fig. 1 below was constructed. Silicon tube was used for the flow path, and pressure gauges (KEYENCE, AP-12S) and a flow meter (KEYENCE, FD-XS8) were installed in the arrangement. The flow meter required calibration, and the relationship between the flow meter value and the actual flow rate was examined in advance. Changing the voltage applied to the pump from the DC variable power supply (Texio, PFR-100L50), the pressure at the inlet of the tube was controlled. The contrast agent was injected using an injector from the branch made immediately after the pump.

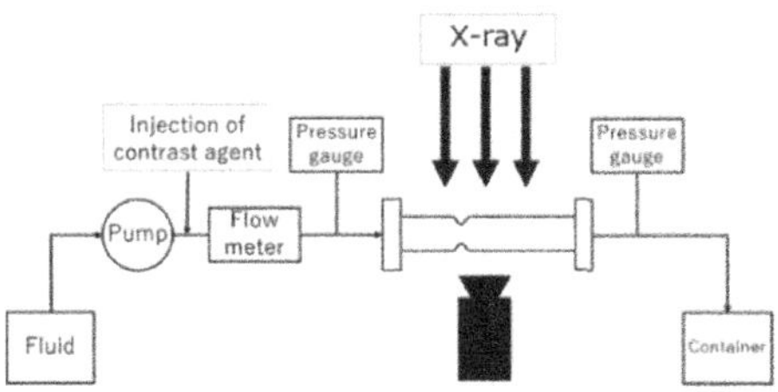

Fig. 1. Schematic image of the experiment equipment

Water and swine blood were used as the fluid. In the case of water, a silicone rubber tube with an inner diameter of 3 mm, an outer diameter of 5 mm, a thickness of 1 mm, and a length of 70 mm was used. A stenosis was created by tying a fishing line at a position of 26 mm from the fluid inlet side. In the case of blood, a tube made of polyvinyl alcohol (PVA) with an inner diameter of 3 mm, an outer diameter of 5 mm, a thickness of 1 mm, and a length of 60 mm was used. A stenosis was created by tying a fishing line to a position 24 mm from the fluid inlet side. At the time of the experiment, the tube made of PVA shrunk overall, and the inner diameter of the part other than the stenotic part was 2.1 mm. The voltage applied to the pump was kept constant and the flow conditions were set for steady flow.

The contrast agent dynamics was projected using An X-ray cardiovascular diagnostic system (Cannon Medical, Infinix Celeve-i INFX-8000C). This system can acquire images at 30 fps, and can reconstruct a 3D image of a stationary object.

2.2 Computational Fluid Simulation and Particle Tracing Simulation

The computational fluid simulation program developed in the previous study [3] was used. For the simulation of contrast agent dynamics, particle tracing simulation was performed. The rubber tube model was 3 mm in inner diameter and 60 mm in length, and the PVA tube model was 2.1 mm and 60 mm. The stenosis shape was represented as a cos curve between 17 mm and 23 mm from the inlet.

3 Result and Discussion

3.1 Result of Experiment

From the 3D images, the inner diameter of the stenosed part was measured. The area stenosis rate of the rubber tube was 62% and that of the PVA tube was 69%.

The result of the case of water and under the condition in which the inlet pressure is 5 kPa is shown. The figure shows 1.5 s after the contrast agent starts flowing. This time was the time when the contrast agent concentration was highest (Fig. 2).

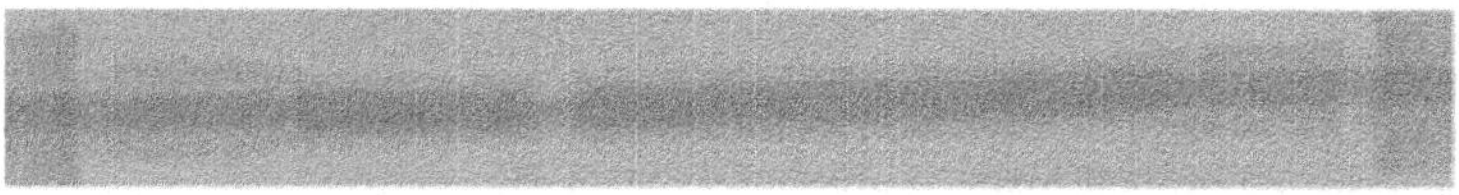

Fig. 2. Contrast agent concentration of water

The result is not clear because X-ray dose of the X-ray cardiovascular diagnostic system is low, so image processing was performed on this result. First, the pixel values of the images for one second before the contrast agent flowed were averaged to create a background image. The reason for averaging the one second images is that one image contains a lot of noise. By taking the difference in pixel value between the background image and each frame, the images that calculated the pixel value change due to the contrast agent were created. In addition, three consecutive frames were averaged to reduce the effect of noise. The following image shows the image whose color depth is tripled for easy viewing (Fig. 3).

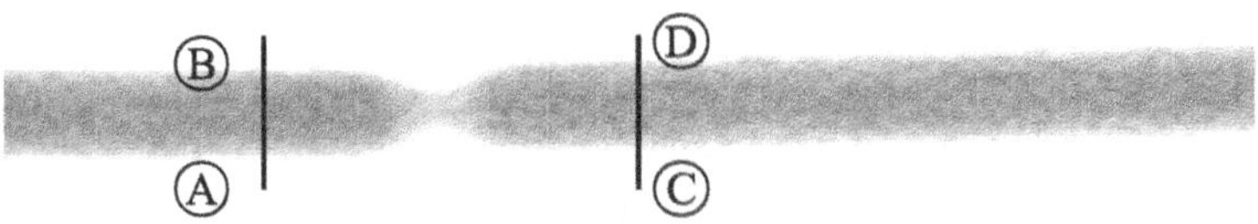

Fig. 3. Contrast agent concentration of water (after image processing)

From this image, it was found that the contrast agent spread evenly throughout the tube before and after the stenosis. The following graphs show the contrast agent concentration distribution in the direction perpendicular to the tube axis. This is obtained by acquiring the pixel values on the AB and CD lines from the imaging results, and calculating the pixel value change per pixel length in the fluid part. "Point" is an axis (mm) with the tube axis set to 0 and the upward direction of the imaging result in the direction perpendicular to the tube axis being positive. Two pixels from the point close to the tube wall are excluded because the effect of the thickness error inside the tube increases. Error bars represent the standard deviation of the pixel values of three consecutive frames used for averaging (Fig. 4).

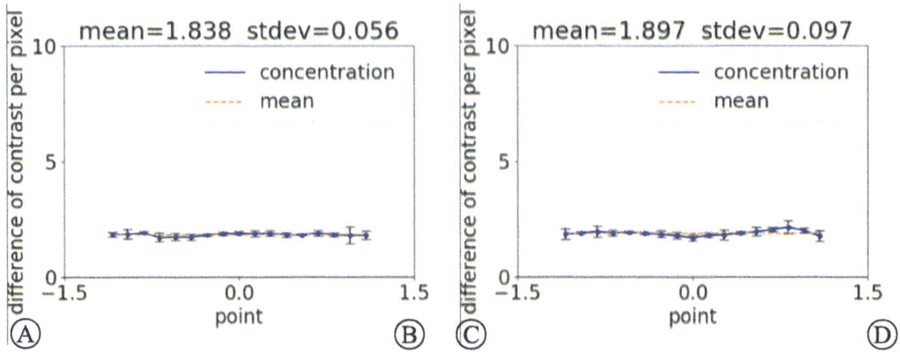

Fig. 4. Contrast agent concentration gradient after stenosis of water (left figure before stenosis, right figure after stenosis)

The average value was almost the same before and after stenosis. The standard deviation before stenosis was 0.056, after stenosis was 0.097. The variability after stenosis was about 1.7 times larger.

The results of the case of blood and under the condition in which the inlet pressure is 20 kPa are shown. The figure shows at 1.0 s after the contrast agent starts flowing (Figs. 5 and 6).

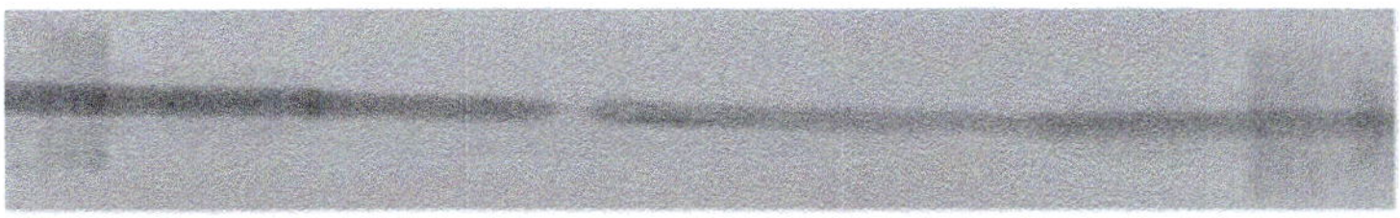

Fig. 5. Contrast agent concentration in flow of blood

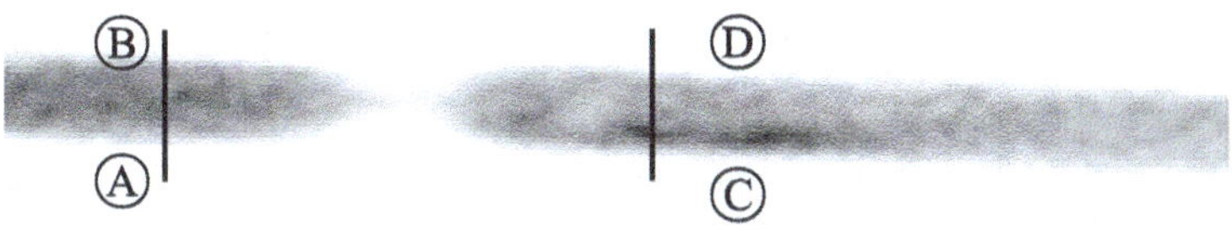

Fig. 6. Contrast agent concentration in flow of blood (after image processing)

The contrast agent concentration distribution in the direction perpendicular to the tube axis with the same calculation for the graph of water is showed.

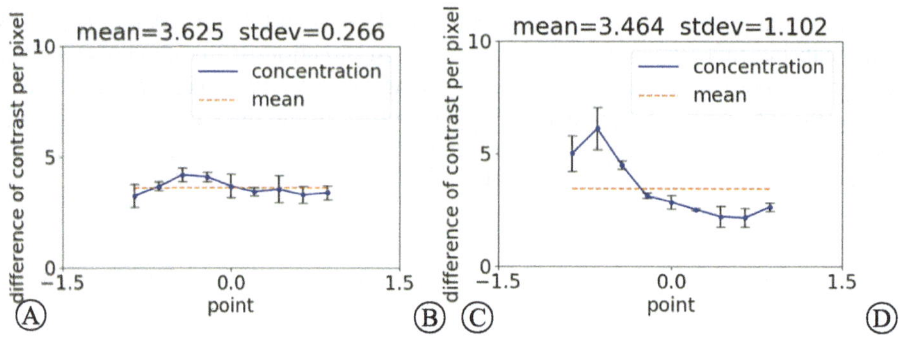

Fig. 7. Contrast agent concentration gradient after stenosis in flow of blood (left figure before stenosis, right figure after stenosis)

The average value was almost the same before and after stenosis. The standard deviation before stenosis was 0.266, and that after stenosis was 1.102. The variation after stenosis was about 4.1 times larger. From the graph, it is considered that the flow of the contrast agent largely biased toward the C side is the cause of the large variation. At this time, it was biased toward C, but at another time it was biased toward D.

The left and right graphs in Fig. 7 are at the same time, but the highest concentration after stenosis was higher than the highest concentration before stenosis even comparing different times. This suggests that the contrast agent may have been concentrated after the stenosis. On the other hand, there is a possibility that the concentrated region of the contrast agent is rotating. Further analysis is needed.

3.2 Result of Fluid Simulation

Figure 8 visualizes the flow velocity in the tube axis direction near the stenosis by color. After the stenosis, the jet flow passed through the tube axis and the surrounding area was a recirculation area.

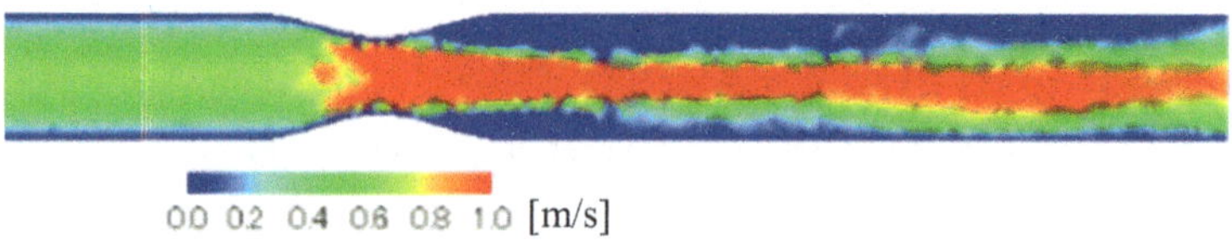

Fig. 8. Visualization of flow velocity near stenosis (water)

In the analysis simulating the experiment with blood, the flow velocity distribution in the tube axis direction was as shown in the figure below (Fig. 9).

The jet flowing down the wall after stenosis may correspond to a region where the contrast agent is concentrated. However, blood cells may be affecting them. So further investigation is needed to identify the cause.

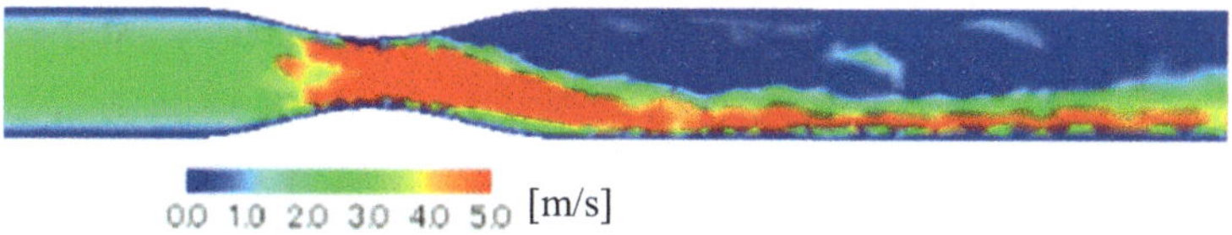

Fig. 9. Visualization of flow velocity near stenosis (water)

3.3 Result of Particle Tracing Simulation

Figure 10 shows the results of the contrast agent dynamics simulation assuming water.

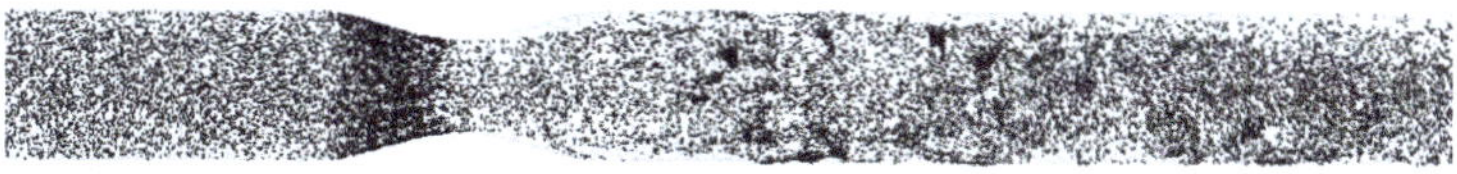

Fig. 10. Particle distribution (water)

The simulation shows that the contrast agent had spread throughout the tube after the stenosis. Even if a recirculation region occurs, the contrast agent flows into the recirculation region and the entire tube is imaged.

Figure 11 shows the results assuming that the fluid is blood.

Fig. 11. Particle distribution (blood)

As in the case of water, even if a recirculation region occurs, the contrast agent flows into the recirculation region and the entire tube is imaged. This is not consistent with the experiment using blood. It shows that Fig. 7 cannot be explained sorely by flow field.

It may be affected by diffusion or blood cell dynamics. Since it dissolves in plasma, the contrast agent has an exclusive relationship with blood cells. Normally, blood cells spread throughout blood. However, since fluid shear stress is large near the tube wall, red blood cells are deformed and move in the tube axis direction. The shear stress is large even at stenosed tube so that blood cells may move in the tube axis direction. Where the proportion of plasma increases, concentration of contrast agent is high and the image is darkened.

4 Conclusion

Experiment was performed in which water and blood flowed through a stenosed tube to project contrast agent dynamics, and a fluid simulation and particle tracing simulation were performed. In the case of water, the contrast agent after stenosis diffused throughout the vessel. In the case of blood, the region with high contrast agent flowed along the vessel wall. According to the simulations, the contrast agent dynamics in blood cannot be explained sorely by the flow field. This study was supported by JSPS KAKENHI 15K17932.

Conflict of Interest Declaration. None

References

1. Tonino, P., Bruyne, B.D., Pijls, N., et al.: Fractional flow reserve versus angiography for guiding percutaneous coronary intervention. N. Engl. J. Med. **360**(3), 213–224 (2009)
2. Yoon, Y.E., Choi, J.-H., Kim, J.-H., et al.: Noninvasive diagnosis of ischemia-causing coronary stenosis using CT angiography. Cardiovasc. Imaging **5**(11), 1088–1097 (2012)
3. 波田野明日可, 住吉谷淳, 鈴木一真 他: 分岐狭窄柔軟ファントム実験とALE 流体構造連成解析による造影剤動態解明. 日本機械学会論文集 **84**(863), 1–12 (2018)

Medical Devices

Development of a Durable Extracorporeal Ventricular Assist Device Consisting of a Centrifugal Pump with Hydrodynamic Bearings

Tomonori Tsukiya(✉), Toshihide Mizuno, Tomohiro Nishinaka, and Eisuke Tatsumi

National Cerebral and Cardiovascular Center, Suita, Osaka 5648565, Japan
tsukiya@ncvc.go.jp

Abstract. An extracorporeal ventricular assist device (VAD) intended for patients with severe heart failure has been developed. The VAD consists of a centrifugal blood pump with a new hydrodynamic bearing system that enables non-contacting rotation of the impeller. Preclinical evaluation of the VAD system was performed by bench tests as well as by a series of chronic animal experiments. The bench tests demonstrated that the impeller successfully rotates without mechanical contact under a wide range of operating conditions. In the 30-day animal experiments, the VAD demonstrated good antithrombogenicity under LVAD conditions as well as RVAD conditions. The system was finally decided to have good biocompatibility for use within 30 days.

Keywords: Ventricular assist device · Centrifugal pump · Heart failure

1 Introduction

The mortality rate of patients with refractory cardiogenic shock (RCS) remains high [1], and for patients who are not responding to pharmacologic agents, advanced therapies with mechanical circulatory support (MCS) devices are used to maintain perfusion. Especially for patients with very severe heart failure, ventricular assist devices (VADs) are used to replace pumping function of the failed hearts. These VADs are used mainly used in two different phases.

1. Extracorporeal device: an initial salvage device to stabilize major organ functions and hemodynamics.
2. Implantable device: a long-term heart replacement device to bridge to heart transplantation.

VADs for an acute use, called bridge-to-decision device [2], must be easy to install, generate sufficient perfusion to the major organs. An acute VAD system composed of extracorporeal pumps and internal cannulae have been widely accepted for this purpose.

Y. Shiraishi et al. (Eds.): APCMBE 2020, IFMBE Proceedings 82, pp. 197–201, 2021.
https://doi.org/10.1007/978-3-030-66169-4_25

Pulsatile blood pumps were used for this purpose for a long time, but currently extracorporeal centrifugal blood pumps that were originally developed for extracorporeal circulation (ECC) are used. However, many of these pumps have poor durability and antithrombogenicity, and therefore require frequent device exchange due to thrombus formation inside the devices. The authors have been developing a centrifugal blood pump with hydrodynamically levitated impeller that eliminated mechanical contacts between the rotating part (impeller) and the stationary part (casing) to acquire superior durability and antithrombogenicity [3, 4]. We designed a new temporary VAD system consisting of this blood pump. This paper summarizes the results of preclinical tests and clinical trials carried out in our facility.

2 Description of the VAD System

2.1 Blood Pump

The newly developed centrifugal pump consisted of a disposable pump head made of polycarbonate, a driving motor unit, and a portable console with a rechargeable battery. The blood pump attached to the motor unit measured 64 mm in diameter and 131 mm in length. The rotating impeller of the blood pump was designed to be levitated hydrodynamically so that it would not come into mechanical contact with other parts of the pump during its rotation. The structure of the centrifugal pump is illustrated in Fig. 1. The impeller contains permanent magnets inside its lower hub, which couple the external permanent magnets attached to the motor shaft across a small blood gap of 100 μm. Bloods filled in this gap act as a hydrodynamic bearing. The impeller has also a protruding inlet port that forms another hydrodynamic bearing with the inner wall of the pump inlet port. These two hydrodynamic bearings contribute to the stable levitation of the impeller. The lower hub also has a bore of 3 mm in diameter that brings a part of leakage flows back to the inlet to enhance good washing out effect throughout the housing.

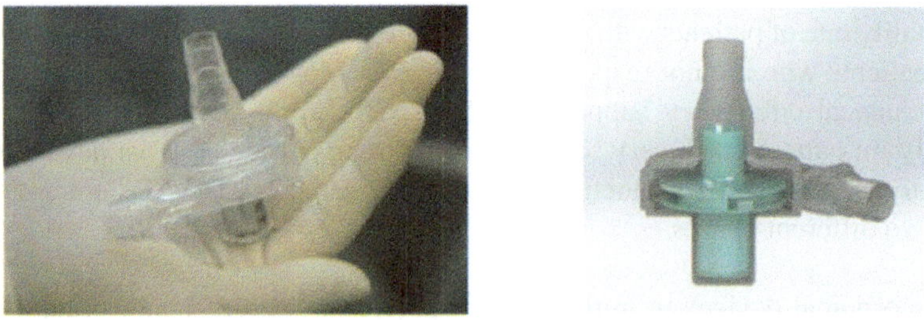

Fig. 1. A centrifugal blood pump with a impeller suspended hydrodynamically inside a housing without mechanical contacts. The impeller is driven by a permanent magnet coupling.

2.2 VAD System

We are currently developing an extracorporeal VAD system mainly for acute uses aiming at immediate recovery of hemodynamics and preserve functions of multiorgan before deciding the following therapeutic strategy. The VAD system consists of an inflow cannula, an outflow cannula, connecting tubes, and metal connectors, as well as the centrifugal pump. The inflow cannula is made of polyurethane and has an inner diameter of 12 mm. The inflow cannula is inserted into the left ventricle via the apex by median thoracotomy. The outflow cannula has a tip made from vascular prosthesis and is anastomosed to the ascending aorta. Both cannulas penetrated the abdominal skin to be connected to the connecting tubes made of polyvinylchloride then to the blood pump. All the blood contacting surfaces of the system are treated with a durable heparin coating that were originally developed by the authors [5].

3 Preclinical Evaluation

Our purpose of the development is to establish a temporary VAD system that can be used continuously for cardiac assist up to 30 days. The system can also be used as a right ventricular assist device. We performed a series of preclinical evaluations including chronic animal experiments for up to 90 days in our laboratory.

3.1 Bench Test

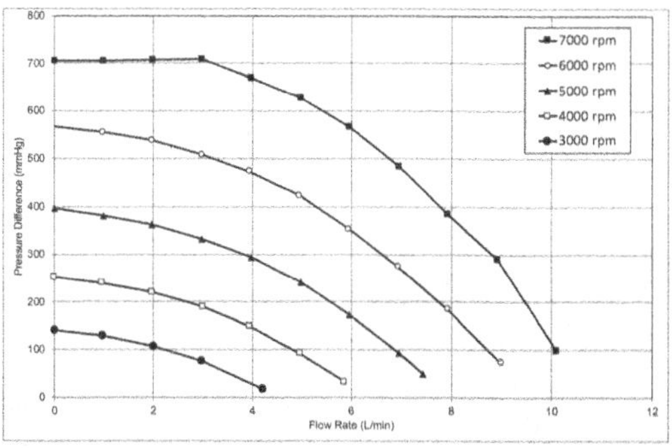

Fig. 2. Hydrodynamic characteristics of the VAD system.

Hydrodynamic characteristics of the VAD system with 2-m connecting tubes were measured in a closed mock circulation loop. The working fluid was aqua glycerol solution with 35% of weight concentration at a room temperature of 25 °C. The flow rates were measured with an ultrasonic flowmeter attached near the outlet of the centrifugal pump. The pressures at the connecting tubes were measure with semiconductor pressure sensors to calculate the pressure differences generated by the VAD system. The hydrodynamic performances of this VAD system are shown the Fig. 2. The VAD system can generate blood flow of 10 L/min against the pressure difference of 100 mmHg at the motor speed of 7000 rpm. Blood pump is estimated to work at around 4000 rpm for an LVAD condition, and at about 5000 rpm for an extracorporeal membrane oxygenation (ECMO) condition.

3.2 Animal Test

We performed chronic animal studies to demonstrate effectiveness and safety of the temporary VAD system for 30 days. Chronic animal experiments were conducted using nine male Holstein calves to evaluate biocompatibility of the VAD system up to 30 days. The experiments were performed in the following three phases;

1. 30-day performance of LVAD system (N = 3)
2. 30-day performance of RVAD system (N = 3)
3. Performance in extended use as LVAD (up to 90 days) (N = 3).

In all the experiments the blood pump was fixed on the harness designed for the calves and kept in the cage. In the LVAD experiments, the inflow cannula was inserted into the apex of the left ventricle and the outflow graft (14 mm) was anastomosed to the descending aorta of the animal. In the RVAD experiments, the inflow cannula was inserted into the right atrium and the outflow graft (7 mm) was anastomosed to the pulmonary artery. Anticoagulant and antiplatelet agents were administered orally to keep prothrombin time (PT) 3 to 4 times as long as the preoperative values.

The average flow rate (L/min) and motor speed (rpm) was 1) 4.1 & 3500, 2) 4.3 & 3900, and 3) 4.1 & 3400, respectively. All the animals were sacrificed after the scheduled periods except for one case in 3), terminated at the 55th day due to inflow cannula occlusion by massive intimal tissues. The blood contacting surfaces of thy pumps were all free of thrombus formation and signs of mechanical scratches. The results so far were promising, and we all hope this device will be approved to rescue the patients requiring acute circulatory supports (Fig. 3 and Table 1).

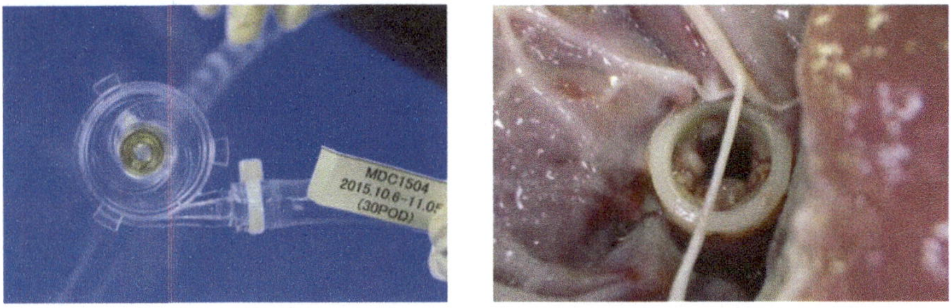

Fig. 3. An example of the explanted centrifugal pumps (left) and the inflow cannula tip obstructed by thrombus at the apex in the experiment #8.

Table 1. Summary of chronic animal experiments

#	LVAD or RVAD	Duration (day)	Motor speed (rpm)	Flow rate (L/min)	Cause of termination
1	LVAD	30	3528 ± 30	4.12 ± 0.66	Scheduled
2	LVAD	30	3510 ± 0	4.20 ± 0.50	Scheduled
3	LVAD	30	3507 ± 5	3.92 ± 0.63	Scheduled
4	RVAD	30	3610 ± 0	3.87 ± 0.17	Scheduled
5	RVAD	30	4000 ± 0	4.60 ± 0.29	Scheduled
6	RVAD	30	4095 ± 185	4.26 ± 0.40	Scheduled
7	LVAD	90	3703 ± 420	3.88 ± 0.68	Scheduled
8	LVAD	55	3562 ± 450	4.30 ± 0.79	Flow obstruction due to thrombus at inlet cannula
9	LVAD	91	3080 ± 103	4.07 ± 0.65	Scheduled

4 Conclusions

The newly developed extracorporeal ventricular assist device system was evaluated in multiple phases including chronic animal experiments and investigator-initiated clinical study. The results so far were quite promising, and we all hope this device will be approved to rescue the patients requiring acute circulatory supports.

References

1. Vahdatpour, C., et al.: Cardiogenic shock. JAHA **8**, e011991 (2019)
2. Takayama, H., et al.: Bridge-to-decision therapy with a continuous-flow external ventricular assist device in refractory cardiogenic shock of various causes. Circ. Heart Fail **7**(5), 799–806 (2014)
3. Tsukiya, T., et al.: Preclinical study of a novel hydrodynamically levitated centrifugal pump for long-term cardiopulmonary support: in vivo performance during percutaneous cardiopulmonary support. J. Artif. Organs **18**(4), 300–306 (2015)
4. Kishimoto, S., et al.: Novel temporary left ventricular assist system with hydrodynamically levitated bearing pump for bridge to decision: initial preclinical assessment in a goat model. J. Artif. Organs **21**(1), 23–30 (2018)
5. Nishinaka, T., et al.: At least thirty-four days of animal continuous perfusion by a newly developed extracorporeal membrane oxygenation system without systemic anticoagulants. Artif. Organs **26**(6), 548–551 (2002)

Sliding Mode Control Based on Nonlinear Disturbance Observer for Pneumatic Artificial Muscle

Thuy-Nguyen Vu[1,2(✉)], Quy-Thinh Dao[2], Van-Thuc Tran[1,2], and Shin-ichiroh Yamamoto[1]

[1] Shibaura Institute of Technology, Tokyo, Japan
nb19503@shibaura-it.ac.jp
[2] Hanoi University of Science and Technology, Hanoi, Vietnam

Abstract. Pneumatic artificial muscles (PAMs) have many advantages such as the natural compliant, lightweight, high ratio of weight to power, thus making them an ideal choice for humanoid robots using in the rehabilitation system. However, control strategies of PAMs are facing difficulties due to their high nonlinear characteristics, uncertainties and response lag. The three-element model [1], which consists of a contractile element, spring element and damping element in parallel, was investigated to describe the dynamics of PAMs. To improve the trajectory tracking performance of PAM, a sliding mode control based on nonlinear disturbance observer (SMCBNDO) was applied. The experimental results show that the identified three-elements model of PAM and the SMCBNDO can achieve good accuracy of trajectory tracking controller. This work proved that this model and non-linear control method can be used for the application of the PAM-based rehabilitation system.

Keywords: Pneumatic muscle actuators · Three-element model · Sliding mode control · Nonlinear disturbance observer

1 Introduction

In recent years, low-cost pneumatic artificial muscles (PAMs) have been widely implemented in human-assistive rehabilitation systems because of their advantages. This actuator is shortened in the longitudinal direction and enlarged in the radial when it is being inflated; when being deflated, it will turn back to its initial form. PAMs are able to create a tensile force to an attached load during the contraction stage. This type of actuator acts in a similar way to the human muscle: shorter muscle length produces smaller contracting force and vice versa. Furthermore, this pneumatic muscle also has an inherently natural compliant attribute which is suitable for human-robotic systems. In addition, it also has about five times higher power to weight ratio in comparison to an electric motor or a hydraulic actuator [2].

The main problem with pneumatic artificial muscle is their inaccuracy and difficult to control, which limit the applicability of PAMs. The first drawback of PAM is the

Y. Shiraishi et al. (Eds.): APCMBE 2020, IFMBE Proceedings 82, pp. 202–214, 2021.
https://doi.org/10.1007/978-3-030-66169-4_26

nonlinear characteristic due to the pressure, and the second one is the hysteresis effect due to its geometric structure.

To cope with the drawbacks of PAMs, the nonlinear mathematical model is built to describe the behavior of PAMs in different input conditions such as amplitude and frequency of actuation pressure or change of load mass. Some nonlinear models are exploited for modeling and control of PAM in literature, i.e. Maxwell slip model [3], Grey-box experimental model [4], Three-element model [1]. Most of the motion dynamic models used in a robot manipulator employ intermediate joint-space variables such length change of PAMs while pressure is being the true input controlled variable. Among the above model, the motion equation of the three-element model describes the relationship between length change and internal pressure, which can be applied to robot actuated by pneumatic muscles easily.

In order to solve the strong nonlinearity and uncertainties, characteristics for trajectory tracking purposes, a variety of advanced control approaches have been developed. Zhao et al. [6] used a neural network method to adjust the parameters of Proportional-Integral-Derivative (PID) controller. Jiang et at. [7] proposed an adaptive fuzzy control algorithm to control humanoid lower limb orthosis actuated by PAM. However, these methods cannot achieve high imprecise tracking control when the external loads change. To obtain imprecise control performance, the external disturbances and uncertainties of the model need to be known. Nonlinear disturbance observer is designed to estimate the lumped disturbance and unmodeled dynamics to improve trajectory tracking quality is proposed [10–13].

In this study, the behavior of self-made PAMs is investigated using the three-element model. The model is a parallel arrangement of a spring element, damping element, and contractile force element. Each element has a coefficient that is a function of the actuation pressure. Based on the identified model of PAM, the sliding mode control based on a nonlinear disturbance observer is designed to prove the effectiveness in tracking control regardless of the modeling uncertainties and disturbance.

The paper is organized as follows. Section 2 is the mathematical model of self-made PAMs and its identification procedure. Section 3 presents the general steps to design SMCBNDO based on the PAM model. Section 4 shows the experimental results of the closed-loop controller. Finally, the conclusion is given in the last section.

2 Modeling of the PAMs

2.1 Mathematical Model

The schematic diagram of PAM and the three-element model of PAM consists of a contractile element (force-generating), spring element and damping element in parallel [1]. The expression for the coefficients of the elements as functions of the input pressure P is determined by the experiment.

In this model, the displacement x of the PAM can be described by the equation of motion:

$$M\ddot{x} + B\dot{x} + Kx = F_{ce} - Mg \tag{1}$$

$$K(P) = K_0 + K_1 P \tag{2}$$

$$B(P) = B_{c0} + B_{c1} P \quad \text{(in contraction state)} \tag{3a}$$

$$B(P) = B_{d0} - B_{d1} P \quad \text{(in deflation state)} \tag{3b}$$

$$F_{ce}(P) = F_0 + F_1 P \tag{4}$$

where M is the mass, g is the acceleration of gravity, F_{ce}, K, B is the contractile (force-generating), spring and damping element.

2.2 Experimental Apparatus

The Pneumatic Muscle (PM) used in the experiment is a self-made McKibben artificial muscle actuator. It is made from an inner bladder made from a bicycle tire tube and enclosed in a nylon sheath used for supporting electrical cables (Clean Cut CCP1.0) is shown in Fig. 1. The PM is 500 mm length and 25.4 mm diameter in an unstretched and uncompressed condition. The input pressure of the PAM is supplied by a proportional electric control valve (ECV) ITV2000/3000 produced by the SMC company. The displacement of the PAM is measured by a laser sensor ZX-SLA11 from OMRON. The

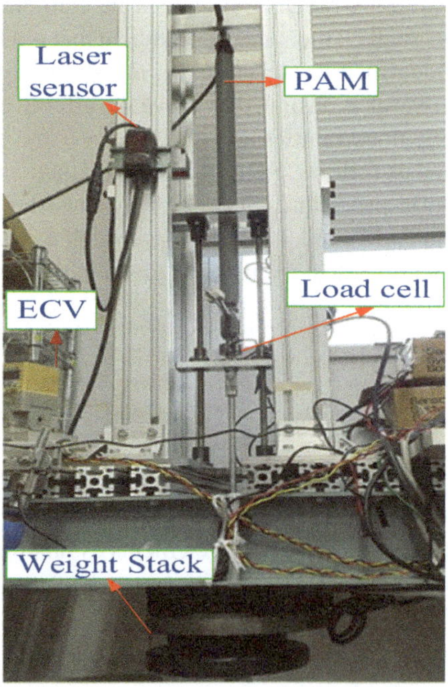

Fig. 1. The PM at 200 kPa is pulling load of 137.8 N.

mass of load is measured by LUR-A-2KNSA1 Load cell from Kyowa Electronic Instrument. The NI cDAQ-9178 controller from National Instrument is employed to regulate the output voltage of ECV as well as acquire data from the laser sensor.

2.3 Identification Procedure

To estimate coefficients as independent functions of pressure, the perturbation method was employed. After pressure reached a steady state, we suddenly reduced weight from the initial weight stack by cutting the nylon rope that connected the removed weight to PAM. Pressures in the range of 200–400 kPa (50 kPa steps) and removed weight in the range of 13.7–52 N. An example of the results at 200 kPa is shown in Fig. 2. The first portion of the response is the displacement of PAM pulling 137.8 (N) load. At 3.279 s, 52 (N) load was suddenly removed and the muscle contracted an additional 22.43 mm.

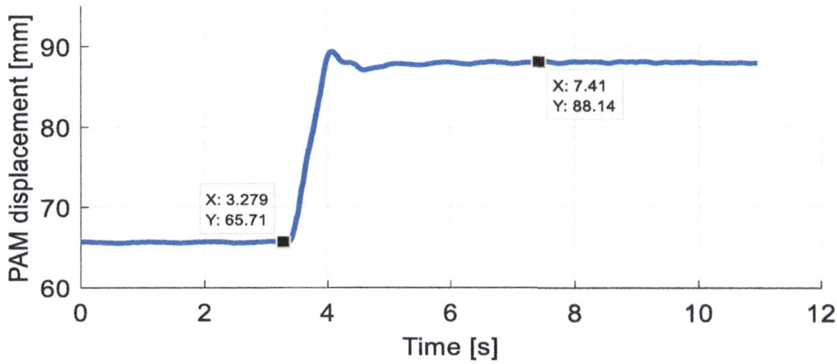

Fig. 2. Response of PAM at 200 kPa was suddenly removed 52 N load.

The experimental process was arranged as follow:

Step 1. Identification of K and B coefficient.
The pressure is held constant so there is no effect to length change of muscle from the contractile force F_{ce}, the length changes after the removed load is depended on spring and damping coefficient K and B, which are estimated by fitting data to Voight Model [5] using (Eq. 5). This experiment was repeated three times at each pressure with different weights of removed load: 13.7 N, 30.4 N, 52 N. Both K and B coefficient increase linearity with pressure represented by the linear functions obtained using least-square fit as shown in Fig. 3 and Fig. 4.

$$x(t) = \frac{1}{K}\left(1 - e^{-\left(\frac{K}{B}\right)t}\right)\Delta\mathrm{F} \tag{5}$$

where ΔF is the external force exerted to PAM (or the differential force when removed load). The spring K and damping B coefficient functions can be determined by perturbating the load at a constant P.

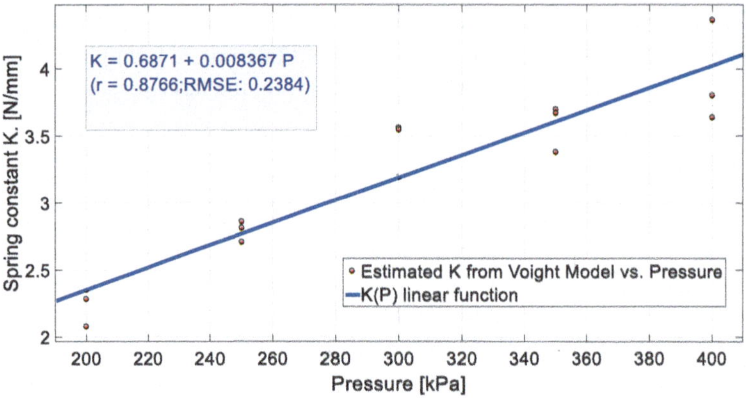

Fig. 3. Relationship between K and P

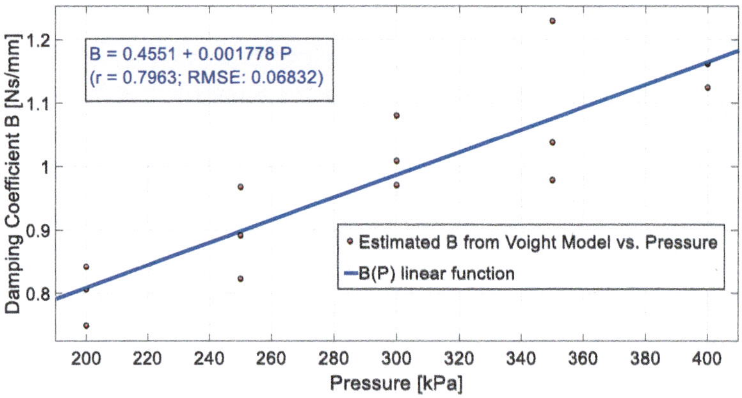

Fig. 4. Relationship between B and P

Step 2. Identification of F_{ce} coefficient.
The contractile force F_{ce} does not depend on the transient portion of the contraction response. It only depends on the steady-state portion. At the steady-state in a defined pressure input condition, K and B coefficients can be calculated from the function of pressure. Displacement of the PAM can be measured by the laser sensor. F_{ce} is the only unknown parameter in the motion equation (Eq. 1) is:

$$F_{ce} = Mg + Ky \tag{6}$$

This experiment was also repeated three times at each pressure with different weights of a removed load: 13.7 N 30.4 N, 52 N. Linear function between F_{ce} and pressure is shown in Fig. 5 with the linear fit and correlation coefficient $r = 0.9443$.

Step 3. Identification of B coefficient in deflation status.
In an operating cycle of PAM, it always works in the loop of contraction and deflation state. The dynamics of deflation needs to be studied further for more accurate behavior model of PAMs.

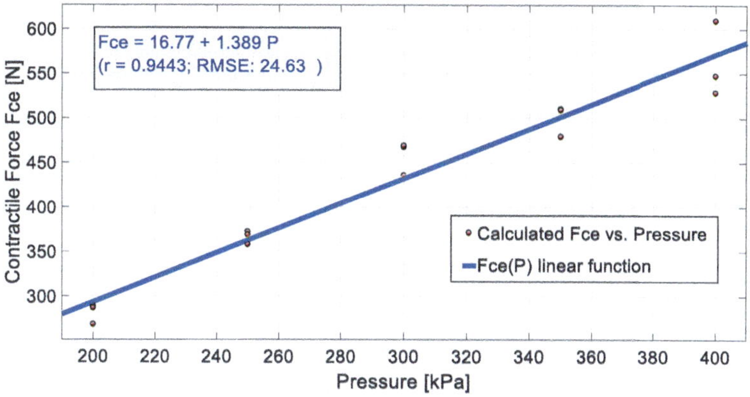

Fig. 5. Relationship between F_{ce} and P

The deflation of a contracted PAM was measured at pressure from 200 kPa to 400 kPa in 50 kPa step. In all experiments, a constant load of 124 N was applied and the PAM was deflated from the steady contracted state to the fully deflated state by suddenly set the output control voltage of ECV to 0 V. An example of the deflation of PAM is indicated in Fig. 6. The PAM was fully deflated from 200 kPa pressure with 124 (N) load. $K(P)$, $F_{ce}(P)$ can be estimated from their functions at each experimental pressure. From (Eq. 1) B coefficient in the deflation status can be calculated, linear function $B(P)$ in deflation state can be found by least-square fit Fig. 7.

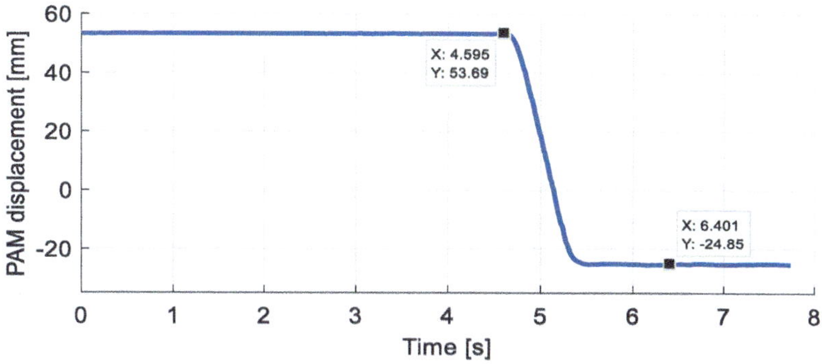

Fig. 6. Response of PAM with 124 (N) load fully deflated from pressure 200 kPa.

2.4 Model Verification

In summary, the model coefficients of PAM are presented in the Table 1.

To evaluate the identified model of PAM, multiple experiment scenarios were set up. The estimated displacement of the PAM was observed online on LabVIEW software. A

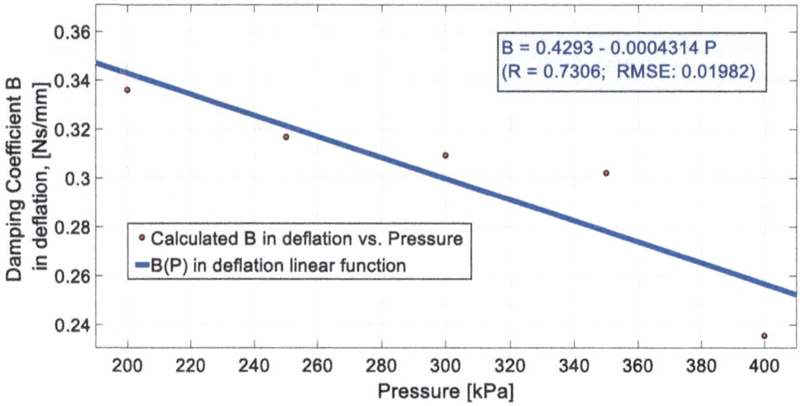

Fig. 7. Relationship between *B* and *P* in deflation status

Table 1. The model coefficients.

K_0	K_1	F_0	F_1
0.6871	0.008367	16.77	1.389
B_{c0}	B_{c1}	B_{d0}	B_{d1}
0.4551	0.001778	0.4293	0.0004314

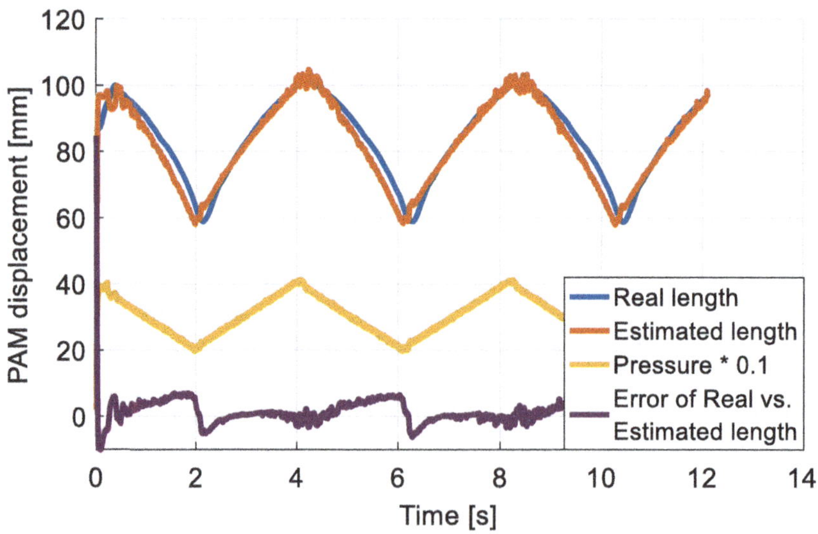

Fig. 8. Estimated and measured displacement of PAM with a triangular input pressure in range of (200–400) kPa at 0.25 Hz – Load of 138 N

triangular input waveform pressure which varied from 200 kPa to 400 kPa, was used to investigate the contraction state and deflation state of PAM.

The maximum error of measured and estimated displacement of PAM when the input pressure frequency is 0.25 Hz, is 7.064 mm (Fig. 8), and 10.07 mm at 0.5 Hz (Fig. 9). The error is higher when the frequency increases and a maximum at the lowest pressure level. The estimated model can predict change length in different load conditions: 138 N in Fig. 8 and 98 N in Fig. 9. The results approve the PAM's nonlinearity characteristic and verify the performance of the identified model.

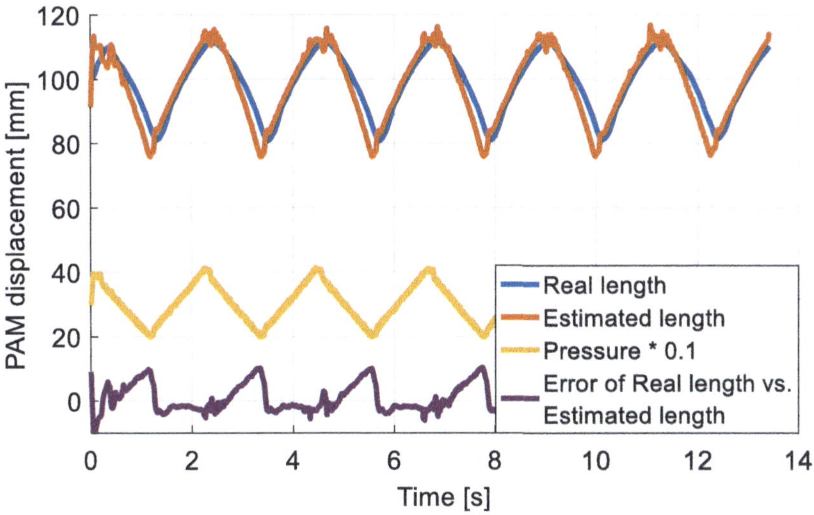

Fig. 9. Estimated and measured displacement of PAM with a triangular input pressure in the range of (200–400) kPa at 0.5 Hz – Load of 98 N

3 SMC Based on Non-linear Disturbance Observer Design

For the class of non-linear systems, sliding mode control is a suitable control approach to the problems of uncertain, time-varying characteristic and unmodeled dynamics. Unmodeled dynamics of PAM-based systems are generated because of the approximation of non-linear PAM parameters to the linear model. A non-linear disturbance observer based on control was proposed by Chen [13] to improves the performance and robustness of the control system. In this section, SMCBNDO is designed to solve the trajectory tracking problem of PAM regardless of the disturbance and unmodeled dynamics.

The structure of SMCBNDO is presented in Fig. 10.

The dynamic equation (Eq. 1) can be rewritten in the state model:

$$\begin{cases} \dot{x}_1 = x_2 \\ \dot{x}_2 = \frac{1}{M}[F_0 + F_1P - Mg - (B_0 + B_1P)x_2 - (K_0 + K_1P)x_1 + d] \end{cases} \tag{7}$$

where $x_1 = x$, $x_2 = \dot{x}$ and d is the disturbance and model errors. So we have the generic nonlinear model

$$\dot{x} = f_1(X) + g_1(X)P + g_2(X)d \tag{8}$$

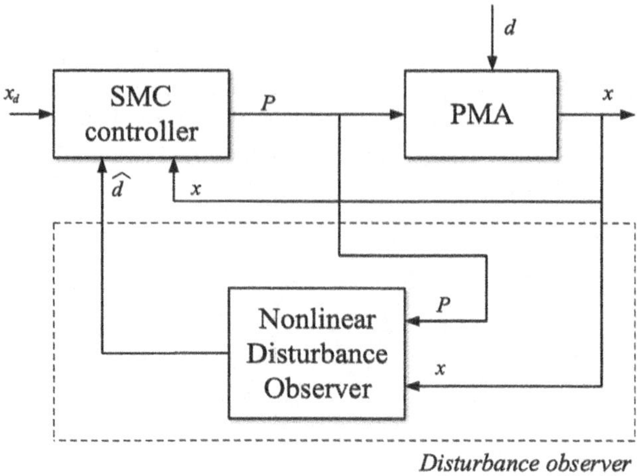

Fig. 10. Structure of nonlinear disturbance observer-based control.

where

$$\dot{X} = [\dot{x}_1 \ \ \dot{x}_2]^T;\ f_1(X) = \left[0 \ \ \frac{1}{M}(F_0 - Mg - B_0x_2 - K_0x_1)\right]^T;$$

$$g_1(X) = \left[0 \ \ \frac{1}{M}(F_1 - B_1x_2 - K_1x_1)\right]^T;\ g_2(X) = \left[0 \ \ \frac{1}{M}\right]^T \tag{9}$$

To estimate the disturbance d a nonlinear observer is proposed as

$$\dot{\hat{d}} = L\left(\dot{X} - f_1(X) - g_1(X)P - g_2(X)\hat{d}\right) \tag{10}$$

with $L = [c_1 \quad c_2]$ where $c_1 > 0,\ c_2 > 0$ is the gain matrix of observer.

Define an auxiliary vector $z = \hat{d} - \rho(X)$ where $z \in R^2$, $\rho(X)$ is a nonlinear function to be designed, and the nonlinear observer gain is defined as$L = (\partial\rho/\partial X)$. Thus

$$\dot{z} = \dot{\hat{d}} - \dot{p} = L(-f_1(X) - g_1(X)P - g_2(X)(z + \rho(X)) \tag{11}$$

Let $\tilde{d} = d - \widehat{d}$, so $\dot{\tilde{d}} = \dot{d} - \dot{\hat{d}}$.

Assume that disturbance changes slowly to the observer dynamics, so $\dot{d} = 0$, so transform (Eq. 10) and (Eq. 11) we have a dynamic equation of disturbance error

$$\dot{\tilde{d}} = -Lg_2(X)\tilde{d} \tag{12}$$

$\rho(X)$ is given by

$$\rho(X) = c_1x_1 + c_2x_2 = c_1y + c_2\dot{y} \tag{13}$$

Substituting (Eq. 12) and (Eq. 13) into (Eq. 11) we have the model of the disturbance observer as

$$\begin{cases} \hat{d} = z + \rho(X) \\ \dot{z} = -(c_1\dot{x} + c_2(f + bP) + \frac{c_2}{M}(z + c_1x + c_2\dot{x})) \\ f = \frac{1}{M}\left[F_0 - Mg - B_0\dot{x} - K_0x\right] \\ b = \frac{1}{M}[F_1 - B_1\dot{x} - K_1x] \end{cases} \quad (14)$$

Because of imperfect estimation of coefficient F, K, B we assume that f and b function are uncertain or imprecise. The extent of the imprecise on f and b can be bound by a known and continuous function of $(x,\ \dot{x})$. The control target is to keep the trajectory $x(t)$ of PAM to track a desired trajectory $x_d(t)$ in the presence of model uncertainties off and b.

Assume we have an estimate $\hat{f}$ of f and $\hat{b}$ of b. Let F^* be a positive function such that

$$\left|\hat{f} - f\right| \leq F^* \quad (15)$$

Assume the control gain b is unknown but we have upper and lower bound for it. Let Δ is the bound band of b, that is satisfies

$$b = (1 + \Delta)\hat{b} \text{ and } |\Delta| \leq B^* \quad (16)$$

Consider the sliding surface as

$$\sigma = \dot{\tilde{x}} + \mu\tilde{x} \quad (17)$$

where $\tilde{x}(t) = x(t) - x_d(t)$ is the tracking error and μ is a scalar design parameter. Based on [8] and (Eq. 7) the SMC law based on disturbance observer can be given as

$$P = \hat{b}^{-1}\left(-\hat{f} + \ddot{x}_d - \mu\dot{\tilde{x}} - \frac{\hat{d}}{M} - k\ \mathrm{sgn}(\sigma)\right) \quad (18)$$

where

$$k \geq \left(F^* + B^*\left|-\hat{f} + \ddot{x}_d - \mu\dot{\tilde{x}} - \frac{\hat{d}}{M}\right| + \frac{D_{max}}{M} + \eta\right)/(1 - B^*) \quad (19)$$

η is a positive constant.

The SMC law (Eq. 18) guarantees that the state trajectory reaches the boundary layer in a finite time regardless of the initial state, inside the boundary layer constrain trajectory to stay inside it for all later time. The global stability of the system and estimate disturbance $\hat{d}$ converge to nonlinear disturbance d was proved in [10, 13].

4 Experimental Results

To verify the effectiveness of the SMCBNDO, the multiple-scenario experiment is carried out with the designed controller. The operating region of the PAM is selected to be suitable for applying as an actuator in the Body Weight Support system [9] using in the rehabilitation robot.

We set the load $M = 14$ kg and set up the initial pressure $P = 200$ kPa.

The control parameters were chosen from trials and errors in the experiment as

$$\mu = 160,\ \eta = 0.2,\ F^* = 0.3\left|\hat{f}\right|,\ B^* = 0.2,\ D_{max} = 0.1\left|\hat{d}\right|$$

After setting the initial condition, the desired trajectory is given by

Case 1.

$$x_d(t) = 20\ \sin(0.195 * 2 * pi\ *\ t) + 80\,(\mathrm{mm}) \tag{20}$$

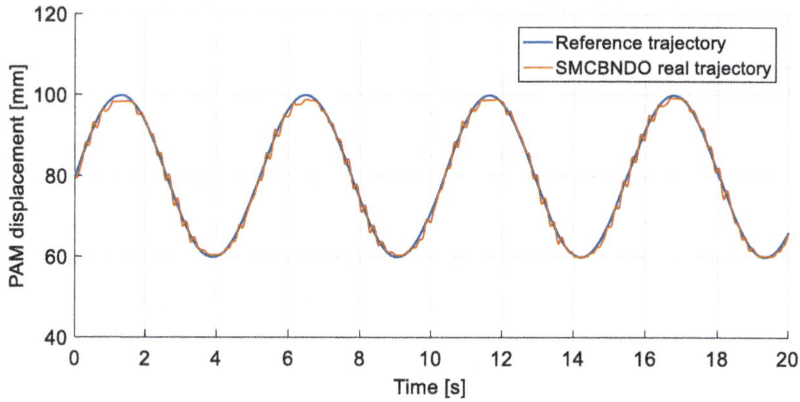

Fig. 11. Sine trajectory tracking with $M = 14$ kg at 0.195 Hz

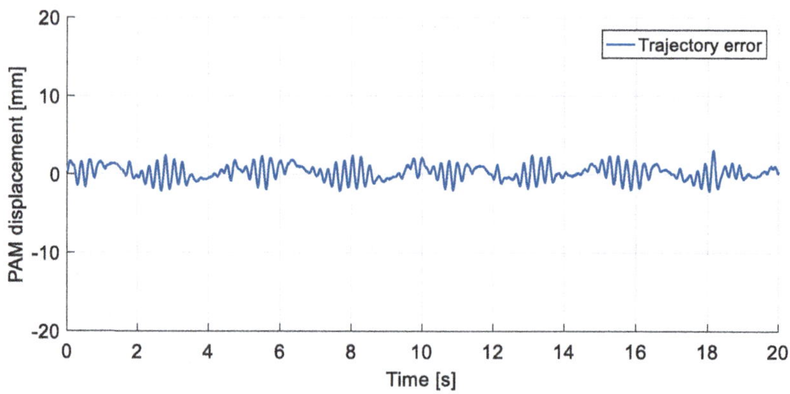

Fig. 12. Sine trajectory error

Case 2.

$$x_d(t) = 20\ \sin(0.5 * 2 * pi\ *\ t) + 80\,(\mathrm{mm}) \tag{21}$$

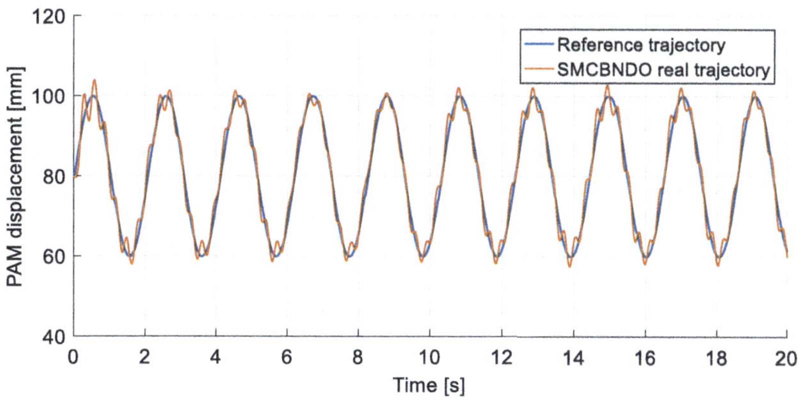

Fig. 13. Sine trajectory tracking with $M = 14$ kg at 0.5 Hz

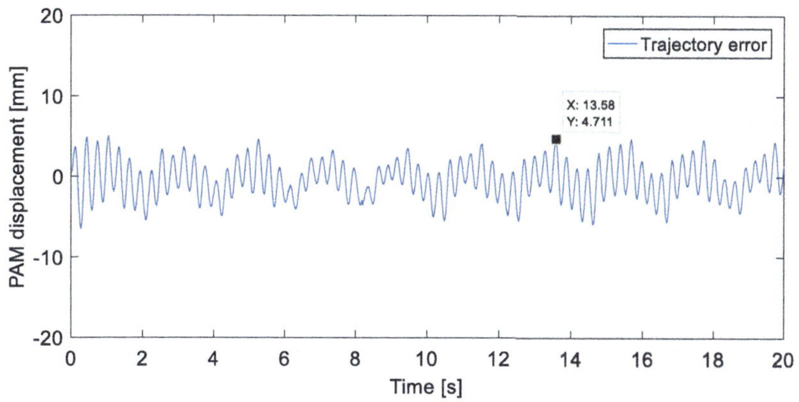

Fig. 14. Sine trajectory error

In all experimental scenarios, the sampling time of the discrete-time control system in LabVIEW was Ts = 5 ms. All the data were recorded for 20 s from the start time of the experiment. The experimental results are shown in Fig. 11, Fig. 12, Fig. 13, Fig. 14. In both Fig. 11 and Fig. 13, the real trajectory of PAM (red line) followed the reference trajectory (blue line). We can see the error trajectory in the Fig. 12 and Fig. 14 oscillates around "zero" due to characteristics of the sliding mode controller. The performance of the system in the lower frequency is better than the higher frequency. The maximum error at 0.195 Hz of frequency in Fig. 12 is 2.7 mm and 4.71 mm at 0.5 Hz of frequency in Fig. 14. Increasing the value F^* can improve the performance of trajectory tracking but it may cause a stronger "chattering" phenomenon. In the control law, the $\text{sgn}(\sigma)$ function should be replaced by $\text{sat}(\sigma)$ function to decrease the "chattering" phenomenon.

Comparing to the other control methods, SMCBNDO proves its effectiveness to apply to trajectory tracking control for PAMs. The proposed control method can overcome the strong nonlinearity and uncertainties characteristics of PAM with very fast response and good tracking performance in different experimental scenarios.

5 Conclusions

The paper focuses on the modeling and control strategy based on the three-element model [1] of a self-made PAM. The identified model demonstrates a good approximation of the nonlinear characteristics. Based on the model, we can understand the dynamics and predict the motion of PAM. The sliding mode control based on a nonlinear disturbance observer was proposed to the identified model. The experimental results show the effectiveness of the method in order to improve the performance and robustness of the PAM-based application. Future work will improve the performance of the system at the higher frequency to increase the feasibility of PAMs in the rehabilitation field. Experiments with real subjects supported by BWS system will be considered.

References

1. Reynolds, D.B., Repperger, D.W., Phillips, C.A., Bandry, G.: Dynamic characteristics of pneumatic muscle. Ann. Biomed. Eng. **31**(3), 310–317 (2003)
2. Dao, Q.-T., Tran, V.-T., Dzahir, M.A.M., Yamamoto, S.-I.: Proceedings of the 13th SEATUC Symposium. ISSN 2186-7631
3. Vo-Minh, T., Tjahjowidodo, T., Ramon, H., Van Brussel, H.: A new approach to modeling hysteresis in a pneumatic artificial muscle using the Maxwell-slip model. IEEE/ASME Trans. Mechatron. **16**(1), 177–186 (2010). https://doi.org/10.1109/TMECH.2009.2038373
4. Ba, D.X., Ahn, K.K.: Indirect sliding mode control based on gray-box identification method for pneumatic artificial muscle. Mechatronics **32**, 1–11 (2015). ISSN 0957-4158
5. Fung, Y.C.: Biomechanics: Mechanical Properties of Living Tissues, 2nd edn, p. 568. Springer, New York (1993)
6. Zhao, J., Zhong, J., Fan, J.: Position control of a pneumatic muscle actuator using RBF neural network tuned PID controller. Math. Probl. Eng. **2015**, 1–6 (2015)
7. Jiang, F., Tao, G., Li, Q.: Analysis and control of a parallel lower limb based on pneumatic artificial muscles. Adv. Mech. Eng. **9**, 1–4 (2016)
8. Slotine, J.J., Li, W.: Applied Nonlinear Control. Prentice Hall, Englewood Cliffs (1991)
9. Van Thuc, T., Yamamoto, S.: Development of a body weight support system using pneumatic muscle actuators: controlling and validation. Adv. Mech. Eng. **8**(12), 1–3 (2016)
10. Xing, K., Huang, J., Wang, Y., Wu, J., Xu, Q., He, J.: Tracking control of pneumatic artificial muscle actuators based on sliding mode and non-linear disturbance observer. IET Control Theory Appl. **4**(10), 2058–2070 (2010)
11. Wu, J., Huang, J., Wang, Y.J., Xing, K.X.: Nonlinear disturbance observer-based dynamic surface control for trajectory tracking of pneumatic muscle system. IEEE Trans. Control Syst. Technol. **22**, 440–455 (2014)
12. Yang, H.J., Yu, Y., Zhang, J.H.: Angle tracking of a pneumatic muscle actuator mechanism under varying load conditions. Control Eng. Pract. **61**, 1–10 (2017)
13. Chen, W.-H.: Disturbance observer based control for nonlinear systems. IEEE/ASME Trans. Mechatron. **9**(4), 706–710 (2004)

A Novel OCT Image-Guided Laser Automatic Ablation Method Based on Non-common Optical Path Structure

Chengquan Hu[1], Wei Chang[1], Yangxi Li[1], Yingwei Fan[2], Xinran Zhang[1], Hongxiang Kang[2], and Hongen Liao[1](✉)

[1] School of Medicine, Tsinghua University, Beijing 100084, China
liao@tsinghua.edu.cn
[2] Beijing Institute of Radiation Medicine, Beijing 100039, China

Abstract. During tumor resection, the boundaries of some types of tumors are difficult to distinguish with the naked eye. Doctors also face problems with tissue deformation and avoidance of important blood vessels and nerves. Although optical biopsy provides a rapid intraoperative diagnosis method, how to effectively combine with laser ablation and other treatment methods to achieve accurate ablation or resection of tumor tissue still faces great challenges. Therefore, we designed a theranostics system with non-common optical path structure and an intelligent control method. By fixing the front-end of optical coherence tomography (OCT) sample arm with the high power laser probe, and intelligently driving the micromirror to control high power laser spot based on the diagnosis result of OCT image, the system can excise the tumorous tissues with high precision. We verified the feasibility of the system and method on the ex vivo porcine brainstem model. The experimental results showed that the system can sensitively recognize and quickly ablate lesions in real time. The diagnostic and planning process takes 26 s. The ablation precision can reach millimeter order. The theoretical removal rate of light spot planning method is about 90%. The ablation cavity's position error is about 0.15 mm.

Keywords: Theranostics system · Optical coherence tomography · Automatic laser ablation

1 Introduction

The tumorous tissues in preoperative images for diagnosis and location may de-form intraoperatively or its boundary may be unclear with naked eyes. During the operation, to identify the suspected lesion area and the scope of metastasis of the tumor, frozen sections are often acquired for rapid pathological diagnosis, while it is invasive, range-limited, relatively time-consuming, and requires a high level of experience of the pathologists. In recent years, optical biopsy technology has begun to play an important role in surgery as an instant diagnostic method. Optical biopsy technology mainly refers to the acquisition and analysis of tissue information in vivo or ex vivo by means of confocal microscopy, fluorescence spectroscopy or other optical imaging methods. Among

Y. Shiraishi et al. (Eds.): APCMBE 2020, IFMBE Proceedings 82, pp. 215–222, 2021.
https://doi.org/10.1007/978-3-030-66169-4_27

them, optical coherence tomography (OCT) is an imaging method that acquires the tissue microstructure information by measuring the interference signal of tissue back-scattering light [1]. With micron order resolution, OCT can display the pathological information of tissue in real time and have great application prospects in cancer diagnosis.

3D OCT images need to be matched with the actual surgical area to realize a better location of the lesions. However, the space limitation and complexity of the actual surgical scene make it difficult to achieve fast and accurate image registration. By combining the front end of OCT with the probe of laser scalpel in a hardware way, the image registration process can be omitted. The hardware combination methods can be sorted as common-path and non-common-path. Common-path combination means that imaging laser and high-power ablation laser will propagate in the same optical path, so as to achieve a more compact structure. The most representative method is based on semi-transparent mirror [2–4], while physical size of light path is large and it is difficult to transmit two kinds of lasers efficiently at the same time. Therefore, some researchers have proposed a combiner-based structure [5] with a compact structure, and it has the potential to achieve high-quality imaging and high-efficiency ablation at the same time. However, due to the limitation of present manufacturing method of the combiner, the imaging quality is poor. Common-path combination requires the common-use of optical components between imaging and ablation lasers. Thus, some researchers have proposed non-common-path methods. A more common method in non-common-path design is to deflect a high-power laser through a larger lens or a prism fixed adjacent to it, so that a single OCT image monitors a single point ablation process [6, 7]. However, this method has a relatively fixed working distance and limited imaging range. Some researchers controlled the high power laser probe with the manipulator, and cooperated with the sample translation table [8]. However, the system structure is not compact and the matching speed between laser spot and images is slow.

In this article, we proposed a theranostics system with a neighboring fixed front-end to achieve automatic ablation. The system can drive the micromirror scanner to deflect the laser beam to ablate the target spot according to the OCT image analysis results, while ablation process is displayed in real time via OCT images.

2 System and Methods

2.1 Theranostics System

System Components. The sweep source OCT system consists of a sweep light source (HSL-20-100-B, Santec, central wavelength: 1310 nm; power: 20 mW; bandwidth: 80 nm), a balanced photodetector (Thorlabs, int-msi-1300b), a data acquisition device (ATS9350, AlazarTech, sampling rate: 500 m/s) and other components. The axial and lateral resolution of the OCT image are about 12 microns and 22 microns, respectively. The depth of field of OCT scanning objective is ± 0.8 mm. The high power laser is a semiconductor laser (Max-ray Photonics, central wavelength 808 nm, power 0.1–4 W) and the fiber probe core diameter is 105 μm with a self-focusing lens (C-lens) and a working distance of 20 mm. The micromirror, a Micro-Electro-Mechanical System (MEMS) scanner, can be deflected up to $\pm 25^\circ$ around two axes.

System Design. The diagram of the theranostics system is shown in Fig. 1 (a). Under the control of OCT scanner, the OCT front-end imagines the target area, and the interference signal is transmitted to the host for processing. Then, the host controls the high-power laser and the MEMS scanner to automatically ablate the tumorous area.

The high-power laser probe and the MEMS scanner are fixed on the OCT probe with an annular mounting in Fig. 1 (b). By adjusting the height of the mounting, the imaging area can coincide with the high-power laser irradiation area. The ablation laser perpendicular to the sample platform irradiates on the sample surface after being reflected by the MEMS scanner. Driving the MEMS scanner to perform two-dimensional deflection can control the laser spot to move within a certain range.

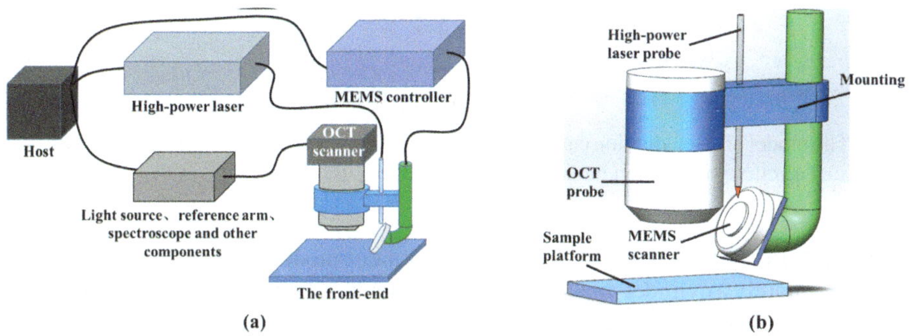

Fig. 1. The diagram of theranostics system and the neighboring-fixed front-end. (a) The diagram of theranostics system; (b) The neighboring-fixed front-end.

2.2 Methods

Ex Vivo Porcine Brainstem Model. In this study, fresh porcine brain-stem and mouse liver slices were used to simulate the normal tissues and tumorous tissues separately. An irregular area approximately $2 \times 2\ mm^2$ square was marked with India ink on the porcine brainstem. The mice liver slice about 1 mm thick with the same size and shape as the marked area was placed on the marked area as a tumor. After a few minutes, the mouse liver slice fits well into the porcine brain-stem. As the mouse liver slice is transparent, the India ink can mark the tumor area well, especially in OCT images. This study was approved by the Tsinghua University Experiment Animal Center-approved animal protocols (No. 18-LHE1).

Model of Laser Ablation Cavity. The interaction between tissue and laser is related to the tissue, laser, environment and other factors. So it is difficult to define the therapeutic effect with a unified model. It is necessary to conduct a pre-experiment to obtain a relatively stable experience-based laser ablation cavity model and to select the most suitable laser parameters under this condition.

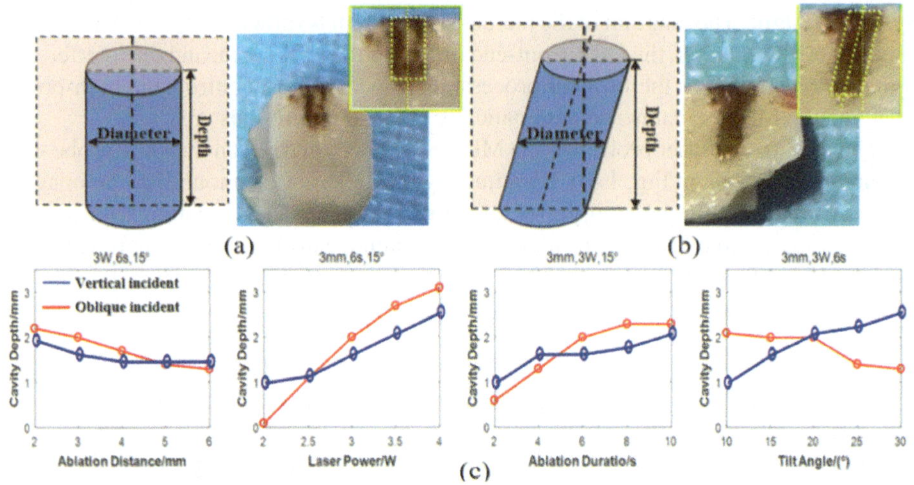

Fig. 2. The model of laser ablation cavity. (a) The vertical incident model; (b) The oblique incident model; (c) The curve of ablation parameters and the size of the cavity.

In Fig. 2 (a)–(b), the vertical and oblique incident models of laser ablation cavities are approximately a cylinder and an oblique cylinder, respectively. The size of the cavity can be calculated by a reference object in white-light images.

The size of the ablation cavity was measured under different laser irradiation distance, output power, duration and the angle between incident direction and vertical direction, respectively in Fig. 2 (c). Considering the effective imaging depth of OCT (about 2 mm), the ablation efficiency and accuracy, we choose 3 W as the laser power and 6 s as the ablation duration. The data in Fig. 2 (c) shows that the cavity size varies slightly under the change of ablation distance and tilt angle. Therefore, it is acceptable to approximate the formed cavity size as 1.5 mm diameter and 1.5 mm depth. These size parameters of ablation cavities will be used in subsequent experiments as the theoretical parameters formed by the 3 W, 6 S laser.

The Function Between the MEMS Scanner Tilt Angles and the Laser Spot Position. The dual-axis MEMS scanner can deflect in both longitudinal and transverse directions to make the laser spot on the sample surface move in two dimensions. The function between the longitudinal tilt angle θ and the transverse tilt angle γ of the MEMS scanner and the spot position coordinates (x, y) is deduced. As shown in Fig. 3 (a)–(b), the plane of the sample platform is assumed to be the xy coordinate plane, and the intersection point of the incident laser line and the xy coordinate plane is the coordinate origin O. When the scanner's longitudinal tilt angle θ keeps unchanged, the laser spot will move along the arc of radius L with the transverse tilt angle γ changing. When the longitudinal tilt angle θ changes, the radius L of the arc will change accordingly. Therefore, the calculation of the function between the tilt angles and the laser spot coordinate position can be divided into two steps. First, let the distance from incident point to the coordinate origin O be d and the angle between the scanner and the horizon be α. Let the initial installation α be α_0. The radius L can be calculated as Eq. 1.

$$L = -d \cdot \tan 2\alpha = -d \cdot \tan 2(\alpha_0 + \theta) \tag{1}$$

As shown in Fig. 4 (c), when the θ keeps unchanged, the normal plane of the scanner will tile γ with the scanner tilting γ in transverse direction. And the tilt angle φ of laser spot on xy coordinate plane can be calculated as Eq. 2.

$$\varphi = 2\sin^{-1}\left(\frac{\sin\frac{\gamma}{2}}{\sin(\alpha_0 + \theta)}\right) \tag{2}$$

With Eq. 1 and 2, the function of MEMS tilt angles θ and γ with the laser spot coordinate position (x, y) is calculated as below.

$$x = L\cos\varphi = -d\tan 2(\alpha_0 + \theta)\cos\left(2\sin^{-1}\left(\frac{\sin\frac{\gamma}{2}}{\sin(\alpha_0 + \theta)}\right)\right) \tag{3}$$

$$y = L\sin\varphi = -d\tan 2(\alpha_0 + \theta)\sin\left(2\sin^{-1}\left(\frac{\sin\frac{\gamma}{2}}{\sin(\alpha_0 + \theta)}\right)\right) \tag{4}$$

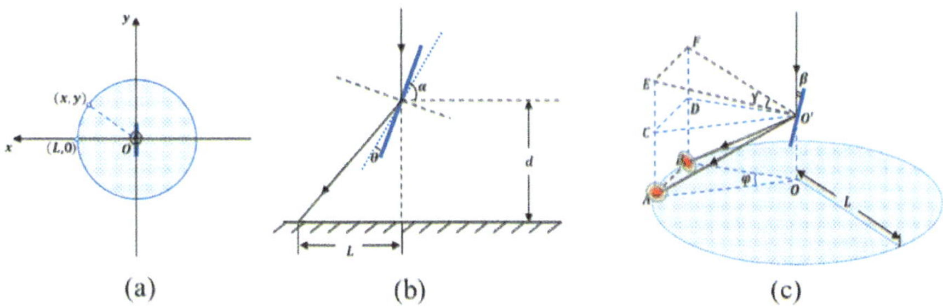

Fig. 3. The function between tilt angles of MEMS scanner and the coordinate position of light spot. (a) The radius L and the coordinate position (x,y); (b) The longitudinal tilt angle θ and the radius L; (c) The transverse tilt angle γ and the radius L.

2D En Face Map Extraction and Automatic Ablation. In order to analyze and show whether there are lesions in target area, we extracted an optical characteristic parameter that was related to the deep tissue structure and simplified the 3D OCT image into a 2D en face map in Fig. 4 (a). The signal intensity I_{OCT} of each A-scan data can be approximately expressed by the Eq. 5,

$$I_{OCT}(z) = A \cdot \mu_{bs} \cdot e^{-2\mu_{att}z} \cdot h^2(z) \tag{5}$$

where z is the depth from tissue surface, A is an overall system constant, μ_{bs} is the backscattering coefficient, μ_{att} is the total attenuation coefficient, $h(z)$ is the point spread function (PSF). The total attenuation coefficients μ_{att} of each A-scan data can be calculated by FD algorithm [9] and be processed to a two-value map by Otsu's method [10]. Based on the two-value map, the coordinate positions of tumorous area can be converted to the location of the OCT image pixels, combined with Eq. (3), (4) to control the MEMS scanner and ablate the lesions by arcs with grading radius in Fig. 4 (b). After automatic ablation according to the flow chart in Fig. 4 (c), OCT re-images the target area to determine whether the newly exposed tissue layer contains lesions. Repeating this process, the lesions can be automatically ablated layer by layer.

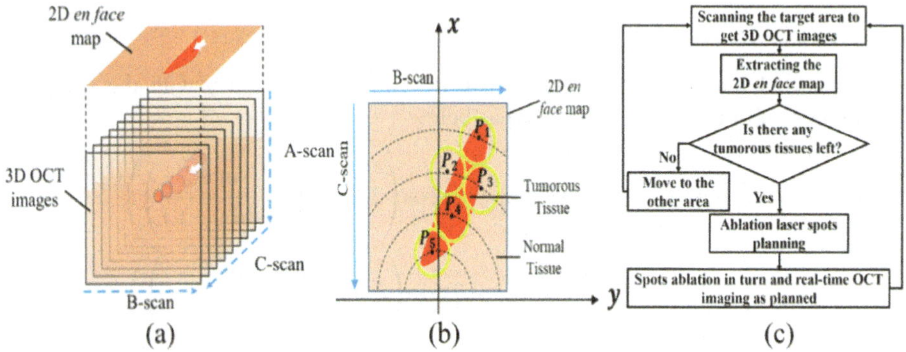

Fig. 4. 2D *en face* map extraction and ablation planning. (a) The 2D *en face* map; (b) Ablation spot planning; (c) The diagnosis and ablation flow chart. The lesion is marked by white arrow. The A-scan, B-scan and C-scan are showed in blue arrow. The red part is tumorous and the orange part is normal. P_1 to P_5 are the centers of ablation spots and the yellow circles are the outlines. The arcs in black dotted line are the moving trace of light spot.

3 Results

3.1 Ablation Effect Assessment

By comparing the 2D en face maps before and after ablation, we can intuitively evaluate the therapeutic effect of this system. The obvious tumorous area marked by white arrow in Fig. 5 (b) is ablated in Fig. 5 (c).

The theoretical ablation rate can be roughly evaluated by comparing the coverage area of lesion area in spot planning process with the area in *en face* map diagnosis. The results of three brainstem tissue model experiments are shown in Table 1. As the simulated lesions will deform during ablation and curl into the laser spot, the actual ablation tissue volume is slightly larger than the theoretical volume in the spot planning process. This leads to the actual ablation rate slightly higher than the theoretical removal rate calculated in Table 1, which can reach about 90%. Through this method, automatic lesion area excision is realized.

3.2 The Cavity Distribution Assessment

As shown in Fig. 6, the positions of laser ablation cavities are basically consistent with the spots planning. By calculating the position errors between the center of each ablated

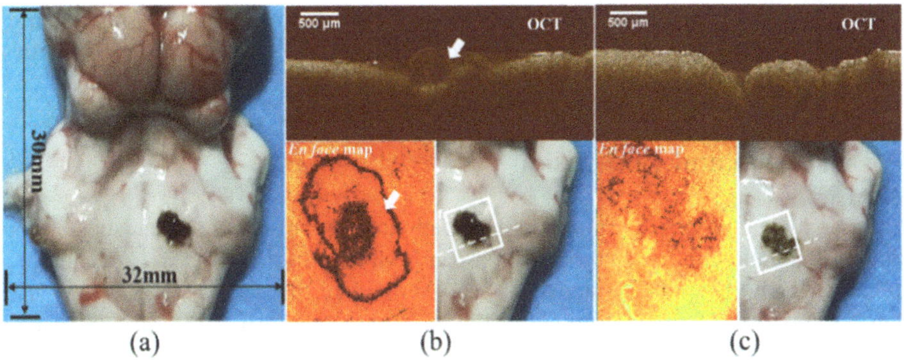

Fig. 5. OCT images and *en face* maps of the model. (a) The size of the brainstem specimen; (b) Before ablation; (c) After ablation. The white squares are imaging areas. OCT images are corresponding to the white dotted areas. The tumor areas are marked by white arrows.

Table 1. The removal rate of the tumorous area.

	Model 1	Model 2	Model 3	Mean	S.D
Theoretical removal rate	89%	94%	91%	91%	0.024

cavity and the center of each planning spot, the accuracy of cavity distribution was evaluated. Considering the change of the center position of the oblique cylinder with the change of depth, the accuracy of the spot position can be quantitatively evaluated. 23 cavities were formed on three brainstem tissue models. The mean error of the positions of these circles was about 0.16 ± 0.092 mm.

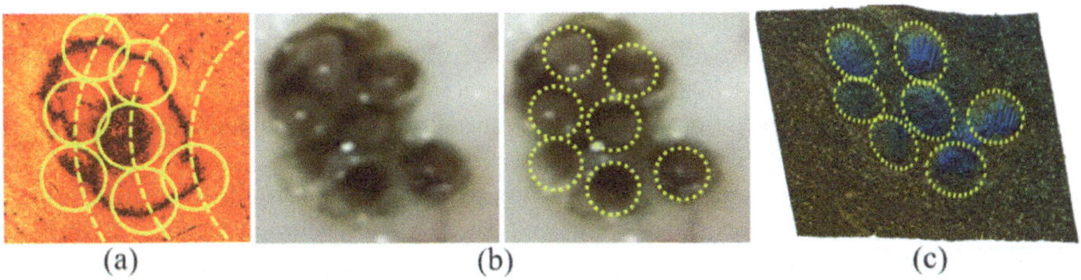

Fig. 6. The laser spot planning and corresponding actual results. (a) The *en face* map and spot planning; (b) The white light image; (c) The surface reconstruction of 3D OCT images. The yellow dotted circles are the outlines of the ablative cavities calculated according to the images' information.

3.3 Time Consuming Assessment

To evaluate the rapidity of the system, the time costs in each step during the whole process were measured. The OCT imaging took about 6 s to collect for a 4 mm × 4 mm square target area. The 2D *en face* maps extraction took about 18 s, and the spot planning took 2 s. The time cost of laser ablation process is affected by the number of cavities. The whole diagnosis and ablation process took about 1 min.

4 Discussion and Conclusion

In this study, we proposed a non-common-path neighboring-fixed theranostics system and its automatic control method, which can omit registration process. And the preliminary feasibility of the system has been verified in this article. The system uses high-resolution OCT image to identify the lesion area and drives the MEMS scanner to control the high-power laser spot to ablate lesion area, while displaying the ablation status in real time. The system can effectively combine intraoperative imaging and treatment, and reduce the dependence on doctors' experience. The structure of the system also has the potential of miniaturization to combine with endoscopic forward-scanning OCT and be used in minimally invasive surgery. Subsequent work will focus on building the corresponding hardware system combined with the control method for overall testing.

Acknowledgements. The authors would like to acknowledge the supports of National Key Research and Development Program of China (2017YFC0108000), National Natural Science Foundation of China (81771940, 81901907), and Beijing Municipal Natural Science Foundation (L172003, 7172122).

References

1. Huang, D.: Optical coherence tomography. Science **254**(5035), 1178–1181 (1991)
2. Katta, N.: Optical coherence tomography image-guided smart laser knife for surgery. Lasers Surg. Med. **50**(3), 202–212 (2018)
3. Palanker D.V.: Femtosecond laser-assisted cataract surgery with integrated optical coherence tomography. Sci. Transl. Med. **2**(58), 58ra85–58ra85 (2010)
4. Boppart, S.A.: High-resolution optical coherence tomography-guided laser ablation of surgical tissue. J. Surg. Res. **82**(2), 275–284 (1999)
5. Jivraj, J.: Coaxial cavity injected OCT and fiber laser ablation system for real-time monitoring of ablative processes. In: Clinical and Translational Neurophotonics, p. 100500C. International Society for Optics and Photonics, San Francisco (2017)
6. Oh, W.Y.: Ultrahigh-speed optical frequency domain imaging and application to laser ablation monitoring. Appl. Phys. Lett. **88**(10), 103902 (2006)
7. Li, Z.: Miniature forward-imaging B-scan optical coherence tomography probe to guide real-time laser ablation. Lasers Surg. Med. **46**(3), 193–202 (2014)
8. Fan, Y.: A novel integration of spectral-domain optical-coherence-tomography and laser-ablation system for precision treatment. Int. J. Comput. Assist. Radiol. Surg. **13**(3), 411–423 (2018)
9. Yuan, W.: Robust and fast characterization of OCT-based optical attenuation using a novel frequency-domain algorithm for brain cancer detection. Sci. Rep. **7**, 44909 (2017)
10. Otsu, N.: A threshold selection method from gray-level histograms. IEEE Trans. Syst. Man Cybern. **9**(1), 62–66 (1979)

Power Transmission Efficiency Evaluation of Spiral Coil with Center Tap for Transcutaneous Energy Transmission System

Tomoki Okinaga, Takahiko Yamamoto(✉), and Kohji Koshiji

Tokyo University of Science, Tokyo, Japan
t_yamamoto@rs.tus.ac.jp

Abstract. The transcutaneous energy transmission system (TETS) is the most promising method for transmitting driving energy to a totally-implantable artificial heart (TIAH) without additional invasive procedures. In addition to a rectifier, TETS also requires a regulator to supply current at constant voltage to the TIAH and the implanted secondary battery in the inner circuit. In this study, we describe a double wire spiral coil with center tap that can perform full-wave rectification with a simpler circuit configuration. Hence, it suggests that the number of components in the internal circuit can be reduced while achieving almost the same power transmission efficiency.

Keywords: Transcutaneous energy transmission system · Totally-implantable artificial heart · Center tap

1 Introduction

In this study, we propose a novel coil shape to reduce the number of components and heat generation. Transcutaneous energy transmission system (TETS) [1] is the most promising method for transmitting driving energy to a totally-implantable artificial heart (TIAH) without additional invasive procedures [2]. The TIAH requires a regulator that supplies a stable voltage to the internal body battery and actuator, and outputs a constant voltage regardless of changes in the input voltage due to load fluctuations. However, the regulator generates heat and may cause burns in the body.

2 Transcutaneous Energy Transmission System (TETS)

In the TETS power generated by a DC power supply outside the body is converted to AC power using an inverter (Fig. 1). This AC power is transmitted non-invasively from the external coil to the internal coil facing the skin using a magnetic coupling. The AC power transmitted to the internal coil is converted back to DC power using a rectifying, smoothing, and stabilizing circuit in the body. It is then supplied to the secondary battery for backup and the control circuit of the TIAH.

Y. Shiraishi et al. (Eds.): APCMBE 2020, IFMBE Proceedings 82, pp. 223–227, 2021.
https://doi.org/10.1007/978-3-030-66169-4_28

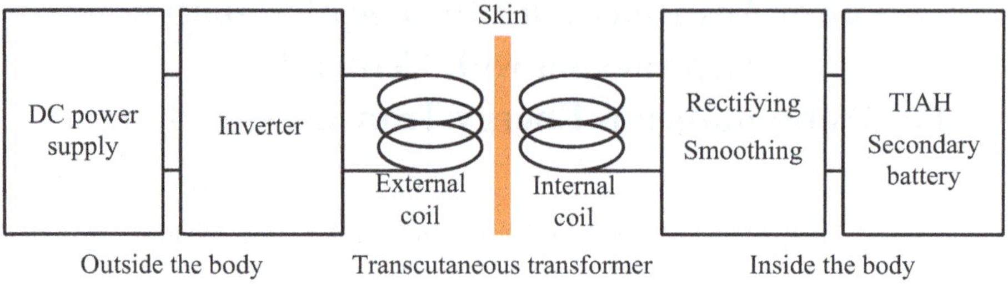

Fig. 1. Block diagram of TETS

3 Spiral Coil with Center Tap

We used Litz wire (4/30/0.05) to create two coil designs: (a) typical spiral coil by winding a single line (Fig. 2), and SCCT by winding two lines (indicated by distinct colors [black and red] in Fig. 3). We assessed the characteristics of both types using an impedance analyzer (Agilent 4294A), a summary of which is presented in Table 1. Furthermore, full-wave rectification using a typical spiral coil requires four diodes. However, only two diodes are needed when SCCT is used.

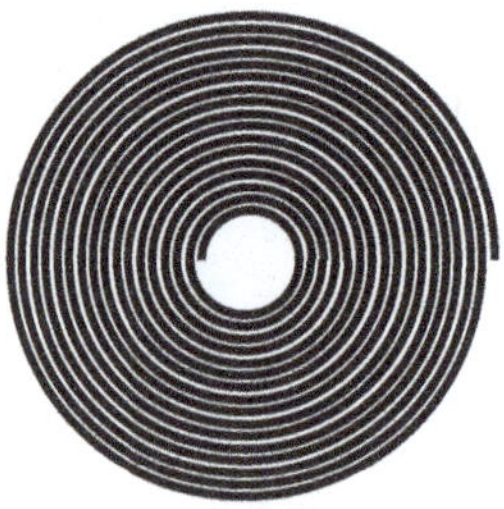

Fig. 2. Typical spiral coil

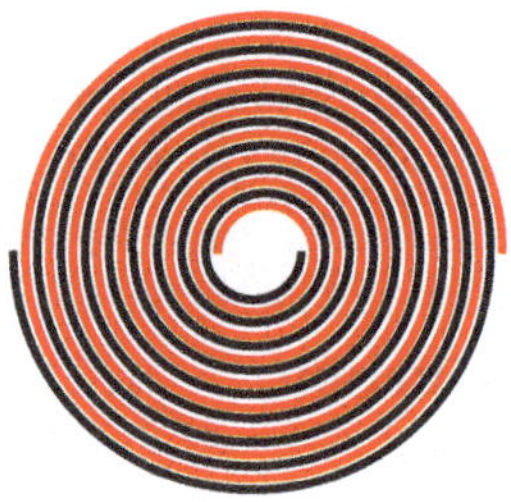

Fig. 3. Spiral coil with center tap

Table 1. Characteristics of coils

Type of coils	Turns	Inductance [μH]	Resistance [mΩ]
Power supply coil	26	15.32	334.4
Reception spiral coil	27	16.68	371.4
SCCT red	13	3.580	97.89
SCCT black	14	3.750	105.4

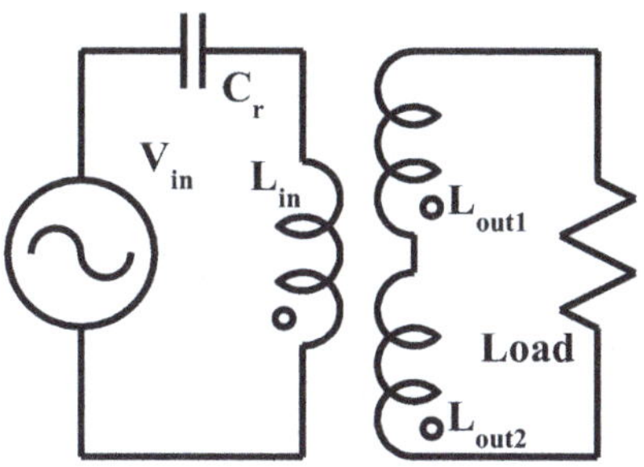

Fig. 4. Measurement circuit

Table 2. Components values

Components	Values
Frequency	300 kHz
C_r	15.7 nF
Load	30 Ω

4 Measurement

The circuit used for measuring the characteristics of each coil is indicated in Fig. 4 and the specifications of other components are listed in Table 2. As the main focus of this paper was on the characteristics of the coils, we simplified the experimental setup by substituting the inverter and the rectifier circuits with a signal generator (Keysight 33500B) and power amplifier (NF Electronic Instruments 4025), and a load resistor.

We investigated the power transfer efficiency between the transmitting coil and the receiving coil by changing the input voltage. The following five conditions were considered for investigation (refer to Fig. 5) and the power transfer efficiency per coil was measured using power meter (Yokogawa PX8000).

1. To receiving spiral coil
2. To SCCT connecting inside of the black line and inside of the red line (Fig. 5a)
3. To SCCT connecting outside of the black line and inside of the red line (Fig. 5b)
4. To SCCT connecting inside of the black line and outside of the red line (Fig. 5c)
5. To SCCT connecting outside of the black line and outside of the red line (Fig. 5d).

Fig. 5a. Connect inside black and red

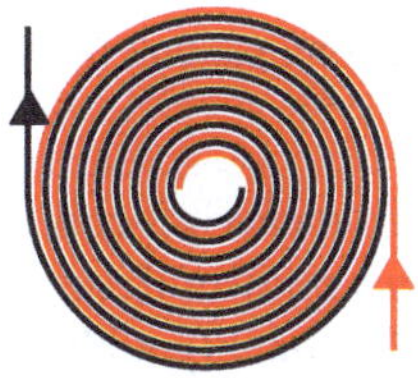

Fig. 5b. Connect outside black and inside red

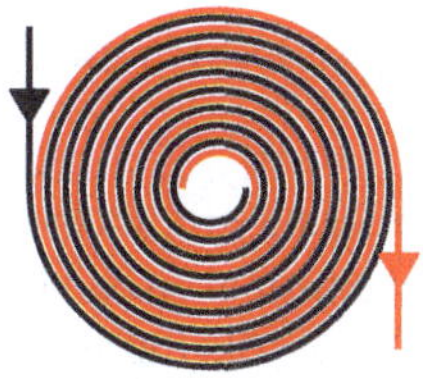

Fig. 5c. Connect inside black and outside red

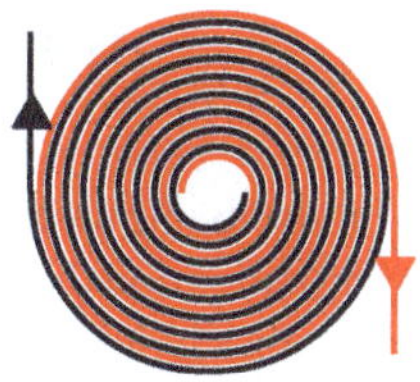

Fig. 5d. Connect outside black and red

Fig. 5. Schematic of connect line SCCT

Table 3 lists the measurement results. The power transfer efficiency between conventional spiral coils was 70.78 ± 0.47%. Under types 3 and 4, the power transfer efficiency

Table 3. Power transfer efficiency

Schematic of coil	Power transfer efficiency [%]
Type.1	70.78 ± 0.47
Type.2	0.767 ± 0.030
Type.3	68.19 ± 0.59
Type.4	69.95 ± 0.20
Type.5	0.560 ± 0.314

on condition 3 was 68.19 ± 0.59%, and on condition 4 was 69.95 ± 0.20%. On the other hand, conditions 2 and 5 were less than 1% in all ranges, suggesting that they were not suitable for power transmission. These results show conditions 3 and 4 receiving almost the same power transmission efficiency as the conventional method.

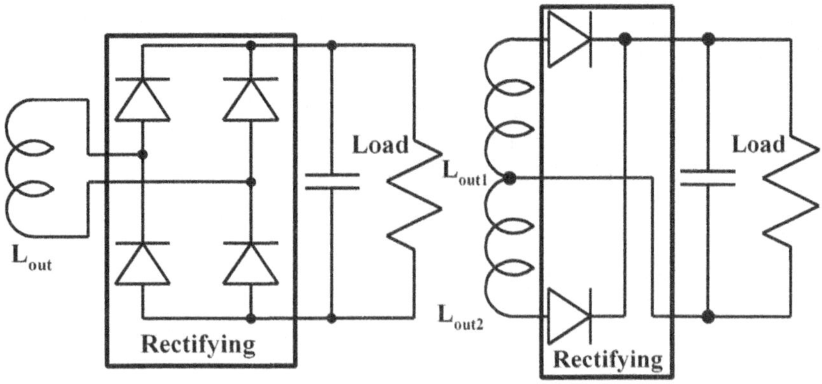

Fig. 6a. Conventional circuit **Fig. 6b.** Proposed circuit

Fig. 6. Internal rectifying and smoothing circuit for each coils

Figure 6 shows the internal circuit for each receiving coil. When performing full-wave rectification the conventional circuit using a spiral coil requires four rectifier diodes. On the other hand, SCCT achieves the same functionality with only two rectifier diodes. Hence, this leads to a reduction in the number of components and in the size of the internal circuit. The power transfer efficiency of the SCCT is slightly less than that of the spiral coil. However, the overall power transfer efficiency differs in Fig. 6, therefore it is necessary to consider the voltage drop that depends on the rectifier diodes.

Figure 7 shows the AC to DC power transfer efficiency from transmitting coil to internal load resistance while changing load resistance and keeping input 15 V amplitude sine wave. The efficiency of the spiral coil drops greatly when the load is heavy, but the SCCT exceeded the power transfer efficiency even at heavy loads by more than 60% and exceeded the power transfer efficiency of the spiral coil in all resistance ranges.

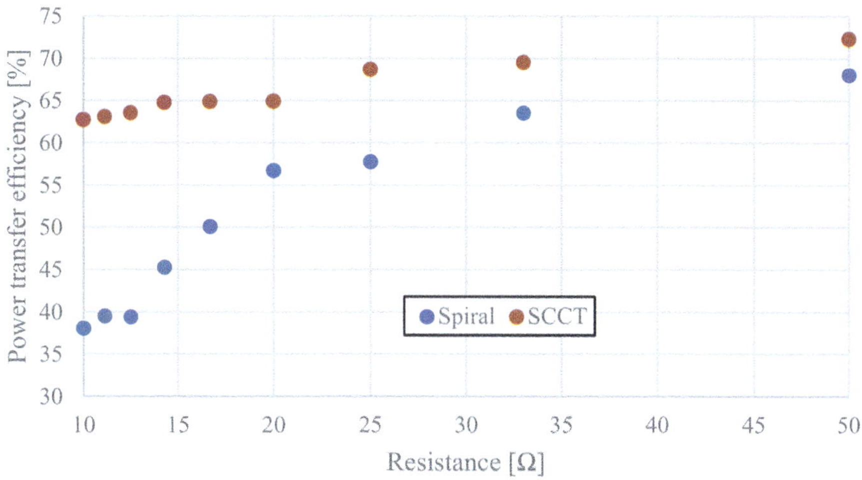

Fig. 7. Resistance vs. AC to DC power transfer efficiency

5 Conclusion

In this study, we proposed using SCCT for the receiving coil of TETS to reduce the number of components in the internal circuit. It was confirmed that the power transmission efficiency remained almost unchanged compared to the power transmission using a spiral coil in the power receiving circuit. Moreover, using SCCT resulted in a reduction in the number of components used in the conventional internal circuit. It was demonstrated that the same functionality could be achieved using SCCT.

References

1. Knecht, O., Bosshard, R., Kolar, J.W.: High-efficiency transcutaneous energy transfer for implantable mechanical heart support systems. IEEE Trans. Power Electron. **30**(11), 6221–6236 (2015)
2. Gunji, D., Imura, T., Fujimoto, H.: Basic study of transmitting power control method without signal communication for wireless in-wheel motor via magnetic resonance coupling. In: Proceedings of IEEE/IES International Conference on Mechatronics (ICM), pp. 313–318 (2015)

Machine Learning

Initial Study of Emotion Recognition Using Brain Rhythm Sequencing Method

Jia Wen Li[1,2,3], Shovan Barma[4], Sio Hang Pun[1,2(✉)], Mang I. Vai[1,2,3], Pan Ke Wang[1,2,3], Xin Liu[1,2,3], Yuan Yu Yu[1,2,5], Jiu Jiang Wang[1,2,5], Shuang Zhang[5], and Peng Un Mak[3]

[1] Institute of Microelectronics, University of Macau, Macau SAR, China
lodgepun@um.edu.mo
[2] State Key Laboratory of Analog and Mixed-Signal VLSI, University of Macau, Macau SAR, China
[3] Department of Electrical and Computer Engineering, Faculty of Science and Technology, University of Macau, Macau SAR, China
[4] Department of Electronics and Communication Engineering, Indian Institute of Information Technology Guwahati (IIITG), Guwahati, India
[5] College of Computer Science and AI, Neijiang Normal University, Neijiang, China

Abstract. A method named brain rhythm sequencing has been proposed to interpret Electroencephalography (EEG) signal for emotion recognition. For this purpose, first, the reassigned smoothed pseudo Wigner-Ville distribution (RSPWVD) has been adopted. Then, all generated sequences have been evaluated for classifying the emotional states based on arousal and valence. Moreover, the classification was by similarity measurements between pairwise of the sequences. After evaluating the accuracies by using the sequences from different channels and time durations, on one hand, the significant channel for classification has been identified so that the recognition can be accomplished by considering single-channel. On the other hand, the particular duration which was valuable for classification has been found so that the size of required data can be reduced. Thus, emotion recognition can be performed by fewer data resource. Results revealed that the accuracy of ~70% has been realized and hence, the proposed method could open a new way to implement data analysis for EEG-based applications such as emotion recognition.

Keywords: Brian rhythm sequencing · Electroencephalography (EEG) · Emotion recognition · Sequence classification

1 Introduction

Emotion is an efficient inner indicator of human psychological states, so to automatically recognize it by employing reliable techniques is valuable. For this purpose, various attempts such as voice [1], gesture [2], facial expression [3], etc. have been proposed in recent years. Nonetheless, the emotional states reflected by above modalities are subjective and could be disguised. Hence, the physiological signals controlled by the nervous

Y. Shiraishi et al. (Eds.): APCMBE 2020, IFMBE Proceedings 82, pp. 231–239, 2021.
https://doi.org/10.1007/978-3-030-66169-4_29

system are considered as the relatively truthful patterns available for emotional study and moreover, since Electroencephalography (EEG) is regulated by neural oscillations and it owns a strongly relationship with emotional reactions [4]. Thereby, the emotion recognition using EEG is presenting a greater potential in this field.

To realize EEG-based emotion recognition, the key is suitably extracting the valuable features from the signal. Because the information at both the time and frequency domains are helpful to represent the key properties of signal, thus, the time-frequency analysis (TFA) are commonly considered. Afterward, a fair amount of features can be mined and among them, the brain rhythms are the representatives. In detail, they are derived from classifying the EEG based on five specific sub-bands in frequency: δ(0–4 Hz), θ (4–8 Hz), α (8–13 Hz), β (13–30 Hz), and γ (30–50 Hz) [5]. The great contribution of them is that the variation of power of certain brain rhythm can be regarded as useful clue to identify emotional states in term of arousal and valence [6–8], in which arousal indicates whether a subject is calming or exciting, and valence means whether a subject has negative or positive feeling. Indeed, the previous works mainly based on the relationship between the powers of the brain rhythms and the levels of arousal or valence. However, the effect of the chronological orders of five brain rhythms and using the time-related occurrences to interpret EEG signal has not been investigated. If the brain rhythms are interpreted as the sequence, the time-frequency characteristics of EEG can be expressed simultaneously. In addition, following the sequence classification by performing the similarity measurement, the brain rhythm sequences can be adopted for classification task and from the evaluations of the accuracies by using the sequences from different channels and durations, on one hand, the significant channel for classification can be found. On the other hand, the particular duration which is more valuable for classification can also be determined. Such properties are beneficial to reduce the size of required data. So, in order to fulfill the aforementioned properties, a method named brain rhythm sequencing has been proposed.

The rest of this work is organized as follows: Sect. 2 introduces the experiment through DEAP database. Section 3 elaborates on the details about EEG-based brain rhythm sequencing method and sequence classification by measuring similarity. Section 4 discusses the results and finally, Sect. 5 makes the conclusions.

2 Experiment Overview

The EEG signals for experiment were from DEAP database [6]. In detail, it offered 32 subjects (15 females and 17 males, from 23 to 37 years) and each subject watched 40 different 60s videos to evoke emotions. After watching each video, the subject rated it (from 1–9) based on arousal (A) and valence (V). Therefore, the emotion of each trial (video) can be originally labeled into different sub-groups: $A \geq 5$, high arousal (HA) and $A < 5$, low arousal (LA); $V \geq 5$, high valence (HV) and $V < 5$, low valence (LV); where the threshold of high and low level was 5 [9]. Besides, the standard 10–20 system consisted of 32 channels (FP1, AF3, F3, F7, FC5, FC1, C3, T7, CP5, CP1, P3, P7, PO3, O1, OZ, PZ, FP2, AF4, FZ, F4, F8, FC6, FC2, CZ, C4, T8, CP6, CP2, P4, P8, PO4 and O2) were employed. Consequently, the size of EEG for each subject was 60s $\times$ 32 channels $\times$ 40 trials. Meanwhile, the sampling rate was 128 Hz and analog passband filtering of 0.01–100 Hz was applied for pre-processing.

As for the experimental steps, first, the pre-processed EEG signals were applied for producing the brain rhythm sequences by TFA. Then, to perform sequence classification, the target sequences correlated with emotional states (HA/LA and HV/LV) were selected according to the ratings of 40 trials. Next, the classification was by measuring similarity between pairwise of the sequences (the target sequence and the query sequence) and dynamic time warping (DTW) was adopted. Besides, since there were 32 channels, so 32 sequences were generated for one trial and the total size for one subject was 32 sequences $\times$ 40 trials. The generated sequences were evaluated to calculate the accuracies and from the results, the channel presenting the highest accuracy was defined as the significant channel for emotion recognition. In addition, as the whole stimulus was 60s, the length of generated sequence was also 60s and in order to investigate the segment which was more appropriate for classification, three durations with 20s length have been tested, i.e., 0–20s (beginning), 20–40s (middle), and 40–60s (end). From the accuracies based on different durations, the appropriate one can be surveyed. In this experiment, the emotion recognition was implemented on two dimensions separately, i.e., HA/LA in arousal and HV/LV in valence.

3 Proposed Methodology

The proposed sequence required the specific brain rhythm at a certain instant, so TFA was operated. Moreover, Wigner-Ville distribution (WVD) in TFA seemed proper for brain rhythm sequencing as it enabled the estimation of power value in the specific sub-band and to localize it in corresponding instant [10]. Nevertheless, WVD could produce unnecessary cross-terms which confine to acquire the frequency information at a shorter timestamp accurately. To improve it, an appropriately smoothing of WVD along the time direction as well as frequency direction was considered and this modified version named as smoothed pseudo WVD (SPWVD) [11]. Meanwhile, in order to obtain the precise time indices of higher power regions, the reassignment operation was recommended [12], as expressed in (1):

$$SPW_x^{(r)}(t', \omega'; g, h) = \int_{-\infty}^{+\infty}\int_{-\infty}^{+\infty} SPW_x(t, \omega; g, h) \delta(t' - \hat{t}(x; t, \omega))\delta(\omega' - \hat{\omega}(x; t, \omega))dtd\omega \tag{1}$$

where

$$\hat{t}(x; t, \omega) = t - \frac{SPW_x(t, \omega; \tau_g, h)}{2\pi SPW_x(t, \omega; g, h)} \tag{2}$$

$$\hat{\omega}(x; t, \omega) = \omega + j\frac{SPW_x(t, \omega; g, D_h)}{2\pi SPW_x(t, \omega; g, h)} \tag{3}$$

with $t_g = tg(t)$ and $D_h(t) = dh(t)/dt$.

After performing the reassigned SPWVD (RSPWVD), a time-frequency plane has been acquired and aiming to produce the sequence data from this plane, on one hand, the whole duration should be divided into some small timestamps (t_1, t_2, t_3 ...), which can be inspired by the average reaction time of neurons. In previous studies, Rey *et al.* [13] reported that the average evoked power from the EEG signal occurred for a timestamp around 0.2s. In [14], Chandra *et al.* claimed that the average reaction time of neurons was between 0.14–0.2s. Thus, 0.2s can be considered as a unit time period of reaction and in this work, the length of timestamp was defined as 0.2s. On the other hand, the frequency of the plane has been divided based on five brain rhythms (r) and therefore, there were five brain rhythms at each 0.2s, as illustrated in Fig. 1.

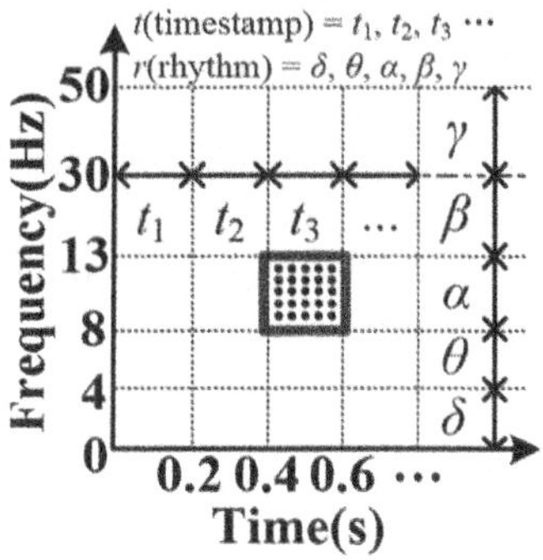

Fig. 1. Power estimations of brain rhythms (r) at each timestamp (t) in time-frequency plane.

Indeed, the sequence needed only one rhythm at each timestamp. Therefore, toward find the significant one for denoting each timestamp, the instantaneous power in each bound of the plane has been considered. For example, in Fig. 1, the power of α at t_3 has been highlighted, which was calculated by the average of all instantaneous values inside that bound. Following such a manner, the powers of remaining brain rhythms at the same timestamp can be calculated. Further, at each timestamp, because the maximum one reflected the crucial sub-band having a relatively great contribution of power, which can be regarded as a significant feature in the frequency. Thereby, the brain rhythm having maximum power contribution at each timestamp has been chosen for sequencing and Fig. 2 depicted the sample of generated brain rhythm sequence data following above steps, in which the EEG signal from FP1 channel presented at the bottom and its part of 5–10s with the corresponding sequence data depicted at the top.

Next, the generated sequences were tested for classification. In sequence classification, a similar structure revealed a similar function, whereas the dissimilarity or mismatch could incur mutation [15]. So, the attribute of the query sequence can be identified through the scores derived from the similarity measurements with the target sequences. Besides, as for selecting target sequences, the ratings have been considered and in detail, for each subject, among 40 trials, the one shown a maximum rating of A or V was correlated with HA or HV. Oppositely, the one indicated a minimum rating of A or V was correlated with LA or LV. As a result, 2 trials were defined as target sequences and the remaining 38 trials were assigned as query sequences. For instance, the maximum and minimum of subject S1 in A were 8.15 (13 trial) and 2.08 (14 trial); and in V were 9.00 (37 trial) and 1.36 (23 trial). Then, the sequences from those corresponding

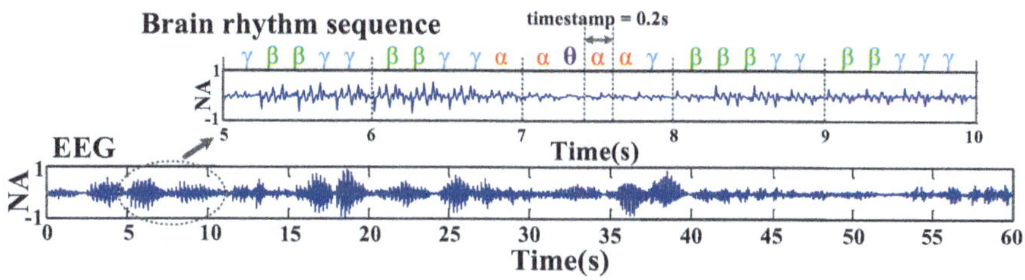

Fig. 2. Sequence data generated by choosing one of the brain rhythms (δ, θ, α, β, and γ) having maximum instantaneous power at each timestamp of 0.2s. (NA: Normalized Amplitude)

trials were acted as the target sequences. Meanwhile, because there were 32 channels, the target sequences were also chosen for all channels based on the same trials so that each channel possessed its particular target sequences for testing. Moreover, DTW, a distance-based method to search an optimal alignment between the sequences, has been applied for measuring the similarity level. Figure 3 illustrated a sample of HA/LA classification by the sequence data from F7 channel for S1. Here, the similarity levels between the query sequence (from one of the 38 tested trials) and the target sequences (HA and LA) were measured by DTW two times. Afterward, from the similarities with the two target sequences, the query sequence was assigned as the same emotion with the one which revealed a higher similarity level and in DTW, the smaller the distance implied the higher the similarity level. Finally, after comparing the testing results from the 38 query sequences (i.e., 38 tested trials) with the original labels, the accuracy (i.e., correct rate) can be calculated. Such procedure was repeated to all channels and durations, and the results were discussed in Sect. 4.

Fig. 3. Arousal classification using the sequences from F7 channel for subject S1 and the classification was by similarity measurements (from DTW) between pairwise of the sequences (the target sequence and the query sequence).

4 Results and Discussion

First, as an example, the results of arousal classification for subject S24 have been depicted in Fig. 4, in which the three scalp maps refer to the accuracies based on three time durations respectively and the deeper the red indicated the higher the accuracy. In Fig. 4, it can be found that the accuracy from the same channel was varying during the whole stimulus. For instance, the accuracies of F3 at 0–20s and 20–40s were relatively higher, whereas a lower result at 40–60s was obtained. It implied that the sequence data of F3 from the beginning and middle segments were more valuable than the end segment for classification. Thus, from the evaluations, the significant channel and particular time duration for recognition can be mined, as summarized in Table 1.

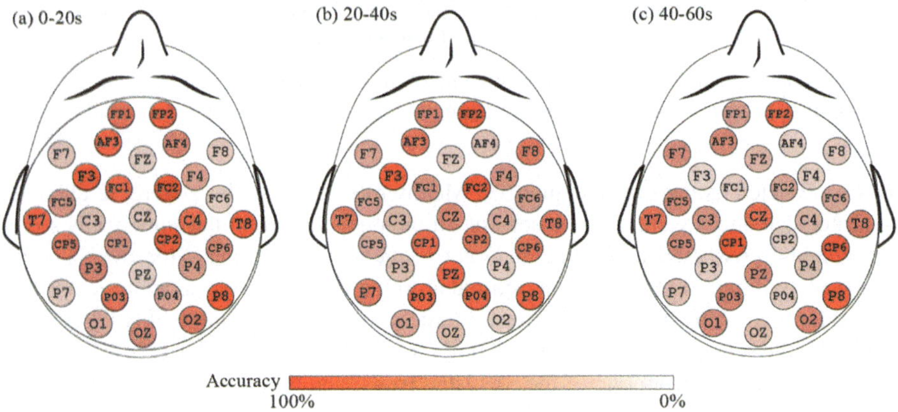

Fig. 4. Arousal classification accuracies based on the sequence data from different channels and durations by using the proposed method. The deeper the red revealed the higher the accuracy. (a) 0–20s; (b) 20–40s; and (c) 40–60s. (From EEG of subject S24 in DEAP)

Table 1. Significant channel and particular time duration of 32 subjects in DEAP database.

	A-SC	A-PTD (s)	A-ACC (%)	V-SC	V-PTD (s)	V-ACC (%)
S1	PZ	40–60	68.42	CP2	40–60	68.42
S2	CP1	20–40	68.42	FC5	20–40	71.05
S3	F4	0–20	81.58	FC5	40–60	63.16
S4	F7	0–20	65.79	F8	20–40	65.79
S5	PO3	20–40	68.42	P3	20–40	78.95
S6	F8	0–20	68.42	T7	40–60	81.58
S7	P3	0–20	68.42	PO4	20–40	76.32
S8	OZ	40–60	76.32	T7	20–40	68.42
S9	F7	0–20	68.42	CP6	20–40	68.42
S10	PO4	0–20	65.79	FC1	0–20	71.05
S11	P3	0–20	68.42	FC5	20–40	71.05
S12	FC2	0–20	86.84	P4	0–20	63.16
S13	PO3	0–20	86.84	CP5	20–40	73.68
S14	F8	40–60	76.32	CZ	0–20	65.79

(*continued*)

Table 1. (*continued*)

	A-SC	A-PTD (s)	A-ACC (%)	V-SC	V-PTD (s)	V-ACC (%)
S15	F3	0–20	65.79	C3	20–40	65.79
S16	C4	0–20	65.79	P7	20–40	71.05
S17	FP2	0–20	73.68	P4	20–40	71.05
S18	FZ	0–20	68.42	C3	20–40	71.05
S19	AF4	40–60	76.32	PO4	40–60	63.16
S20	F8	0–20	81.58	F7	0–20	65.79
S21	PZ	20–40	81.58	C4	0–20	65.79
S22	PO4	0–20	71.05	F7	20–40	73.68
S23	P4	0–20	73.68	CP2	20–40	71.05
S24	CP2	40–60	81.58	AF4	0–20	71.05
S25	F7	40–60	81.58	O2	0–20	65.79
S26	CP5	40–60	71.05	PZ	40–60	71.05
S27	FC1	0–20	76.32	C4	20–40	73.68
S28	AF4	20–40	68.42	F7	20–40	73.68
S29	CP1	20–40	76.32	PO4	40–60	68.42
S30	PZ	20–40	65.79	FC2	0–20	73.68
S31	FC5	0–20	68.42	CP1	0–20	73.68
S32	PO4	0–20	71.05	FC6	20–40	65.79

Note: A-Arousal, V-Valence, SC-Significant Channel, PTD-Particular Time Duration, ACC-Accuracy

In Table 1, regarding the significant channel, 9 subjects were found from the same lobe, while for the rest 23 subjects, their channels were from different lobes, i.e., for ~72% subjects, their significant channels for arousal and valence classifications were different. As known, different lobes dealt with different functions [16] and here, the significant channel could imply the personal active location correlated with the arousal and valence factors separately. In addition, as for particular time duration, in arousal classification, 19 have 0–20s, 6 shown 20–40s, and 7 presented 40–60s, so the prominent duration was 0–20s (beginning); in valence classification, 9 have 0–20s, 17 shown 20–40s, and 6 presented 40–60s, so the prominent duration was 20–40s (middle). In this regard, the appropriate moment for recognizing the levels of arousal and valence were related to different durations, which revealed different emotional factors could be evoked by considering different parts of the stimulus. Meanwhile, the differences in Table 1 also demonstrated the individual characteristics in emotions and thus, to provide more insights for recognition, the subject-dependent model was preferred.

Moreover, the comparative study has been performed and it has been described in Table 2, in which the first column presented the work, and the rest listed the number of channels used, the applied time length of EEG data, methodology, and the accuracy correspondingly. Using the proposed method, several properties of emotion recognition can be investigated, including the significant channel provided a single-channel solution for classification and the particular time duration helped to reduce the size of applied data. Further, regarding the accuracy, compared with the previous works, the proposed method produced the impressive results by applying a 20s time length and single-channel data only. As a result, it can be said that the proposed method provided a novel path for EEG-based emotion recognition through fewer data resource compared with the previous works.

Table 2. Comparative study of previous EEG-based emotion recognition works

	No. of channels	Time length (s)	Methodology	Accuracy (%) Valence-ACC	
				A	V
Chao *et al.* [17]	32	60	Capsule network	68.28	66.73
Chen *et al.* [18]	32	60	C4.5 decision Tree	69.09	67.89
Atkinson *et al.* [19]	14	60	MRMRM	73.06	73.14
Kumar *et al.* [20]	2	30	Bispectral analysis	64.84	61.17
This work	1	20	BRS	73.03	70.07

Note: A-Arousal, V-Valence, MRMRM-Maximum Relevance Minimum Redundancy Method, BRS-Brain Rhythm Sequencing

5 Conclusion

In this work, brain rhythm sequencing for emotion recognition has been proposed and 32 subjects from DEAP database have been applied for evaluations. Results indicated that the accuracy of ~70% can be accomplished by using single-channel data with a 20s particular time duration, which offered superior result through fewer data resource compared with previous works. Besides, during the evaluations, the target sequences were chosen based on individual ratings and thereby, if the ratings have the errors due to misunderstandings by the subjects, the selections could be influenced. In the future, the experiments for evoking strong and valid emotional reactions will be performed so that the proper target sequences can be selected accordingly for sequence classification.

Acknowledgment. This work was funded by The Science and Technology Development Fund (FDCT), Macau SAR (093/2015/A3, 088/2016/A2, and SKL-AMSV(FDCT-funded)), and the Research Committee of the University of Macau under Grant MYRG2018–00146-AMSV and Grant MYRG2019–00056-AMSV. The authors express faithful thanks to the financial supports listed above.

Conflict of Interest. The authors declare that they have no conflict of interest.

References

1. El Ayadi, M., Kamel, M.S., Karray, F.: Survey on speech emotion recognition: features, classification schemes, and databases. Pattern Recogn. **44**(3), 572–587 (2011)
2. Metcalfe, D., McKenzie, K., McCarty, K., et al.: Emotion recognition from body movement and gesture in children with Autism Spectrum Disorder is improved by situational cues. Res. Dev. Disabil. **86**, 1–10 (2019)
3. Ko, B.C.: A brief review of facial emotion recognition based on visual information. Sensors **18** (2018). Article 401
4. Li, X., Song, D., Zhang, P., et al.: Exploring EEG features in cross-subject emotion recognition. Front. Neurosci. **12** (2018). Article 162
5. Choi, S.J., Kang, B.G.: Prototype design and implementation of an automatic control system based on a BCI. Wireless Pers. Commun. **79**(4), 2551–2563 (2014)
6. Koelstra, S., Mühl, C., Soleymani, M., et al.: DEAP: a database for emotion analysis using physiological signals. IEEE Trans. Affect. Comput. **3**(1), 18–31 (2012)
7. Kim, M.K., Kim, M., Oh, E., et al.: A review on the computational methods for emotional state estimation from the human EEG. Comput. Math. Methods Medi. **2013** (2013). Article 573734
8. Onton J., Makeig S.: High-frequency broadband modulation of electroencephalographic spectra. Front. Hum. Neurosci. **3** (2009). Article 61
9. Mohammadi, Z., Frounchi, J., Amiri, M.: Wavelet-based emotion recognition system using EEG signal. Neural Comput. Appl. **28**(8), 1985–1990 (2017)
10. Tzallas A. T., Tsipouras M. G., and Fotiadis D. I.: Automatic seizure detection based on time-frequency analysis and artificial neural networks. Comput. Intell. Neurosci. **2007** (2007). Article 80510
11. Barma, S., Chen, B.W., Ji, W., et al.: Detection of the third heart sound based on nonlinear signal decomposition and time-frequency localization. IEEE Trans. Biomed. Eng. **63**(8), 1718–1727 (2016)
12. Li, J.W., Barma, S., Mak, P.U., et al.: Brain rhythm sequencing using EEG signals: a case study on seizure detection. IEEE Access **7**, 160112–160124 (2019)
13. Rey, H.G., Fried, I., Quiroga, R.Q.: Timing of single-neuron and local field potential responses in the human medial temporal lobe. Curr. Biol. **24**(3), 299–304 (2014)
14. Chandra, A.M., Ghosh, S., Barman, S., et al.: Effect of exercise and heat-load on simple reaction time of University Students. Int. J. Occup. Saf. Ergon. **16**(4), 497–505 (2010)
15. Barne, S.L., Belliveau, N.M., Ireland, W.T., et al.: Mapping DNA sequence to transcription factor binding energy in vivo. PLoS Comput. Biol. **15**(2), 1–29 (2019)
16. Dai, Y., Wang, X., Li, X., et al.: Sparse EEG compressive sensing for web-enabled person identification. Measurement **74**, 11–20 (2015)
17. Chao, H., Dong, L., Liu, Y., et al.: Emotion recognition from multiband EEG signals. Sensors **19** (2019). Article 2212
18. Chen, J., Hu, B., Moore, P., et al.: Electroencephalogram-based emotion assessment system using ontology and data mining techniques. Appl. Soft Comput. **30**, 663–674 (2015)
19. Atkinson, J., Campos, D.: Improving BCI-based emotion recognition by combining EEG feature selection and kernel classifiers. Expert Syst. Appl. **47**, 35–41 (2016)
20. Kumar, N., Khaund, K., Hazarika, S.M.: Bispectral analysis of EEG for emotion recognition. Procedia Comput. Sci. **84**, 31–35 (2016)

Fast Learning from Imperfect Labels to Segment Brain Based on Active Contour Model and 3D U-Net

Honglei Ji[1], Yi Lv[1], and Junchen Wang[1,2(✉)]

[1] School of Mechanical Engineering and Automation, Beihang University, Beijing 100191, China
wangjunchen@buaa.edu.cn

[2] Beijing Advanced Innovation Center for Biomedical Engineering, Beihang University, Beijing 100083, China

Abstract. Brain extraction is an essential segmentation step in many neuroimage analysis studies. Convolutional neural network approaches have been proposed in the literature and widely used for medical image segmentation, outperforming traditional segmentation approaches. However, these learning-based methods highly rely on ground truth data which needs significant time and labor inputs from experienced experts. In addition, the independent identical distribution assumption is difficult to comply with in practice, leading to the overfitting problem across different datasets. This paper introduces a 3D U-Net based self-fine-tuning network for brain volumetric segmentation that learns from quickly available but imperfect labeled images generated by a model-based approach. We outline two typical application scenarios of our method: (1) In a ground truth absence setup, an improved brain extraction tool (IBET) algorithm is used to generate brain label images fully automatically. The self-fine-tuning network learns from these imperfect labels and yields more accurate predictions. (2) In a transfer learning setup, the IBET algorithm is used to fast extract dataset-specific labels for fine tuning the pretrained network (overfitting prevention). We have evaluated the performance of the proposed method on four datasets and achieved satisfactory results.

Keywords: Brain extraction · 3D U-Net · Self-fine-tuning · Active contour model

1 Introduction

Brain extraction, a.k.a. skull stripping, from brain magnetic resonance (MR) images, is an essential step in many neuroimage analysis studies. It refers to the segmentation of brain tissue from non-brain tissues in the brain MR images. Applications in brain disease prediction, diagnosis, and therapy such as brain volumetric measurement, cortical surface reconstruction, assessing Alzheimer's disease require stripped MRI scans. Nevertheless, automatic accurate brain segmentation is generally a difficult task due to the obscure brain boundaries, low contrast, and absence of intensity standardization [1]. To this end, the

Y. Shiraishi et al. (Eds.): APCMBE 2020, IFMBE Proceedings 82, pp. 240–248, 2021.
https://doi.org/10.1007/978-3-030-66169-4_30

research community has produced a number of methods to automate this procedure. The existing methods can be generally categorized into four types: intensity and morphology based [2], atlas based [3], deformable model based [4, 5] and learning based [6–9] approaches. After the pioneering work of Krizhevsky et al. [10], deep learning especially convolutional neural network (CNN) approaches, have been actively investigated and widely used for medical image segmentation tasks in recent years, and in most cases deep learning based methods outperform traditional segmentation approaches.

However, these learning-based methods highly rely on "ground truth" data which needs significant time and labor inputs from experienced experts. In addition, the generalization ability of a CNN trained on a specific training dataset depends on the independent identical distribution assumption which is difficult to comply with in practice, leading to the overfitting problem when applied in a different dataset. This motivates us to think two questions: 1. How to train a CNN from scratch with quickly available imperfect training data generated from a model-based approach that yields better segmentation accuracy; and 2. How to fine tune a pretrained CNN on a different dataset with the least human input. Recent research [11] shows that recent progress on handling label noise in deep learning has been largely omitted by the medical image analysis community. Fedorov et al. [12] explore the segmentation of brain tissue from noisy label learning and prove that the network with "dropout" can generalize from imperfect training data. However, no quantitative evaluations are presented. Heller et al. [13] analyze the effect how label errors influence the performance of segmentation, and they find that for each network architecture, performance declines with boundary-localized errors but robust to non-boundary-localized errors. It shows that learning from boundary noisy labels is a very challenging problem.

In this paper we propose a 3D U-Net based self-fine-tuning network (**learner**) for 3D brain segmentation that learns from quickly available but imperfect labeled images generated from a model-based approach (**generator**). We show two typical application scenarios that answer the aforementioned two questions: (1) In a ground truth absence setup, an improved brain extraction tool (IBET) algorithm proposed in our previous work is used to generate brain label images fully automatically. The self-fine-tuning network learns from these imperfect labels and yields more accurate predictions. (2) In a transfer learning setup, the IBET algorithm is used to fast extract dataset-specific labels for fine tuning the pretrained network so as to prevent overfitting. We have evaluated the performance of the proposed method on four datasets and achieved satisfactory results.

2 Methods

2.1 Label Generator

Annotation of volumetric data such as CT or MR with segmentation labels in a slice-by-slice manner generally takes up to hours for one volume, which needs significant human inputs. In our previous work, we propose an improved brain extraction tool (IBET) to fast segment the 3D brain in an MR volume with instant 3D visualization. The algorithm evolves an icosphere to fit the brain surface in the MR volume as shown in Fig. 1. For the details of our IBET method, we refer interested readers to the original paper [5].

In this paper, the IBET algorithm serves as a label generator to fast produce labeled training data. As the algorithm runs quickly (appr. 500 ms per volume of 200 × 200 × 200), we can write a script to fully automatically segment a whole dataset. Note that the parameters of IBET for the whole dataset are fixed so the segmentation results are generally not optimal for an individual volume (i.e., imperfect labels).

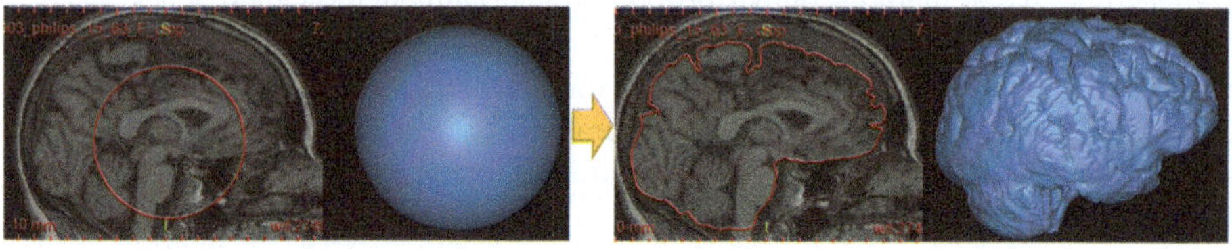

Fig. 1. Skull stripping using the improved brain extraction algorithm.

2.2 Self-fine-Tuning CNN Learner

Architecture. The self-fine-tuning CNN learner architecture is based on the 3D U-Net [14], which consists of an encoder part to extract features and a decoder part to obtain precise localization. It is built of convolution layers, pooling layers, transposed convolution layers, skip connections and self-fine-tuning mechanisms (see Fig. 2).

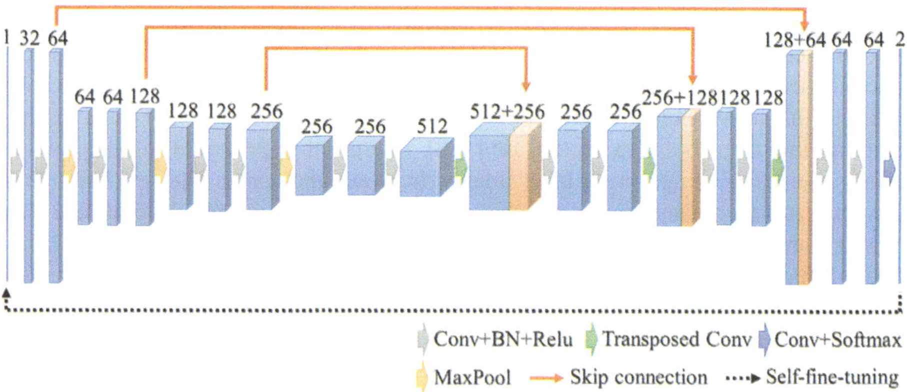

Fig. 2. Architecture of the self-fine-tuning segmentation network. Blue boxes represent feature maps. The number of channels is denoted above each feature map.

Inspired by the fact that CNNs can learn and generalize from data, we assume that the CNN model could correct the labels when trained on the data with imperfect labels, and the corrected labels can be utilized in the next training iteration. The pipeline of our proposed CNN learner based on imperfect labels is illustrated in Fig. 3. Different with the original 3D U-Net which is supposed to be trained on pure ground truth, our architecture is supposed to be trained on imperfect training data, therefore a self-fine-tuning mechanism is used to encourage the model to correct imperfect labels when entering the next training iteration. As soon as the self-fine-tuning mechanism takes effect (controlled by a hyper-parameter), the next training iteration will start and use the prediction results from the previous training iteration as the new training labels. A hyper-parameter l_t is introduced to control the commence of the self-fine-tuning iteration. When the training loss and validation loss are both below l_t, the self-fine-tuning mechanism works (entering the next training iteration) and l_t is updated to the current validation loss.

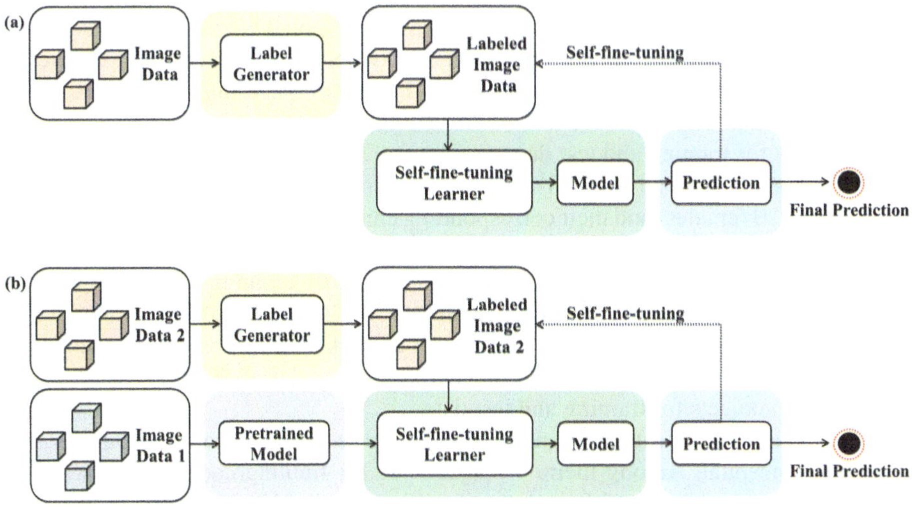

Fig. 3. The pipeline of our proposed CNN learner based on imperfect labels. (a) Ground truth absence segmentation: IBET is used to generate brain label images fully automatically. The self-fine-tuning network learns from these imperfect labels and yields more accurate predictions. (b) Transfer learning segmentation: the IBET algorithm is used to fast extract dataset-specific labels for fine tuning the pretrained network so as to prevent overfitting.

Training. For training the network, Adam optimizer is employed with an initial learning rate of 0.001, and learning rate decreasing with the epoch is formulated as follows:

$$Lr_{\text{current\_epoch}} = Lr_{\text{initial}} \times 0.97^{\text{epoch}} \tag{1}$$

where $Lr_{\text{current\_epoch}}$ is the learning rate of the current epoch, and Lr_{initial} is the initial learning rate. The batch size is set to 2. Data augmentation including random rotation, translation and scaling is performed on-the-fly. One minus Dice coefficient is used as the loss function:

$$loss = 1 - \frac{2|X \cap Y|}{|X| + |Y|} \tag{2}$$

where X and Y are the ground truth mask and predicted mask.

3 Experiments and Results

3.1 Datasets

Four different datasets with ground truth are used in this study. Note that the ground truth is only used to evaluate the accuracy of the proposed method in the experiments, not for network training since we assume the ground truth is not always available in practice. The details of the datasets are as follows:

- NFBS skull-stripped images dataset [15] is obtained from the Neurofeedback Skull-stripped (NFBS) repository. It consists of 125 T1-weighted scans from subjects with a variety of clinical and subclinical psychiatric symptoms (ages from 15 to 45 years) as well as their corresponding annotations. The size of the volumes is $192 \times 256 \times 256$ voxels, and the voxel size is $1 \times 1 \times 1$ mm. The dataset is divided by 100 and 25 subjects for training and test data.
- LPBA40 skull-stripped images dataset [16] consists of 40 T1-weighted scans (20 males, 20 females) and their corresponding annotations. Among them, 12 subjects are for training and the others are for testing. The size of the volumes is $256 \times 124 \times 256$ voxels. Coronal slices are acquired 1.5 mm apart with an in-plane resolution of 0.86 mm (38 subjects) or 0.78 mm (2 subjects).
- CC359(12) skull-stripped images dataset [17] consists of 12 T1-weighted scans from healthy subjects with hand labeling brain masks as ground truth. The dataset is divided by 5 and 7 subjects for training and test data.
- Anonymous clinical data contains of 8 T1-weighted scans obtained from our collaborated hospital. Among them, 4 subjects are for training and the others are for testing. The average size of the volumes is $168 \times 264 \times 218$ voxels. Brain masks are delineated by professional neurosurgeons as the ground truth.

3.2 Implementation

We implemented the self-fine-tuning network using Pytorch. All experiments were performed on a computer workstation with Intel(R) Xeon(R) Sliver 4110 CPUs (2.10 GHz $\times$ 2) with 32GB RAM and two NVIDIA Quadro RTX 8000 GPUs with 48 GB RAM.

3.3 Ground Truth Absence Segmentation Experiment

For the ground truth absence segmentation task, we use the brain images from the NFBS dataset with imperfect labels generated by IBET, and the ground truth masks are used to evaluate the segmentation accuracy metrics. The self-fine-tuning 3D U-Net and the original 3D U-net were trained on the imperfect labels for 200 epochs, respectively. During the training of the self-fine-tuning 3D U-Net, the training labels were corrected by four times (4 training iterations) in total.

Table 1. Evaluation metrics compared with the ground truth on the NFBS dataset.

Methods	Dice	Jaccard	VS	HD (mm)	ASSD (mm)
IBET	0.9474	0.9005	0.9163	16.2021	1.7408
3D U-net	0.9508	0.9064	0.9174	15.4071	1.5468
Self-fine-tuning network	0.9647	0.9319	0.9733	15.9505	1.0760

Dice coefficient (DC), Jaccard Index (IoU), and volumetric similarity (VS) are used as evaluation metrics, which are adopted to calculate the similarity of two masks X and Y as follows, respectively:

$$DC = \frac{2|X \cap Y|}{|X| + |Y|} \tag{3}$$

$$J(X, Y) = \frac{|X \cap Y|}{|X \cup Y|} \tag{4}$$

$$VS = 1 - \frac{||X| - |Y||}{|X| + |Y|} \tag{5}$$

In addition to the voxel overlap evaluation, Hausdorff distance (HD) and average symmetric surface distance (ASSD) are also calculated. Table 1 summarizes the overall analysis of individual metrics. Note that the results of IBET are different with that of our previous work because the parameters of IBET for the whole NFBS dataset were fixed in our experiments rather than being optimized for individual brain volumes.

The results show that the Dice score, Jaccard index, volumetric similarity, and average symmetric surface distance of the self-fine-tuning network are all higher than those of IBET and the original 3D U-Net. The Hausdorff distance score is statistically similar to that of the original 3D U-Net.

3.4 Transfer Learning Segmentation Experiment

For transfer learning segmentation task, a standard 3D U-Net model was pretrained on the NFBS training dataset with the ground truth. The pretrained 3D U-Net model on the NFBS test dataset attained the Dice coefficient of 99%. For other three datasets LPBA40, CC359(12), and anonymous clinical data, we performed the segmentations in the following three ways: 1. Using the pretrained model; 2. Using the IBET algorithm; 3. Using the IBET results as the training data to fine tune the pretrained model. The results are compared against the ground truth masks to calculate evaluation metrics. Table 2 summarizes the overall analysis of individual metrics. Moreover, during the training of the self-fine-tuning 3D U-Net, the labels were corrected by four, nine and two times for LPBA40, CC359(12), and anonymous clinical data, respectively. The statistical results are graphically represented in Fig. 4.

The results show that our proposed self-fine-tuning CNN learner with the IBET label generator achieved the highest mean scores on all the five metrics, outperforming all the other methods.

Table 2. Evaluation metrics compared with the ground truth for the other three datasets.

Methods	Metrics				
	Dice	Jaccard	VS	HD (mm)	ASSD (mm)
LPBA40					
IBET	0.9625	0.9277	0.9583	15.8579	1.1676
Pretrained model	0.9367	0.8819	0.9346	52.3304	2.7003
Self-fine-tuning network	0.9662	0.9348	0.9730	13.6495	0.9876
CC359(12)					
IBET	0.9662	0.9346	0.9861	13.3656	1.5496
Pretrained model	0.9606	0.9243	0.9840	14.9913	1.5961
Self-fine-tuning network	0.9700	0.9418	0.9912	9.5783	1.4183
Clinical data					
IBET	0.9859	0.9722	0.9938	8.5796	0.4097
Pretrained model	0.9750	0.9513	0.9849	16.2524	0.7526
Self-fine-tuning network	0.9870	0.9743	0.9971	7.5819	0.3531

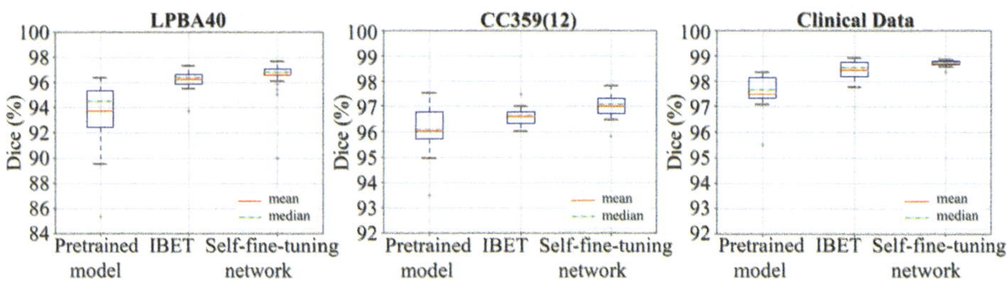

Fig. 4. Performance of our self-fine-tuning network compared with other methods.

4 Conclusion and Discussion

We have introduced a self-fine-tuning network for brain volumetric segmentation that learns from quickly available imperfect labeled images and offers more accurate labels. This mechanism combines both the advantages of model and learning based methods and has two typical application scenarios, i.e., ground truth absence learning and model transfer learning. We have tested the performance of the proposed network on four datasets and achieved satisfactory results, which proves that the self-fine-tuning CNN architecture with the IBET label generator has the ability to fast learn, generalize and correct from imperfect labels.

However, the introduction of self-fine-tuning mechanisms will slightly increase the training time. Additionally, these experiments were all run on datasets with binary labels. In clinical diagnosis or investigations, more work must be done to study whether the method generalizes to different problems, and problems with many class labels.

Acknowledgement. This work was supported in part by the National Natural Science Foundation of China under Grant 61701014 and Grant 61911540075, and in part by the National Key Research and Development Program of China under Grant 2017YFB1303004.

Conflict of Interest Declaration. The authors declare that they have no conflict of interest.

References

1. Iglesias, J.E., Liu, C.Y., Thompson, P.M., et al.: Robust brain extraction across datasets and comparison with publicly available methods. IEEE Trans. Med. Imaging **30**(9), 1617–1634 (2011)
2. Subudhi, A., Jena, J., Sabut, S.: Extraction of brain from MRI images by skull stripping using histogram partitioning with maximum entropy divergence. In: 2016 International Conference on Communication and Signal Processing (ICCSP), pp. 0931–0935. IEEE, Beijing (2016)
3. Ledig, C., Heckemann, R.A., Hammers, A., et al.: Robust whole-brain segmentation: application to traumatic brain injury. Med. Image Anal. **21**(1), 40–58 (2015)
4. Smith, S.M.: Fast robust automated brain extraction. Hum. Brain Mapp. **17**(3), 143–155 (2002)
5. Wang, J., Sun, Z., Ji, H., et al.: A fast 3D brain extraction and visualization frame-work using active contour and modern OpenGL pipelines. IEEE Access **7**, 156097–156109 (2019)
6. Kleesiek, J., Urban, G., Hubert, A., et al.: Deep MRI brain extraction: a 3D convolutional neural network for skull stripping. Neuroimage **129**, 460–469 (2016)
7. Dey, R., Hong, Y.: CompNet: complementary segmentation network for brain MRI extraction. In: International Conference on Medical Image Computing and Computer-Assisted Intervention, pp. 628–636. Springer, Granada (2018)
8. Chen, K., Shen, J., Scalzo, F.: Skull stripping using confidence segmentation convolution neural network. In: International Symposium on Visual Computing, pp. 15–24. Springer, Nevada (2018)
9. Hwang, H., Rehman, H.Z.U., Lee, S.: 3D U-Net for skull stripping in brain MRI. Appl. Sci. **9**(3), 569 (2019)
10. Krizhevsky, A., Sutskever, I., Hinton, G.E.: Imagenet classification with deep convolutional neural networks. In: Advances in Neural Information Processing Systems, Nevada, pp. 1097–1105 (2012)
11. Karimi, D., Dou, H., Warfield, S.K., et al.: Deep learning with noisy labels: exploring techniques and remedies in medical image analysis. arXiv preprint arXiv:1912.02911 (2019)
12. Fedorov, A., Johnson, J., Damaraju, E., et al.: End-to-end learning of brain tissue segmentation from imperfect labeling. In: 2017 International Joint Conference on Neural Networks (IJCNN), pp. 3785–3792. IEEE, Alaska (2017)
13. Heller, N., Dean, J., Papanikolopoulos, N.: Imperfect segmentation labels: how much do they matter? In: Intravascular Imaging and Computer Assisted Stenting and Large-Scale Annotation of Biomedical Data and Expert Label Synthesis, pp. 112–120. Springer, Quebec (2018)
14. Çiçek, Ö., Abdulkadir, A., Lienkamp, S. S., et al.: 3D U-Net: learning dense volumetric segmentation from sparse annotation. In: International Conference on Medical Image Computing and Computer-Assisted Intervention, pp. 424–432. Springer, Athens (2016)
15. Puccio, B., Pooley, J.P., Pellman, J.S., et al.: The preprocessed connectomes project repository of manually corrected skull-stripped T1-weighted anatomical MRI data. Gigascience **5**(1), s13742–s14016 (2016)

16. Shattuck, D.W., Mirza, M., Adisetiyo, V., et al.: Construction of a 3D probabilistic at-las of human cortical structures. Neuroimage **39**(3), 1064–1080 (2008)
17. Souza, R., Lucena, O., Garrafa, J., et al.: An open, multi-vendor, multi-field-strength brain MR dataset and analysis of publicly available skull stripping methods agreement. Neuroimage **170**, 482–494 (2018)

Comparison of Reconstruction Performances Between Deep Learning and Iterative Methods in MR Compressed Sensing

Shohei Ouchi and Satoshi Ito(✉)

Department of Information and Systems Sciences, Graduate School of Engineering, Utsunomiya University, 7-1-2 Yoto, Utsunomiya, Tochigi, Japan
ultuchiboy@gmail.com, itohst@is.utsunomiya-u.ac.jp

Abstract. A deep residual learning convolutional neural network (CNN) is applied to improve image quality and speed up the reconstruction of compressed sensing magnetic resonance imaging. The reconstruction performances of proposed method are compared with iterative reconstruction methods.

The proposed method adopts a deep residual learning CNN to learn the residual component between the input and output images (i.e., aliasing artifacts) for image reconstruction. The CNN-based reconstruction is compared with iterative reconstruction methods to clarify its characteristics. The peak-signal-to-noise ratio (PSNR) and the structural similarity index (SSIM) are examined for various numbers of sampling rates.

The experimental results demonstrate that reconstruction time is drastically reduced to 0.022 s per image compared with that for conventional iterative reconstruction. Compared with iterative reconstruction, superior quality was obtained for low sampling rates (30% or 40%). These results suggest that CNN can learn the rules of the occurrence pattern of aliasing artifact. With the proposed method, the structure and small contrast in the original image are well preserved in the reconstructed image and aliasing artifacts are mostly removed.

A deep residual learning CNN can recognize and predict aliasing artifacts. It is demonstrated that reconstruction time is significantly reduced.

Keywords: Compressed sensing · Reconstruction · Deep learning

1 Introduction

Magnetic resonance imaging (MRI) is commonly used medical diagnostic tools. MRI scanners produce high-quality images. However, the scan time of MRI is much longer than that of other tools such as X-ray computed tomography. The application of compressed sensing (CS) [1, 2] to MRI image acquisition (CS-MRI) [3] has attracted much interest in recent years. CS is a signal recovery theory that allows image reconstruction from fewer sampled signals than those required by the Nyquist-Shannon sampling theorem.

Y. Shiraishi et al. (Eds.): APCMBE 2020, IFMBE Proceedings 82, pp. 249–254, 2021.
https://doi.org/10.1007/978-3-030-66169-4_31

With the application of CS, scan time can be shortened changing the imaging sequence and algorithm for image reconstruction. However, CS-MRI has the problems of image quality degradation and a long reconstruction time. Image reconstruction is performed by solving the L1-L2 minimization problem with some constraints. An iterative reconstruction method is used to solve the minimization problem. The reconstruction time required per image is relatively long compared to that for standard Fourier transform-based methods.

Deep learning has received much attention because of its excellent performance in speech and image recognition [4, 5]. Deep learning has been applied to MRI image reconstruction [6–8]. Because deep learning involves the computation of a function that maps some inputs to their corresponding outputs using a huge number of training images, it takes a lot of time to train a convolutional neural network (CNN). However, once learning has been completed, rapid reconstruction with no iterative processing can be achieved.

In this study, we applied deep residual learning in image domain learning using a CNN to the image reconstruction problem using an undersampled MRI signal. Zhang et al. proposed a CNN in which DnCNN is adopted for removing noise from noisy images [9]. Inspired by Zhang et al.'s network, we propose a CNN reconstruction method that uses DnCNN with a modified version of Zhang et al.'s network (DnCNN-CS). Our method predicts aliased components caused by undersampling using patch-based learning and then obtains an alias-free image by subtracting the predicted alias components from the input alias-superimposed image. To the best of our knowledge, no previous work has directly predicted aliasing components using patch based learning for CS image reconstruction.

To clarify the characteristics of images obtained using the proposed CNN-based reconstruction, a comparison is conducted with reconstructed images obtained with conventional iterative reconstruction methods, namely the Split Bregman method [10].

2 Material and Methods

2.1 Network Structure of DnCNN-CS

The proposed DnCNN-CS is designed to predict the residual image, i.e., the difference between the alias-superimposed image and the ground truth image, which is different from a typical CNN that predict the de-aliased image directly. Our proposed CNN is a modified version of Zhang et al.'s denoising CNN. The input layer has a convolutional (Conv) layer and a leaky ReLU. The middle layers have a Conv layer, batch normalization, and a leaky ReLU. The output layer has a Conv layer. CNN architecture and reconstruction flow are shown in Fig. 1.

2.2 Image Reconstruction in DnCNN-CS

Let the original image (ground truth) vector be $y \in \mathbb{C}^N$, $x \in \mathbb{R}^N$, respectively and the full Fourier encoding matrix be $F \in \mathbb{C}^{N \times N}$, then Fx represents the full k-space data. Consider a diagonal matrix representing the under-sampling mask as $U \in \mathbb{R}^{N \times N}$, then

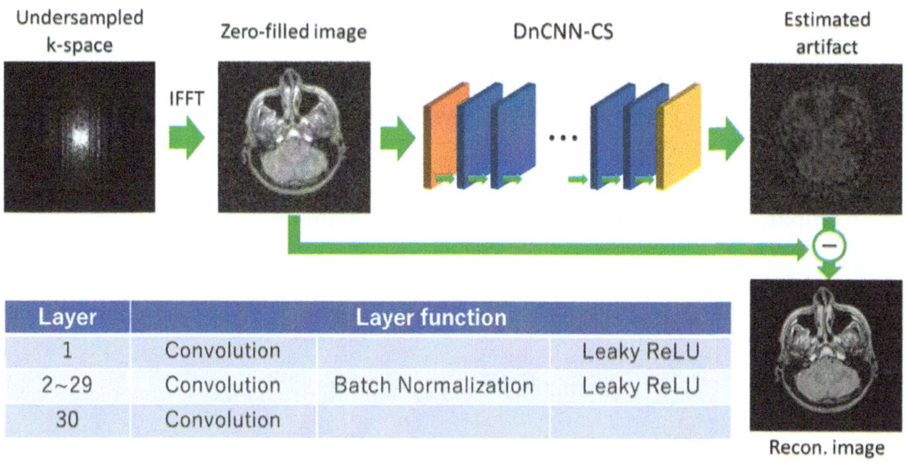

Layer	Layer function		
1	Convolution		Leaky ReLU
2~29	Convolution	Batch Normalization	Leaky ReLU
30	Convolution		

Fig. 1. The network structure of DnCNN-CS and image reconstruction procedure. CNN is composed of convolution, Batch Normalization and ReLU. DnCNN-CS is used to estimate the aliasing artifacts and reconstructed image is obtained by subtracting the estimated artifact from the zero-filled image.

the under-sampled k-space signal can be expressed as UFx and zero-filled image x_z is obtained by Eq. (1),

$$x_z = F^H UFx \tag{1}$$

where H represents the Hermitian transpose operation. Aliasing artifacts v caused by under-sampling will superimposed on the zero-filled image x_z. In this time, x_z can be expressed as follows:

$$x_z = x + v. \tag{2}$$

When the original mapping is more like an identity mapping, the residual mapping will be much easier to be optimized [11]. Zero-filled image x_z is much more like the fully scanned image x and close to identity mapping. Therefore, residual learning formulation is more suitable for image de-aliasing in DnCNN-CS. Let $R(x_z)$ be the residual mapping to predict v ($R(x_z) \approx v$), reconstructed image x' is obtained by Eq. (3):

$$x' = x_z - R(x_z). \tag{3}$$

2.3 Datasets

MR images used in this study are proton-density-weighted images (PDW) of the head included in the IXI dataset [12]. For this study, we used 2500 images (100 images per person) for training and 40 images (one person) for testing, that are sampled from various sequences captured using a 1.5T MRI system. If the size of images is smaller than 256 × 256 pixels, then and zero data were filled in the both end of the image to be 256 × 256 pixels image.

2.4 Sampling Pattern

We consider 1D random under-sampling, where k-space signals are randomly under-sampled only the phase-encoding direction as shown in Fig. 1. Sampling rates of 30, 40, and 50% were examined.

2.5 Comparison with Iterative Reconstruction Methods

Generally, CS-MRI is formulated as a penalized inverse problem including the L1-L2 norm. The solution for the object function shown as Eq. (4) is taken as the reconstructed MR image.

$$\underset{x}{\operatorname{argmin}} \| \mathrm{UFx} - \mathrm{y} \|_2^2 + \lambda \| \Psi \mathrm{x} \|_1 \tag{4}$$

According to the CS theory, signal sparsity is an important prior to remove the aliasing artifacts due to under-sampling in k-space, therefore sparsity is introduced by applying sparsifying transform function Ψ to the image. Since L1-minimization problem is non-smooth and non-differentiable function, many algorithms have been proposed so far. Split Bregman algorithm [10] is utilized as variable splitting algorithms that came to be used to improve the quality of reconstructed images in the development period of CS-MRI. This algorithm is a concept from functional analysis for finding extrema of convex functional. We evaluate Split Bregman to compare the results with proposed method. Obtained images are compared in term of the peak signal to noise ratio (PSNR) and SSIM.

2.6 Computing Environment and Training Time

Training and testing were performed using the MATLAB R2017b on a computer equipped with an Intel Core i7-9700K (3.60 GHz) CPU and an NVIDIA GeForce RTX 2080 Ti GPU. The training required 8 h for 2500 images.

3 Result and Discussion

The relationship between PSNR, SSIM, and sampling rates of DnCNN-CS and Split Bregman is summarized in Table 1. The PSNR and SSIM values are the averages of 40 images. Table 1 shows that the DnCNN-CS has higher PSNR, SSIM values than Split Bregman for sampling rates lower than 50%. A comparison of reconstructed images obtained with each method are shown in Fig. 2. Each image includes an enlarged view of the area indicated by the red box. Figure 2 shows that many more aliasing artifacts remain and details are lost for the Split Bregman. Because sampling density is low at middle and higher spatial frequencies, as shown in Fig. 1. The aliasing artifacts are likely to appear in the reconstructed image. In contrast, the structure and small contrast are well preserved with the DnCNN-CS. These results indicate that the DnCNN-CS can learn the rules of the occurrence pattern of aliasing artifact.

The reconstruction time for the DnCNN-CS was 1.78 s using CPU only and 0.022 s using the CPU and GPU. The reconstruction time for Split Bregman was 13.95 s.

Table 1. The PSNR and SSIM results of DnCNN-CS and Split Bregman for the different sampling rate.

Sampling rate [%]	DnCNN-CS		Split Bregman	
	PSNR [dB]	SSIM	PSNR [dB]	SSIM
30	31.73	0.9642	28.11	0.9354
40	33.25	0.9716	31.09	0.9592
50	34.55	0.9777	35.06	0.9809

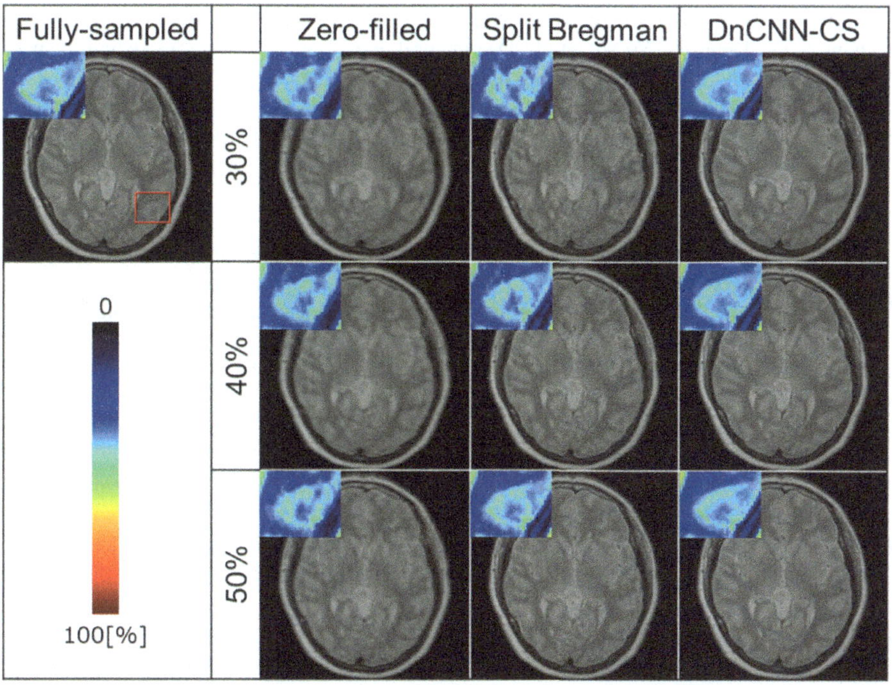

Fig. 2. Comparison of reconstructed images for 1D random sampling. Sampling rate is set as 30, 40, and 50%. Each image includes an enlarged view of the area indicated by the red box.

4 Conclusion

We presented a DnCNN-CS approach for image reconstruction using an undersampled MRI signal in which aliasing artifacts caused by undersampling are learned by a patch-based CNN. A comparison of the proposed DnCNN-CS with conventional iterative reconstruction methods (Split Bregman) shows that the PSNR and SSIM values for the DnCNN-CS are superior in 1D random undersampling. The DnCNN-CS has a drastically reduced reconstruction time.

Acknowledgements. This work was partially supported by JSPS KAKENHI grants 19K04423 and Kayamori Foundation of Informational Science Advancement. We acknowledge https://brain-development.org/ixi-dataset/ as the source of the IXI data used in this paper.

References

1. Donoho, D.L.: Compressed sensing. IEEE Trans. Inf. Theory **52**, 1289–1306 (2006)
2. Candès, E.J., Wakin, M.B.: An introduction to compressive sampling. IEEE Signal Process. Mag. **25**, 21–30 (2008)
3. Lustig, M., Donoho, D.L., Pauly, J.M.: Sparse MRI: the application of compressed sensing for rapid MR imaging. Magn. Reson. Med. **58**, 1182–1195 (2007)
4. Hinton, G., Deng, L., Yu, D., et al.: Deep neural networks for acoustic modeling in speech recognition: the shared views of four research groups. IEEE Signal Process. Mag. **29**, 82–97 (2012)
5. Krizhevsky, A., Sutskever, I., Hinton, G.: ImageNet classification with deep convolutional neural networks. In: 25th International Conference on Neural Information Processing Systems of NIPS 2012, Lake Tahoe, Nevada, pp. 1097–1105 (2012)
6. Wang, S., Su, Z., Ying, L., et al.: Accelerating magnetic resonance imaging via deep learning. In: IEEE 13th International Symposium on Biomedical Imaging, Prague, Czech Republic, pp. 514–517 (2016)
7. Lee, D., Yoo, J., Tak, S., et al.: Deep residual learning for accelerated MRI using magnitude and phase networks. IEEE Trans. Biomed. Eng. **65**, 1985–1995 (2018)
8. Hyun, CM., Kim, HP., Lee, SM., et al.: Deep learning for undersampled MRI reconstruction. Phys. Med. Biol. 63, 135007 (2018)
9. Zhang, K., Zuo, W., Chen, Y., et al.: Beyond a Gaussian denoiser: residual learning of deep CNN for image denoising. IEEE Trans. Image Process. **26**, 3142–3155 (2017)
10. Goldstein, T., Osher, S.: The split Bregman method for L1-regularized problems. SIAM J. Imag. Sci. **2**, 323–343 (2009)
11. He, K., Zhang, X., Ren, S., et al.: Deep residual learning for image recognition. In: IEEE Conference on Computer Vision and Pattern Recognition, Las Vegas, USA, pp. 770–778 (2016)
12. IXI dataset. https://brain-development.org/ixi-dataset/. Accessed 8 Jan 2020

Tissue Engineering

Ceramide: An Extravagant Prospective for Treatment of Burn Patients

Md. Ashrafuzzaman(✉), Faiza Ahmed, Afsana Mimi, Farahdiba Zarin, and Sabrina Shetu

Military Institute of Science and Technology, Dhaka, Bangladesh
ashezaman@gmail.com, faiza.ahmed123@gmail.com, afsanamimiasa@gmail.com, farahdiba.zarin96@gmail.com, sabrinashetu23@gmail.com

Abstract. Progressions in wound controlling to encourage rapid and successful healing of wound by providing a warm humid environment at the site of application of treatment aims to facilitate patient immediate rescue. Ceramide has an extravagant prospective in providing structural support and preserving the water-retaining properties of the epidermis. Hydrogel and Composite sponges based on chitosan/gelatin, with and without incorporating ceramide, was synthesized by chemical crosslinking and evaluated as wound dressing material. Physical characteristics of the prepared Hydrogels such as water vapor transmission rate (WVTR) and composite sponges (water absorption and weight loss) were inspected. The lowest WVTR was showcased by the sample containing chitosan and hydrogel in 1:1 ratio, with ceramide included in them. Addition of ceramide led to lower WVTR. FTIR of hydrogel membranes and composite sponges authenticated the existence of free hydroxyl groups. N-H and O-H stretching and the intramolecular hydrogen bonds were indicated by presence of bands for the composite material in the region $3272.6 - 1028.7\ cm^{-1}$ and $3280.1 - 1028.7\ cm^{-1}$ in hydrogel membrane. Stretching of $C = O$ bonding due to the formation of hydrogen bonds are detected in amide I mode, which implies the presence of hydrogen bonding. The SEM images revealed that the surface of hydrogel membranes were compact in nature, which prevents microbes from entering. The porous nature of the internal structure of the composite sponges allows for improved water retention. Increasing the amount of chitosan by threefold that of hydrogen improved the porosity of the structure.

Keywords: Ceramide · Burn patients · Wound healing · Wound dressing · Biopolymers

1 Introduction

Ceramides are components of sphingolipids and form their main supporting structure. In human skin, the epidermal permeability barrier is formed by them residing stratum corneum [1]. They perform the crucial functions of providing a layer of protection, shedding and maintaining moisture of the skin, and functioning as a protective barrier

Y. Shiraishi et al. (Eds.): APCMBE 2020, IFMBE Proceedings 82, pp. 257–271, 2021.
https://doi.org/10.1007/978-3-030-66169-4_32

against external pathogens and environmental contaminants [2, 3]. Wound management with ceramide would have favorable effects mitigating the results of barrier disruption that tape stripping causes [4]. In contrary to the role of ceramide in forming the epidermal protection layer, the signaling characteristics of ceramide promote proliferation, differentiation, cell growth and apoptosis in epidermal keratinocytes [5, 6].

A wound is a disturbance of typical anatomic structure and function (defect or a break), from damage or due to an underlying medical or physiological condition [7, 8]. Wound healing is an evolutionarily conserved, complex, multicellular process of regenerating tissues of dermis and epidermis, whereby the body aims to restore the injured site to its normal condition as successfully as possible (in skin, aims at barrier restoration) [9]. Recently, it has been proved that a warm humid environment at the site of wound dressing allowing the fluids to remain bound to the wound and preventing them from spreading to unaffected areas, accomplish more rapid and successful wound healing whereas principle function of the conventional wound dressing (such as natural or synthetic bandages, cotton wool, lint, and gauzes) was keeping the wound dry by permitting the wound exudates to leave via evaporation and averting entrance of harmful microorganism [10, 11].

Hydrogels are the hydrophilic, insoluble, 3D cross-linked polymeric structures that exhibit the capability to imbibe enormous water or biological fluids by swelling extensively, while nevertheless conserving their structural integrity [12, 13]. Due to their cooling and hydrating effect, they play a dominant role in burn treatment [14]. This makes Hydrogel membranes favorable as a wound management product capable of absorbing water and other exudates, maintaining the moist environment, giving cushioning and cooling effects, easy removal, and reducing malodor by allowing the exudates to be absorbed and retained in the gel [15, 16].

Biomaterials are natural biodegradable polymers [17]. The dermal and the epidermal regions of skin may utilize biopolymers in prompting the wound healing process. Chitin, chitosan, PVA and gelatin are among extensively used natural polymers. Hydrogels of natural polymers are especially of interest given their advantages of polysaccharides being naturally derived, inexpensive, absorbent, non-toxic, non-immunogenic, biocompatible, environmental responsiveness via degradation, cost effective, and highly abundant [18–21].

Chitosan is formed by D-glucosamine and N-acetyl-D-glucosamine in a linear polysaccharide where the components are connected by β(1–4) glycosidic linkages. The components that make up chitosan comprises of results of deacetylation of a naturally occurring poly β(1–4) N-acetyl D-glucosamine called chitin, found abundant in the exoskeleton of crustaceans [22, 23]. The properties of chitosan, deem it beneficial in wound healing, such as: biocompatibility, biodegradability [24], hemostatic activity [25, 26], healing acceleration [27], non-toxicity, adsorption properties, anti-microbial and anti-infection properties [28–30]. These unique set of biological properties make chitosan is an enchanting material for wound healing applications and as a wound healing accelerator. Gelatin is an assortment of proteins soluble in water, which constitutes of collagen extracted from variety of animal by-products (mainly availed from pork skins, pork bones, cattle bones, or split cattle hides) [31]. It enables fluid retention despite exudation with ease of application in the form of films [32, 33]. Its low cost, biocompatible nature [34], and tendency to be a biodegradable [35] biomaterial having hydrophilic, non-toxic, non-antigenic, non-immunogenic properties, make it a promising candidate in wound healing applications. The material has the potential of enhancement via chemical modification and the ability to enable covalent drug delivery by attachment of the drug covalently [36]. The usage of PVA as a hydrogel membrane is major. PVA hydrogels are transparent (allowing continuous monitoring of the progress of healing), malleable, bio-inert, biocompatible, biodegradable (minimizing the repetition of dressing changes), and possess the ability to absorb water in significant amounts (keeping a moisturized wound environment) [37–39]. Crosslinking agents (glutaraldehyde) can be used to chemically bind PVA and other polymers with a covalent bond connection [40].

2 Materials and Methods

2.1 Materials

ORYZA Ceramide was procured from ORYZA OIL & FAT CHEMICAL CO., LTD., Japan. Chitosan, Polyvinyl alcohol (PVA), Glutaraldehyde (GA), and Glycerin were procured from Sigma Aldrich Chemical Co., Germany. Gelatin was brought from OMC. Ethanol and HCl were supplied by Changshu Hongsheng Fine Chemical Co., Ltd.

2.2 Preparation of Hydrogel Membranes

Chitosan (0.2 g) was solvated in 10% aqueous acetic acid (20 mL) to make a 1% chitosan solution (w/v). Temperature was maintained at 70 °C for 10 min with constant stirring in a hotplate magnetic stirrer. To form gelatin (10%) solution, gelatin powder (1 g) was dissolved in deionized water (10 mL) at 70 °C for 5 min with constant stirring by using hotplate magnetic stirrer. Then the chitosan solution was poured in gelatin solution in various proportions (Chitosan: Gelatin = 3:1, 1:1, and 1:3) named as C3G, CG and CG3, respectively and mixed in the magnetic stirrer vigorously until the aqueous solutions turned opaque. 1g of PVA was dissolved in 10 ml water (10% w/v) and the resulting mixture was heated at 70 °C for 5 min with continuous stirring. For ceramide encapsulated in the Hydrogel, 0.1 g of ceramide was mixed with the PVA solution. PVA solution, both with and without ceramide was added to the previously prepared solution. The ceramide encapsulated solution were named CGW, C3GW and CG3W. 0.5 ml of GA and 0.05 ml of HCl was added in 10 ml of ethanol to prepare crosslinking agent, which was then added to the mixture with constant stirring. Then 2 ml of glycerin was added with continuous stirring, and the resulting solution was poured into separated petridishes. The resulting solution mixture left for 24 h at room temperature at an aseptic environment (Fig. 1).

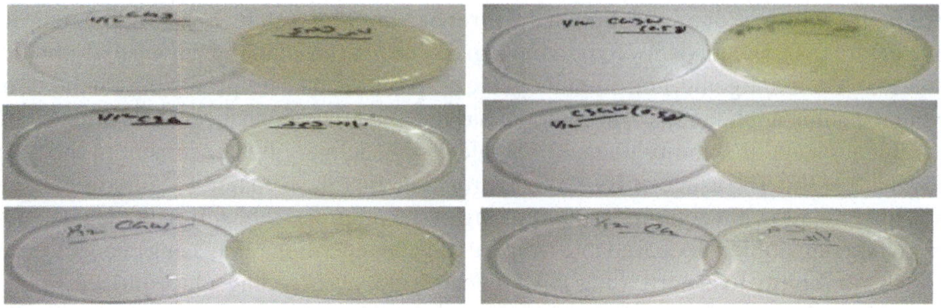

Fig. 1. Prepared hydrogel membranes

2.3 Preparation of Composite Sponge

Chitosan and gelatin solutions for composite sponges were prepared using the process which was used in their preparation for the hydrogel membranes. The composite materials were prepared by mixing chitosan solution with gelatin solution in various proportions (Chitosan: Gelatin = 3:1, 1:1, and 1:3) named as C3G, CG and CG3, respectively by vigorous stirring in a magnetic stirrer until the solutions turned opaque with even consistency. Then the solutions were transferred into distinct petridishes and allowed to cool undisturbed in room temperature. For ceramide encapsulated in the composite sponge, 0.1 g ceramide was added to 5 mL of water. The resulting ceramide solution was added to the opaque solutions of chitosin and gelatin previously prepared, with the new solutions containing ceramide named C3GW, CGW and CG3W. Theceramide composite sponges were obtained after the solutions containing chitosin, gelatin, and ceramide were poured into separated petridishes and cooled at room temperature (Fig. 2).

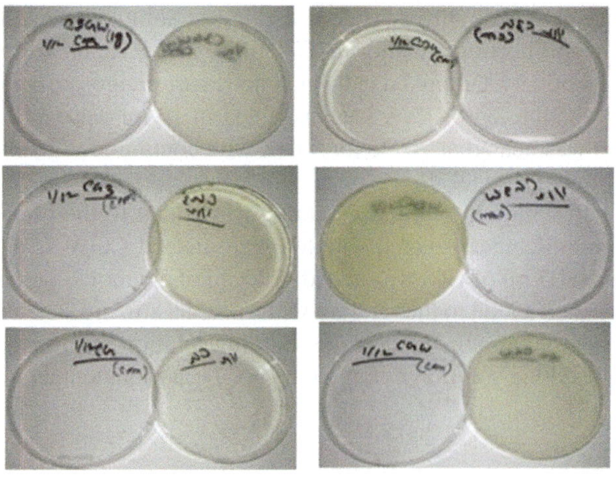

Fig. 2. Prepared composite sponges

2.4 Water Vapor Transmission Rate (WVT) of Hydrogel Membranes

Water vapor transmission rate (WVTR) measurement was calculated as stated by the European Pharmacopoeia monograph pertaining to this analysis [41]. For WVT measurements, test tube of 1.95 cm mouth diameter was filled with 10 ml deionized water. Test tube mouth was closed with our prepared hydrogel membrane and Teflon tape was used to generate an air tight seal. After weighing the test tube, it was placed in an oven for 24 h at 50 °C. The test tube was weighed again after taking out of the oven. The measurement of water vapor transmission rate was done by using formula:

$$\mathrm{WVTR} = \frac{\mathrm{Wi} - \mathrm{Wt}}{\mathrm{A} \times 24} \times 10^6\ \mathrm{gm}^{-2}\mathrm{h}^{-1} \quad (1)$$

The area of the test tube mouth was denoted by A; the initial and final weight (g) of test tube were denoted by W_i and W_t, respectively.

2.5 Water Absorption Test of Composite Sponges

To determine the water absorption capacity, the incubation of the composite sponge (which was freeze-dried for 6 h) in phosphate-buffered saline (PBS) was done at pH 7.4 at 37 °C for 2 h. The weight of the swamped sponge was measured immediately after soaking up the excess moisture of the surface with filter paper. The composite sponge was reweighed after the freeze drying of swolled sponges. The water absorption capacity of the sponge was calculated by using the formula:

$$E\mathrm{ad} = \left[\frac{\mathrm{We} - \mathrm{Wo}}{\mathrm{Wo}}\right] \times 100 \quad (2)$$

Ead is the water adsorption percentage of composite sponge at equilibrium. *We* and W_o **represent** the weight of the sponge(g) at equilibrium and initially, respectively.

2.6 Weight Loss of Prepared Composite Materials

The prepared sponges (freeze-dried for 6 h) were submerged in PBS buffer solution at the temperature of 37 °C for 24 h. Then the sponges were reweighed after taking out and dried in the oven. Weight loss of sponges was measured by using the following equation:

$$\%W = \frac{(\mathrm{W0} - \mathrm{W}e)}{\mathrm{W0}} \times 100 \quad (3)$$

where, W_o and *We* denote the initial and final weight of sponge (g), respectively.

2.7 Fourier Transform Infrared Spectroscopy (FT-IR)

FT-IR was carried out to assess the structural transformation in hydrogel membranes and composite sponges. FT-IR analysis was accomplished on the dried samples in their pure form. FT-IR spectrometer used an attenuated total reflection (ATR) unit ranging from 450 to 4000 cm^{-1} in the recording of the spectra. Regarding the hydrogel membranes, samples were placed directly into FT-IR machine. All analysis were performed at ambient room temperature.

2.8 Scanning Electron Microscope (SEM)

The surface morphology of the hydrogel membrane and composite materials were examined using SEM (JSM-7600F). Samples were covered with a thin gold conductive layer before analysis. Observation of sample surfaces was conducted by a secondary electron detector applying 5 kV accelerating voltage.

3 Result and Discussion

3.1 Water Vapor Transmission (WVT) Rate of Hydrogel Membrane

The hydrogel used as wound dressing is required to have the capability to reduce the loss of liquid fluid from the wound by conserving a moist domain. WVT lesser in value makes the hydrogel beneficial for wound dressing. From the Fig. 3 we can see that, presence of ceramide minimized the water vapor transmission rate. Incorporation of ceramide in the hydrogel membrane can thus preserve moist environment under the wound with better success and shall minimize loss of fluid from the wound, leading to better wound management.

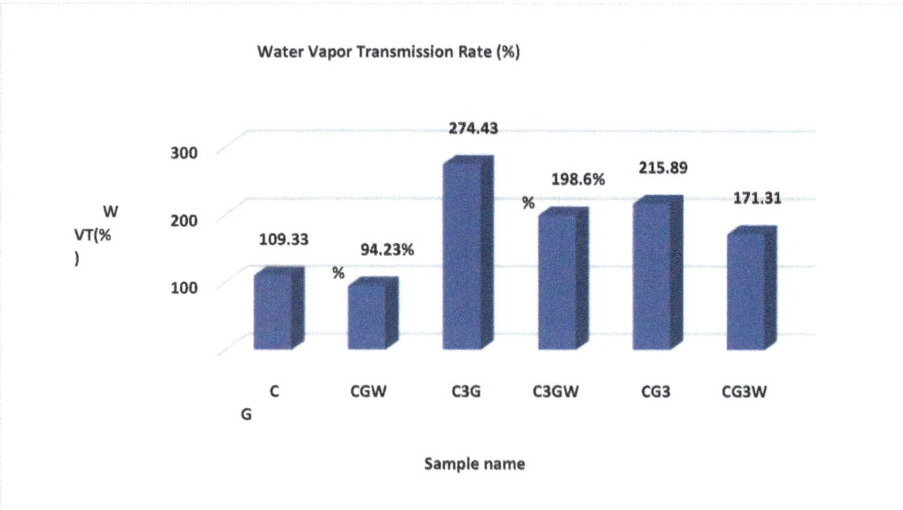

Fig. 3. Water vapor transmission rate of hydrogel membrane.

3.2 Water Absorption and Weight Loss of Composite Sponges

Water absorption capacity is of an utmost importance for biological applications and wound healing. It shows the ability to imbibe wound exudates. For higher gelatin content in the composite materials, the water absorption abilities were significantly reduced. Lower water content in the materials during preparation made way for smaller pore sizes, and as a result, the water uptake was decreased. The increased water uptake by the prepared composite materials can be justified by the hydrophilic and swelling characteristics of gelatin. Of the prepared composited, C3GW showed highest water absorption ability, as it can retain water in the pores surrounding the structure. Additionally, water absorption decreased about 2-fold for CG and 30-fold for CG3, which may refer to the swelling and dissolving of gelatin that could cause the pore sizes to decrease and allow to less water absorption and retention within their network structure. Furthermore, in the water absorption capacity of ceramide loaded composite materials and blank as reported in the Fig. 5, the water uptake has been observed to relate directly to the pore size. An increasing amount of gelatin is retained after some is lost by dissolution with an increase in gelatin content (Fig. 4).

The weight loss is decreased by adding ceramide as the sponges consisting ceramide can resist the dissolution in water.

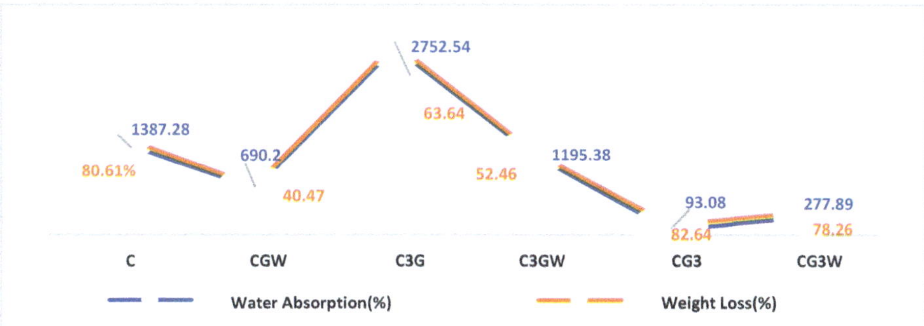

Fig. 4. Water absorption and weight loss of composite sponge

3.3 FT-IR of Hydrogel Membrane and Composite Sponges

The FTIR analysis was performed to evaluate the functional groups present in the composite sponge and hydrogel membrane, as well as to determine the interactions between polymeric chains. Focus of the study was to identify the existence of free hydroxyl groups responsible for the moisture retaining capacity of the hydrogel and composite sponges (Figs. 6, 7 and 8).

The FT-IR of composite sponges depicted a mixture of characteristic absorptions as a result of amine groups of chitosan and the carboxylic acid groups of gelatin. The peaks of C = O of amide I for chitosan at 1650 cm^{-1} shifted to 1632.6 cm^{-1} in composite sponge and 1640.0 cm^{-1} in hydrogel membrane. These refer to the interaction of $-NH^{3+}$ of chitosan and the COO^{-} of gelatin in all composites. A strong band in the region 3272.6 – 1028.7 cm^{-1} in composite material and 3280.1 – 1028.7 cm^{-1} in hydrogel membrane resembles N-H and O-H stretching and the intramolecular hydrogen bonds. The absorption bands at around 2922.2 and 2875 cm^{-1} can be referred to C-H stretching which are symmetric and asymmetric respectively [42]. The residual N-acetyl groups presence was confirmed by the bands at around 1632.6 cm^{-1} in composite sponge and

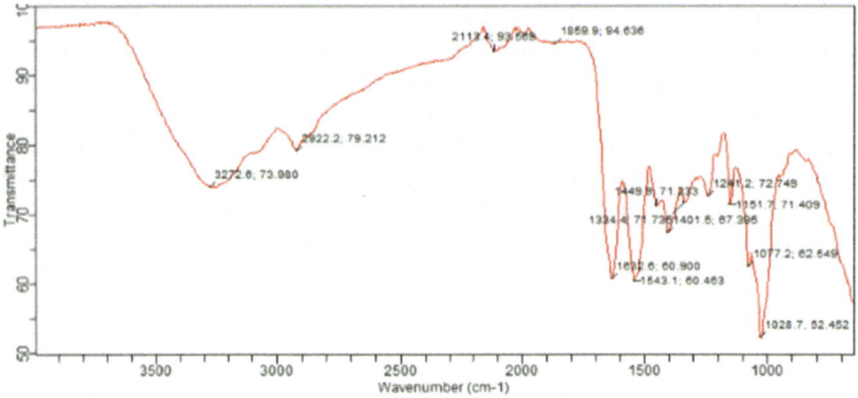

Fig. 5. FT-IR spectra of prepared composite materials (C3GW) with 3:1 rations of chitosan and gelatin

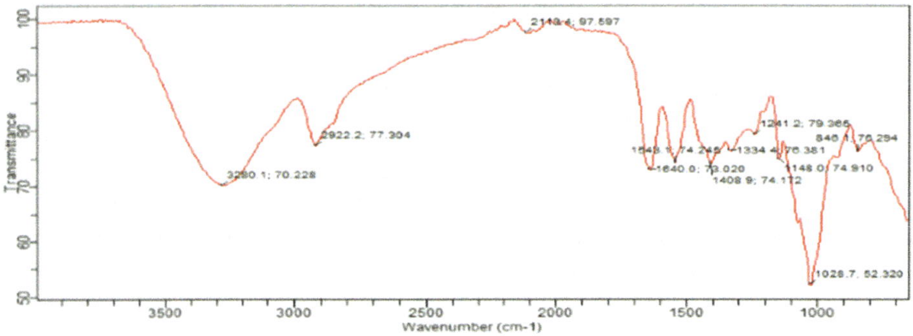

Fig. 6. FT-IR spectra of prepared hydrogel membrane (C3GW) with 3:1 rations of chitosan and gelatin.

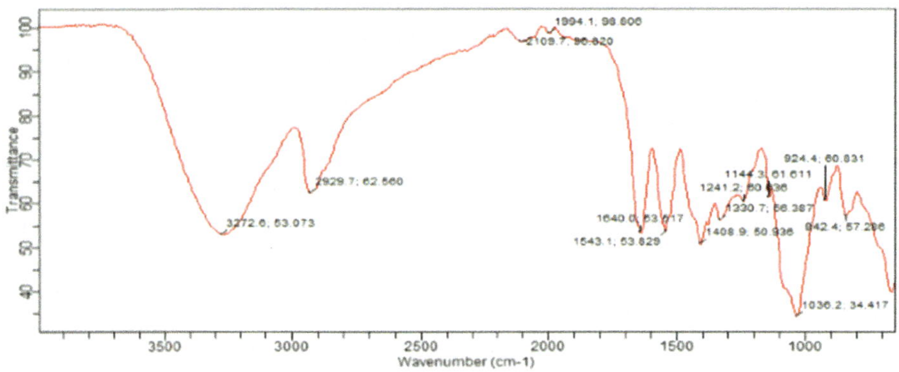

Fig. 7. FT-IR spectra of prepared hydrogel membrane (C3G) with 3:1 rations of chitosan and gelatin.

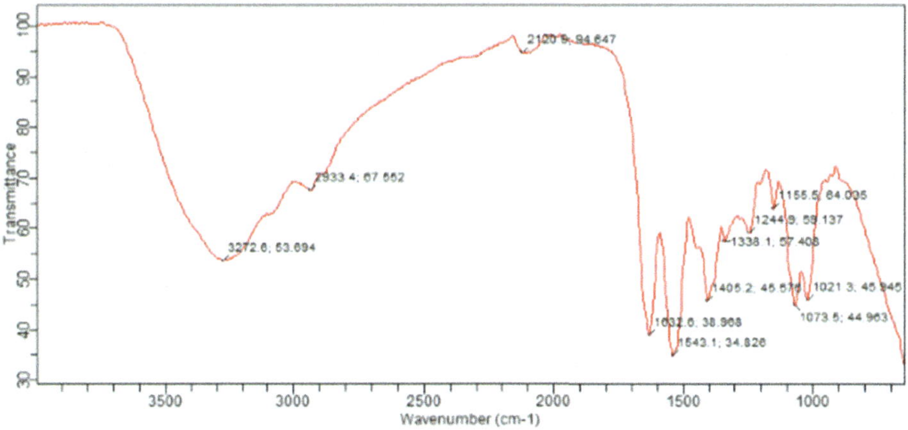

Fig. 8. FT-IR spectra of prepared composite sponge (C3G) with 3:1 rations of chitosan and gelatin.

1640.0 in hydrogel membrane (C = O stretching of amide I) and 1334.4 cm^{-1} in both hydrogel membrane and composite sponge (C-N stretching of amide III), respectively. We found the miniature band at 1543.1 cm^{-1} related to N-H bending of amide II in both hydrogel membrane and composite sponges, which is the third band characteristic of typical N-acetyl group. A band at 1543 cm^{-1} conforms to the N-H bending of the primary amine [43]. The CH2 bending and CH3 symmetrical deformations were established by the existence of bands at around 1449.9 in composite sponge whereas 1408.9 in hydrogel membrane and 1334.4 cm^{-1} in both composite sponge and hydrogel membrane, respectively. The absorption band at 1151.7 cm^{-1} in composite sponge and 1148.0 in hydrogel membrane can be referred to asymmetric stretching of the C-O-C bridge. The bands at 1077.2 and 1028.7 cm^{-1} attribute to C-O stretching in both hydrogel membrane and composite sponge [44, 45]. Stretching of C = O bonding due to the formation of hydrogen bonds are detected in amide I mode, which implies hydrogen bonding. C-N stretching and the plane bending of N-H generate the amide II mode. The water H-O-H bending mode at 1632.6 cm^{-1} interferes with the amide I band.

Some differences observed when ceramide is incorporated in contrast to neat hydrogel and composite sponge. A strong band in the region 3272.6 – 1021.3 cm^{-1} in composite sponges and 3272.6 – 842.4 cm^{-1} in hydrogel membrane attributes to N-H and O-H stretching, as well as the intramolecular hydrogen bonds, whereas the region 3272.6 – 1028.7 cm^{-1} in neat composite sponge and 3280.1 – 1028.7 cm^{-1} in neat hydrogel membrane are observed.

The bands at around 1330.7 cm^{-1} in hydrogel membrane and 1338.1 cm^{-1} in composite sponge supported the existence of residual N-acetyl groups (C-N stretching of amide III), whereas 1334.4 cm^{-1} was observed in both hydrogel membrane and composite sponge without ceramide.

The absorption band at 1155.5 cm^{-1} in composite sponge and 1144.3 in hydrogel membrane can be attributed to asymmetric stretching of the C-O-C bridge, whereas the absorption band at 1151.7 cm^{-1} in composite sponge and 1148.0 in hydrogel membrane are observed without ceramide.

The bands at 1073.5 cm^{-1} in composite sponge and 1036.2 cm^{-1} in hydrogel membrane correspond to C-O stretching whereas, bands at 1077.2 and 1028.7 cm^{-1} were observed in both hydrogel membrane and composite sponge without ceramide (Figs. 9 and 10).

3.4 SEM of Hydrogel Membranes and Composite Materials

The surface morphology of hydrogel membrane was examined by SEM as shown in Fig. 11 and 12. The surface of the hydrogel membrane was found to be impenetrable. This surface will be favorable for wound dressing; it will not grant the penetration of microorganism that can induce wound infection. The composite materials appeared as extremely porous structures. By increasing the percentage of ceramide, the number of pores increase and porous structure were more prominent. The porousity of the structures make them favorable for absorbing wound exudates.

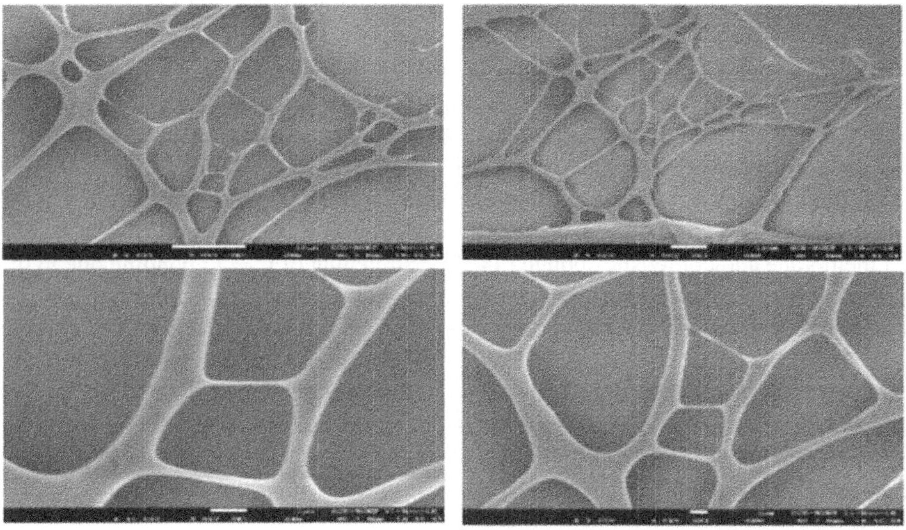

Fig. 9. SEM images of composite materials containing 10% (w/v) ceramide.

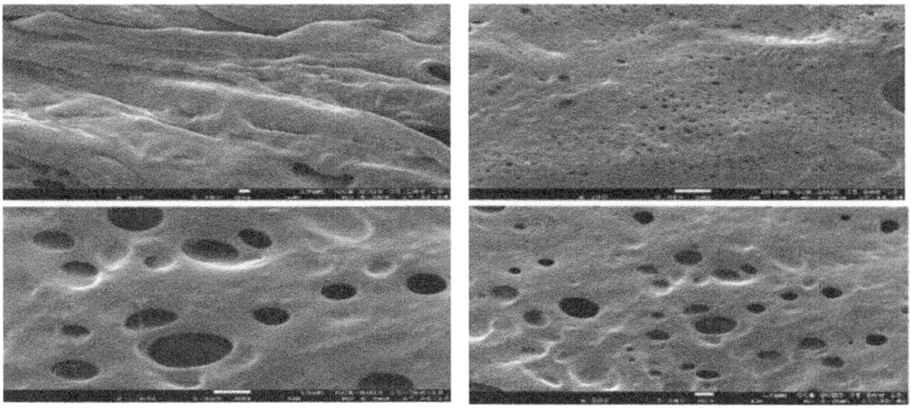

Fig. 10. SEM images of composite materials containing 20% (w/v) ceramide.

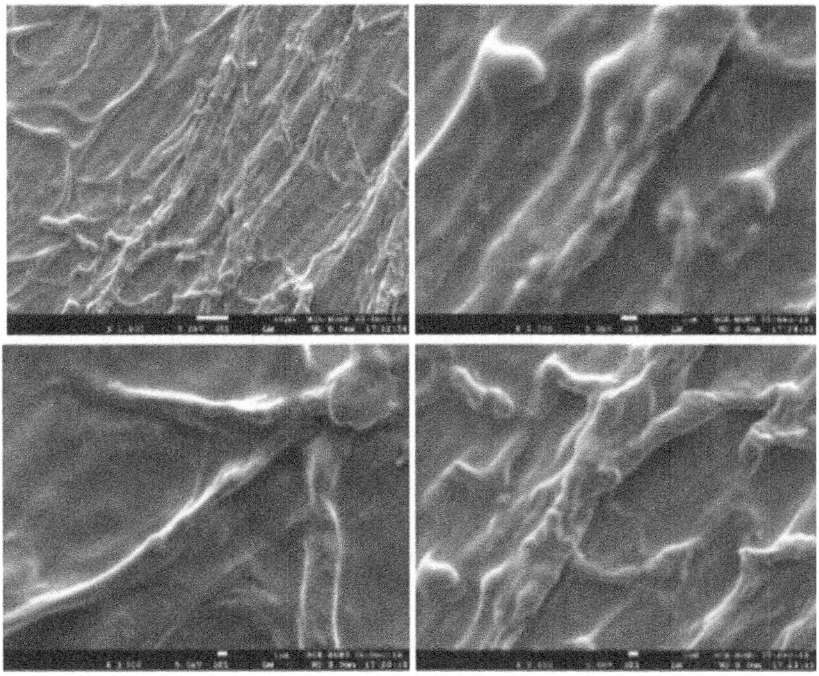

Fig. 11. SEM images of hydrogel membrane containing 10% (w/v) ceramide.

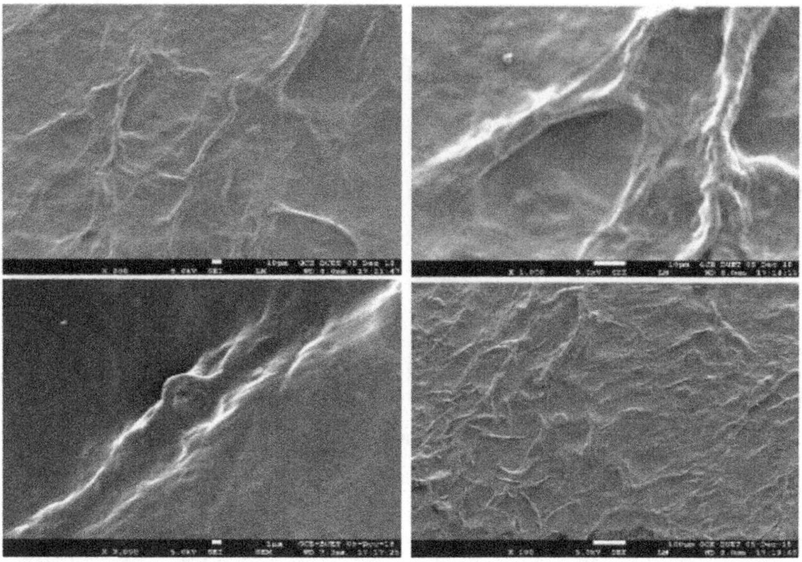

Fig. 12. SEM images of hydrogel membrane containing 20% (w/v) ceramide.

4 Conclusion

A lot of people around the world face difficulties in wound healing and are in danger of developing wound infection. We have chosen ceramide for burn and wound healing application, by virtue of its potential in structuring and maintaining the water-retaining properties of the epidermis. Our focus is to formulate a wound dressing material incorporating ceramide, which can retain a warm and moist environment to benefit the normal wound healing process of dermis and epidermis by providing an optimum microenvironment for healing. In our paper, PVA/chitosan/gelatin has been used in preparation of hydrogel membrane for wound dressing. We have also prepared composite sponge with the same biomaterials without binding agent and PVA. Hydrogel membranes and composite sponges from biopolymers incorporating ceramide, which possess the capability of holding moisture, holds great potential in tackling the concerns posed by the miserable and deplorable condition of wound management. The hydrogel materials produced will act as a barrier against bacteria and environmental pollutants while having the added benefit of being biodegradable. Addition of ceramide improved the water vapor transmission rate (WVT) of hydrogel membranes and water absorption of composite sponges. For chemical structure and morphology testing, the hydrogel membrane and composite sponges without ceramide (neat hydrogel) and with ceramide (having chitosan: gelatin = 3:1) were selected. FT-IR of hydrogel membranes and composite sponges confirmed the presence of free hydroxyl groups. The SEM images evolved that hydrogel membranes were impermeable in nature and the composite sponges appeared as very porous structures.

Acknowledgments. Authors would like to acknowledge the support from the Biomedical Engineering Department, Military Institute of Science and Technology (MIST) and Bangladesh University of Engineering and Technology (BUET), Bangladesh.

Conflict of Interest. The Authors declare that they have no conflict of interest.

References

1. Uchida, Y.: Ceramide signaling in mammalian epidermis. Biochimica et Biophysica Acta (BBA) Mol. Cell Biol. Lipids **1841**(3), 453–462 (2014)
2. Uchida, Y., Holleran, W.M.: Omega-O-acyl ceramide, a lipid essential for mammalian survival. J. Dermatol. Sci. **51**(2), 77–87 (2008)
3. Elias, P.M., Menon, G.K.: Structural and lipid biochemical correlates of the epidermal permeability barrier. Adv. Lipid Res. **24**, 1–26 (1991)
4. Yang, L., Mao-Qiang, M., Taljebini, M., Elias, P.M., Feingold, K.R.: Topical stratum corneum lipids accelerate barrier repair after tape stripping, solvent treatment and some but not all types of detergent treatment. Br. J. Dermatol. **133**(5), 679–685 (1995)
5. Lahiri, S., Futerman, A.H.: The metabolism and function of Sphingolipids and Glycosphingolipids. Cell. Mol. Life Sci. **64**, 2270–2284 (2007)
6. Kleuser, B., Maceyka, M., Milstien, S., Spiegel, S.: Stimulation of nuclear sphingosine kinase activity by plateletderived growth factor. FEBS Lett. **503**(1), 85–90 (2001)
7. Lazarus, G.S.: Definitions and guidelines for assessment of wounds and evaluation of healing. Arch. Dermatol. **130**(4), 489 (1994)
8. Boateng, J.S., Matthews, K.H., Stevens, H.N., Eccleston, G.M.: Wound healing dressing and drug delivery systems. J. Pharm. Sci. **97**(8), 2892–2923 (2008)
9. Bindu, T.H., Vidyavathi, M., Kavitha, K., Sastry, T.P., Kumar, R.S.: Preparation and evaluation of chitosan gelatin composite films for wound healing activity. Trends Biomater. Artif. Organs **24**(3), 123–130 (2010)
10. Sweeney, I.R., Miraftab, M., Collyer, G.: A critical review of modern and emerging absorbent dressings used to treat exuding wounds. Int. Wound J. **9**(6), 601–612 (2012)
11. Bruin, P., Jonkman, M.F., Meijer, H.J., Pennings, A.J.: A new porous polyetherurethane wound covering. J. Biomed. Mater. Res. **24**(2), 217–226 (1990)
12. Peppas, N.: Hydrogels in pharmaceutical formulations. Eur. J. Pharm. Biopharm. **50**(1), 2746 (2000)
13. Kirschner, C.M., Anseth, K.S.: Hydrogels in healthcare: from static to dynamic material microenvironments. Acta Mater. **61**(3), 931–944 (2013)
14. Osti, E., Osti, F.: Treatment of cutaneous burns with burnisheld (hydrogel) and a semi-permeable adhesive film. Ann. Burns Fire Disasters **3**, 137–141 (2004)
15. Edwards, J.: Hydrogels and their potential uses in burn wound management. Br. J. Nurs. **19**, 14–16 (2010)
16. Pal, K., Banthia, A.K., Majumdar, D.K.: Preparation of transparent starch based hydrogel membrane with potential application as wound dressing. Trends Biomater. Artif. Organs **20**, 59–67 (2006)
17. Bindu, H., Vidyavathi, M., Kavitha, K., Sastry, T.P.: Preparation and evaluation of gentamicin loaded chitosan-gelatin composite films for wound healing activity. Int. J. Appl. Biol. Pharm. Technol. **2**, 453–463 (2011)
18. Mohamed, N.A., Fahmy, M.M.: Synthesis and antimicrobial activity of some novel cross-linked chitosan hydrogels. Int. J. Mol. Sci. **13**(9), 11194–11209 (2012)

19. Coviello, T., Matricardi, P., Marianecci, C., Alhaique, F.: Polysaccharide hydrogels for modified release formulations. J. Control. Release **119**, 5–24 (2007)
20. Reis, R.L.: Natural Based Polymers for Biomedical Applications. Woodhead Publishing, Cambridge (2008)
21. Dang, J., Leong, K.: Natural polymers for gene delivery and tissue engineering. Adv. Drug Deliv. Rev. **58**(4), 487–499 (2006)
22. Shahidi, F., Abuzaytoun, R.: Chitin, chitosan, and co-products: chemistry, production, applications, and health effects. Adv. Food Nutr. Res. **49**, 93–135 (2005)
23. Roberts, G.A.F.: Chitin Chemistry, p. 350. Macmillan Press Ltd., Hong Kong xvi, London (1992)
24. Tomihata, K., Ikada, Y.: In vitro and in vivo degradation of films of chitin and its deacetylated derivatives. Biomaterials **18**(7), 567–575 (1997)
25. Wedmore, I., McManus, J.G., Pusateri, A.E., Holcomb, J.B.: A special report on the chitosan-based hemostatic dressing: experience in current combat operations. J. Trauma **60**, 655–658 (2006)
26. Pusateri, A.E., McCarthy, S.J., Gregory, K.W., Harris, R.A., Cardenas, L., McManus, A.T., Goodwin Jr, C.W.: Effect of a chitosan-based hemostatic dressing on blood loss and survival in a model of severe venous hemorrhage and hepatic injury in swine. J. Trauma **54**, 177–182 (2003)
27. Baldrick, P.: The safety of chitosan as a pharmaceutical excipient. Regul. Toxicol. Pharmacol. **56**, 290–299 (2010)
28. Ueno, H., Mori, T., Fujinaga, T.: Topical formulations and wound healing applications of chitosan. Adv. Drug Deliv. Rev. **52**(2), 105–115 (2001)
29. Shelma, R., Paul, W., Sharma, C.P.: Chitin Nanofibre reinforced thin chitosan films for wound healing application. Trends Biomater. Artif. Organs **22**(2), 111–115 (2008)
30. Lim, S.H., Hudson, S.M.: Review of chitosan and its derivatives as antimicrobial agents and their uses as textile chemicals. J. Macromol. Sci. C – Polym. Rev. **43**, 223–269 (2003)
31. Singh, S., Rama Rao, K.V., Venugopal, K., Manikandan, R.: Alteration in dissolution characteristics of gelatin containing formulations: a review of the problem, test methods, and solutions. Pharm. Technol. **26**, 36–58 (2002)
32. Tanaka, A., Nagate, T., Matsuda, H.: Acceleration of wound healing by gelatin film dressings with epidermal growth factor. J. Vet. Med. Sci. **67**, 909–913 (2005)
33. Sinha, R.N., Varma, D.K., Mada,n P.: Burn treatment with gelatin sheets. J. Plastic Reconstr. Surg. **5**, 56 (1972)
34. Ikada, Y., Tabata, Y.: Protein release from gelatin matrices. Adv. Drug Deliv. Rev. **31**, 287–301 (1998)
35. Kuijpers, A.J., van Wachem, P.B., van Luyn, M.J., Plantinga, J.A., Engbers, G.H., Krijgsveld, J., Zaat, S.A., Dankert, J., Feijen, J.: In vivo compatibility and degradation of crosslinkedgelatin gels incorporated in knitted. Dacron. J. Biomed. Mater. Res. **51**, 136–145 (2000)
36. Sahoo, R.K., Biswas, N., Guha, A., Kuotsu, K.: Recent advancement of gelatin nanoparticles in drug and vaccine delivery. Int. J. Biol. Macromol. **81**, 317–331 (2015)
37. Stasko, J., Kalniņš, M., Dzene, A., Tupureina, V.: Poly (vinyl alcohol) hydrogels. Proc. Estonian Acad. Sci. **58**(1), 63 (2009)
38. Kenawy, E.-R., Kamoun, E.A., MohyEldin, M.S., El-Meligy, M.A.: Physically crosslinked-poly(vinyl alcohol)hydroxyethyl starch blend hydrogel membranes: Synthesis and characterization for biomedical applications. Arab. J. Chem. **7**(3), 372–380 (2014)
39. Hassan, C.M., Peppas, N.A.: Structure and applications of poly(vinyl alcohol) hydrogels produced by conventional crosslinking or by freezing/thawing methods. In: Biopolymers. PVA hydrogels, anionic polymerisation nanocomposites, pp. 37–65. Springer, Heidelberg (2000)

40. Oliveira, R.N.: Polyvinyl Alcohol Hydrogel Loaded With Antibacterial Constituents For Burn Heaing Appications. D.Sc thesis, The Alberto Luiz Coimbra Institute for Graduate Studies and Research in Engineering, Rio de Janerio, Brazil (2014)
41. Boonkaew, B., Suwanpreuksa, P., Cuttle, L., Barber, P.M., Supaphol, P.: Hydrogels containing silver nanoparticles for burn wounds show antimicrobial activity without cytotoxicity. J. Appl. Polym. Sci. **131**(9) (2013)
42. Silva, F.R.F., Dore, C.M.P.G., Marques, C.T., Nascimento, M.S., Benevides, N.M.B., Rocha, H.A.O., Chavante, S.F., Leite, E.L.: Anticoagulant activity, paw edema and pleurisy induced carrageenan: action of major types of commercial carrageenans. Carbohydr. Polym. **79**, 26–33 (2010)
43. Lim, S.H., Hudson, S.M.: Synthesis and antimicrobial activity of a water soluble chitosan derivative with a fiber-reactive group. Carbohydrate **339**, 313–319 (2004)
44. Vino, A.B., Ramasamy, P., Shanmugam, V., Shanmugam, A.: Extraction, characterization and in vitro antioxidative potential of chitosan and sulfated chitosan from Cuttlebone of Sepia aculeataOrbigny 1848. Asian. Pac. J. Trop. Biomed. **2**, 334–341 (2012)
45. Song, C., Yu, H., Zhang, M., Yang, Y., Zhang, G.: Physicochemical properties and antioxidant activity of chitosan from the blowfly Chrysomyamegacephala larvae. Int. J. Biol. Macromol. **60**, 347–354 (2013)

Development of Elemental Technologies for Magnetic Hyperthermia in Cancer Treatment

Loi Tonthat[1(✉)], Fumitaka Aki[2], Kazutaka Mitobe[3], Shin Yabukami[1], and Yoshiyuki Yamamoto[3]

[1] Tohoku University, Sendai 980-8579, Japan
loi@ecei.tohoku.ac.jp
[2] Tokai University, Tokyo 151-8677, Japan
[3] Akita University, Akita 010-8502, Japan

Abstract. Hyperthermia utilizing magnetic particles has been widely studied as a possible method for cancer treatment in recent years. This method induces apoptosis (cell death) using heat generation from magnetic particles which are injected into a tumor region and subjected to an external high-frequency magnetic field. In previous studies, we succeeded in developing microsize thermosensitive ferromagnetic particles with a low Curie temperature of around 40–45 °C as a self-controlled heating element that possibly prevents the local overheating surrounding healthy tissues. Further, by utilizing the change of its permeability around its Curie temperature resulting in the change in the magnetic flux density around tumor region, we also developed a wireless temperature measurement. In addition, to localize magnetic particles which cannot be seen from the body surface, we also developed an automatic localization system by using a robot arm equipped with three detection coils symmetrically installed in heating coil. To make our proposed elemental technologies feasible in clinical settings, we have also been developing a magnetic hyperthermia system that can treat deep-seated tumors up to 5 cm. In this paper, we report our elemental technologies which have been developed so far such as heating elements, temperature and position monitoring technologies.

Keywords: Hyperthermia · Magnetic particles · Position and temperature detection

1 Introduction

Cancer is second cause of death globally behind ischemic heart disease and stroke, accounting for approximately10 million deaths annually according to World Health Organization [1]. There are different types of cancer treatment such as surgery, chemotherapy, and radiation therapy. Recently, hyperthermia (also called thermotherapy) utilizing heat generation from magnetic particles to kill cancer cells has been gaining more attention as a promising cancer treatment owing to fewer side effects compared to chemotherapy and radiotherapy, and less invasive than surgery [2, 3]. Cancer cells are

Y. Shiraishi et al. (Eds.): APCMBE 2020, IFMBE Proceedings 82, pp. 272–277, 2021.
https://doi.org/10.1007/978-3-030-66169-4_33

more sensitive to heat than healthy cells, and antitumor effect occurs when the tumor is heated within the therapeutic temperature range of 40–45 °C. This therapy can also be used in combination with chemotherapy or radiotherapy to enhance its therapeutic effectiveness. Heat generation mainly ascribed to hysteresis and/or relaxation loss from magnetic particles subjected to a high-frequency magnetic field applied outside of patient body [4, 5]. Over the past few years, many researchers have been studying functional magnetic particles, and also approaching various biomedical applications utilizing magnetic particles [6–8] such as magnetic hyperthermia, magnetic labeling, MRI contrast agents, and magnetic drug delivery. In magnetic hyperthermia, core technologies include magnetic particles [9–11], its position [12–15] and temperature monitoring [16–18] technologies, and heating control technology, etc. [19–21].

To date clinical applications capable of heating affected parts at a constant temperature while being minimally invasive in detecting the position as well as temperature of heating elements are not established, thus the development of such system is of paramount importance. In this paper, we report our developed elemental technologies such as multifunctional heating elements, non-invasive wireless position and temperature detection, and heating control technologies for magnetic hyperthermia.

2 Development of Wireless Temperature Measurement for Thermosensitive Magnetic Particles

Heat generation from heating elements that continues to increase even after its temperature exceeds therapeutic temperature may damage surrounding healthy tissues. Hence, to monitor its temperature we proposed a wireless temperature measurement using thermosensitive ferromagnetic implant with Low Curie temperature (FILCT) [17, 18]. As shown in Fig. 1, FILCT with high magnetic permeability attracts magnetic flux density around it. If its temperature exceeds the Curie point, then its magnetic permeability will significantly decrease. As a result, the material cannot attract magnetic flux density around it. The shift in the magnetic flux density vector can be measured by the induced voltage in a pickup coil located orthogonally in drive coil.

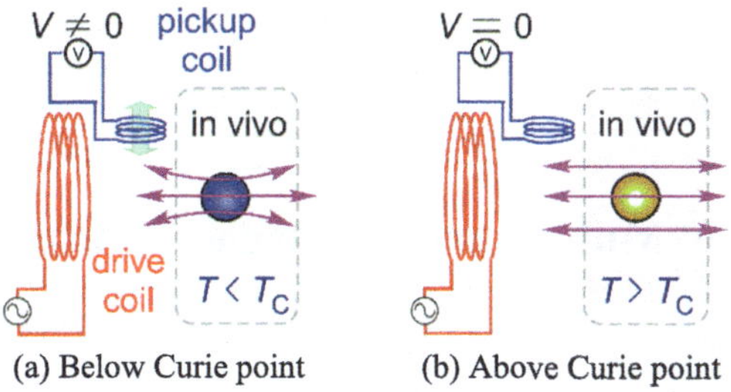

Fig. 1. Fundamental concept of wireless temperature measurement [17, 18].

Using the proposed thermometry, we have also been developing a hyperthermia system for clinical settings as shown in Fig. 2 [22, 23]. The frequency of applied magnetic field frequency was at 129 kHz and the magnetic flux density applied at 1, 2, 3, 4, 5, and 6 cm from the surface of drive coil were 13.7, 10.9, 8.6, 6.4, 4.4, 2.2, and 1.2 mT, respectively. The sample used consisted of 1.0 g of FILCT dispersed in 1.0 mL of a commercial nanofluid named Resovist® (MRI contrast agent, Fuji Film RI Pharma) [24]. Our results indicate that it was possible to heat the sample up to therapeutic temperature and monitoring its temperature via the proposed thermometry up to 5 cm.

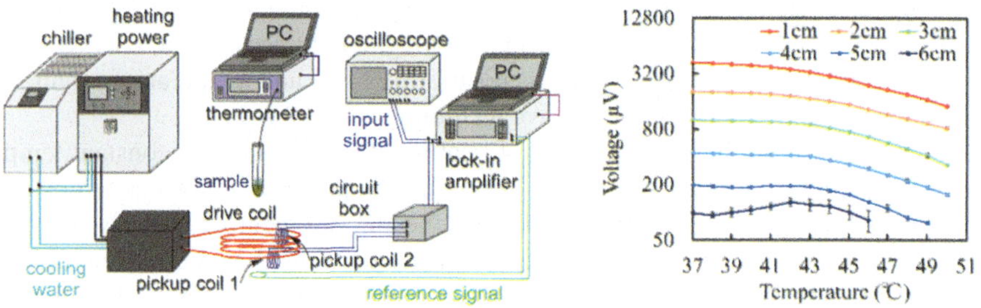

Fig. 2. Diagram of developed system and relationship between sample temperature and detection voltage when there was a change in distance between sample and drive coil [23].

3 Development of Automatic Localization System for Magnetic Particles

When the implant which cannot be seen from the body surface deviates from the central axis of the magnetic field supply and detection unit composed of drive coil and pickup coil (MFSD unit), the magnetic flux density applied on the implant decreases, resulting in a decrease in its heating efficiency and the thermometry accuracy. To solve this problem, we proposed a localization method (see Fig. 3) in which the central axis is aligned directly above the implant by referring to three voltages induced in three pickup coils symmetrically installed inside drive coil [15]. Recently, we also constructed a localization system based on the proposed method. The results obtained indicate that it was possible to automatically locate the position of the implant with accuracy below 1 mm by operating MFSD unit in two modes of coarse adjustment (rotary scanning) and fine adjustment (linear scanning).

4 Development of Multifunctional Magnetic Particles

Our previously developed gold-coated FILCT (hereafter Au-FILCT) improved significantly the heating efficiency of FILCT (8 times) [9]. However, part of the applied magnetic field was shielded ascribed to the conductive coating around FILCT. As a result, the change in the detected voltage induced in pickup coil (hereafter pickup voltage) using Au-FILCT was reduced by half compared to FILCT, thereby the accuracy of our wireless thermometry using Au-FILCT was significantly lowered compared to that of FILCT (see Fig. 4). As an alternative approach to the gold coating, we proposed

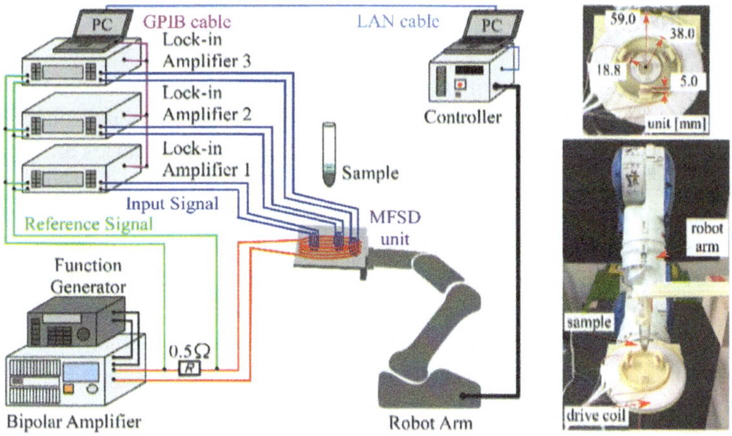

Fig. 3. Diagram and experimental landscape of developed automatic localization system [15].

to mix FILCT with a high heating-efficient magnetic nanofluid. In the case of using a commercial nanofluid named Resovist® [24], the heating efficiency of the proposed mixture of micro/nano-magnetic particles was improved 4.3 times (see Fig. 4e), and the accuracy of our wireless thermometry was improved 1.3 times (see Fig. 4f) compared to that of FILCT under a magnetic field of 500 kHz, 4.95 kA/m [25]. A similar tendency was confirmed when using a lab-made nanofluid Fe_3O_4 [26].

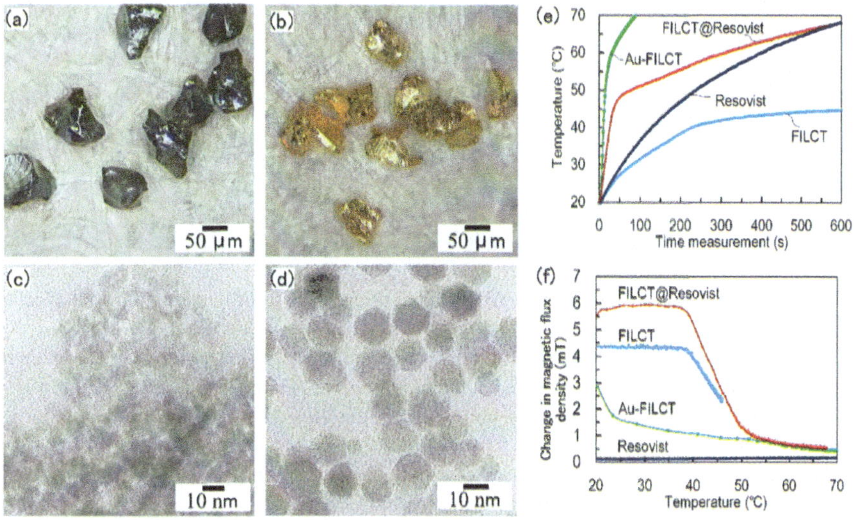

Fig. 4. Heating elements for hyperthermia (a) lab-made FILCT and (b) Au-FILCT, (c) commercial nanofluid Resovist®, (d) lab-made nanofluid Fe_3O_4, (e) temperature rise in sample, (f) relationship between magnetic flux density change in pickup coil and sample temperature [25, 26].

5 Conclusions

Magnetic hyperthermia is a promising cancer therapy and widely studied in the literature in recent years. So far, we developed multifunctional high-performance magnetic material for hyperthermia, and by utilizing its permeability that varies around its Curie temperature that results in the change in magnetic field, we proposed noninvasive wireless temperature and position monitoring methods. At present, using our constructed treatment system based on the proposed methods, it is possible to treat deep-seated tumors up to 5 cm from the body surface. Future work is to extend the treatable distance for deeper tumor by optimizing design of magnetic field supply and detection unit, clarifying heating mechanism in micro/nanomagnetic particles, and constructing automatic heating control technology for magnetic particles.

Acknowledgment. This work was supported by Grant-in-Aid for Scientific Research (KAKENHI) under Grant 19K23597 and 18H03545, all from Japan Society for the Promotion of Science.

References

1. World Health Organization Homepage. https://www.who.int/news-room/fact-sheets/detail/cancer. Accessed 28 Jan 2020
2. Eduardo, M.: Physics of Thermal Therapy: Fundamentals and Clinical Applications, pp. 139–158, 1st edn. CRC Press, Boca Raton, FL, USA (2012)
3. Ihab, M.O., Venkatesha, N., Sulaiman, A., Sangaraju, S., Chandu, V.V.M.G.: Principles of magnetic hyperthermia: a focus on using multifunctional hybrid magnetic nanoparticles. Magnetochemistry **5**, 67–106 (2019)
4. Dutz, S., Hergt, R.: Magnetic particle hyperthermia – a promising tumour therapy? Nanotechnology **25**(45), 452001 (2014)
5. Minkowycz, W.J., Sparrow, E.M., Abraham, J.P.: Nanoparticle Heat Transfer and Fluid Flow, pp. 97–122. 1st edn. CRC Press, Boca Raton, FL, USA (2012)
6. Vanessa, F.C., Antonio, F., Clarisse, R., Manuel, B., Pedro, M., Senentxu, L.: Advances in magnetic nanoparticles for biomedical applications. Adv. Healthcare Mater. **7**, 1700845 (2018)
7. Jose, L.C., Antonio, V.: Biomedical applications of distally controlled magnetic nanoparticles. Trends Biotechnol. **27**(8), 468–476 (2009)
8. Loi, T., Takahashi, S., Onodera, H., Okita, K., Yabukami, S., Yokota, K., Furuya, M., Kanetaka, H., Miura, Y., Takahashi, H., Watanabe, Y., Akiyama, R.: A simple and rapid detection system for oral bacteria in liquid phase for point-of-care diagnostics using magnetic nanoparticles. AIP Adv. **9**(12), 125325 (2019)
9. Mitobe, K., Yoshimura, N.: Low-invasive heating and temperature measurement method for hyperthermia treatment using the metal coated ferromagnetic implant with low curie temperature. Proc. Biodev. **2011**, 341–344 (2011)
10. Yamamoto, Y., Itoh, T., Irieda, T.: A heat dissipation study of iron oxide nanoparticles embedded an agar phantom for the purpose of magnetic fluid hyperthermia. J. Nanosci. Nanotechnol. **19**(9), 5469–5475 (2019)
11. Tong, S., Quinto, C.A., Zhang, L., Mohindra, P., Bao, G.: Size-dependent heating of magnetic iron oxide nanoparticles. ACS Nano **11**(7), 6808–6816 (2017)

12. Knopp, T., Sattel, T.F., Biederer, S., Rahmer, J., Weizenecker, J., Gleich, B., Borgert, J., Buzug, T.M.: Model-based reconstruction for magnetic particle imaging. IEEE Trans. Med. Imaging **29**, 12–18 (2010)
13. Murase, K., Mimura, A., Banura, N., Nishimoto, K., Takata, H.: Visualization of magnetic nanofibers using magnetic particle imaging. Open J. Med. Imag. **5**, 56–65 (2015)
14. Hamanaga, S., Yoshida, T., Sasayama, T., Elrefai, A.L., Enpuku, K.: Three-dimensional detection of magnetic nanoparticles using a field-free line with weak field gradient and multiple pickup coils. Jpn. J. Appl. Phys. **58**, 061001 (2019)
15. Loi, T., Aki, F., Matsuda, E., Saito, H., Yoshimura, N., Mitobe, K.: Position adjustment method and distance estimation method of magnetic field supply and detection unit for magnetic hyperthermia. IEEJ Trans. Elect. Electron. Eng. **12**(S2), S3–S9 (2017)
16. Garaio, E., Collantes, J.M., Garcia, J.A., Plazaola, F., Sandre, O.: Harmonic phases of the nanoparticle magnetization: An intrinsic temperature probe. Appl. Phys. Lett. **107**(12), 123103 (2015)
17. Loi, T., Saito, H., Miyamoto, R., Suzuki, M., Yoshimura, N., Mitobe, K.: Rotary scanning wireless temperature measurement method for hyperthermia using ferromagnetic implants. IEEJ Trans. Electr. Electron. Eng. **10**, S1–S6 (2015)
18. Loi, T., Aki, F., Matsuda, E., Saito, H., Yoshimura, N., Mitobe, K.: Body motion artifact reduction method using rotary scanning for accuracy improvement of wireless temperature measurement. IEEJ Trans. Fundam. Mater. **136**(8), 529–534 (2016)
19. Aoto, T., Takahashi, K., Hoshiyama, H., Yoshioka, Y., Yamada, T., Ota, S., Ikehata, Y., Yamada, S., Takemura, Y.: specific loss power of magnetic particles for hyperthermia excited by pancake-type applicator. IEEJ Trans. Fund. Mater. **137**(8), 476–480 (2017)
20. Aki, F., Loi, T., Saito, H., Mitobe, K.: Examination of the influence on precision of the wireless temperature measurement induction heating system by 37 °C constant temperature environment. IEEJ Trans. Fund. Mater. **54**(6), 2800303 (2018)
21. Yamada, S., Ikehata, Y., Hayashi, R.: Double Pancake Exciting Coils With Back Yoke For Magnetics Field Generator To Medical Treatment. In: Asia-Pacific Symposium on Applied Electromagnetics and Mechanics (APSAEM2014) **23**(3), 591–594 (2015)
22. Aki, F., Loi, T., Saito, H., Yoshimura, N., Mitobe, K.: Study of wireless temperature measurement induction heating system using magnetic properties of au-coated ferromagnetic implant with low curie temperature. Electron Comm Jpn. **101**, 58–66 (2018)
23. Aki, F., Loi, T., Saito, H., Yoshimura, N., Mitobe, K.: Study on wireless temperature measurement induction heating system using magnetic properties of mixture of Resovist® and ferromagnetic implant with low curie temperature. IEEJ Trans. Fund. Mater. **139**(1), 38–44 (2019)
24. Loi, T., Yamamoto, Y., Aki, F., Saito, H., Mitobe, K.: Magnetic field dependence of heating property of resovist® for magnetic hyperthermia. IEEJ Trans. Electr. Electron. Eng. **14**(4), 648–649 (2019)
25. Loi, T., Yamamoto, Y., Aki, F., Saito, H., Mitobe, K.: Thermosensitive ferromagnetic implant for hyperthermia using a mixture of magnetic micro-/nanoparticles. IEEE Trans. Magn. **54**(7), 5400506 (2018)
26. Loi, T., Yamamoto, Y., Aki, F., Saito, H., Mitobe, K.: Thermosensitive implant for magnetic hyperthermia by mixing micro-magnetic and nano-magnetic particles. IEEE Trans. Magn. **54**(6), 5400104 (2018)

Immunogenic Cell Death Induced by Selective Intracellular Vaporization *in vitro*

Ayumu Ishijima[1,2](✉), Satoshi Yamaguchi[3], Takashi Azuma[4], Etsuko Kobayashi[1], Kazuhiro Kakimi[5], Yoshikazu Shibasaki[3], Teruyuki Nagamune[1], and Ichiro Sakuma[1]

[1] Graduate School of Engineering, The University of Tokyo, Tokyo, Japan
ishijima@bmpe.t.u-tokyo.ac.jp
[2] JST-PRESTO, Saitama, Japan
[3] Research Center for Advanced Science and Technology, The University of Tokyo, Tokyo, Japan
[4] Graduate School of Medicine, The University of Tokyo, Tokyo, Japan
[5] Department of Immunotherapeutics, The University of Tokyo Hospital, Tokyo, Japan

Abstract. Recent advances in targeting cancer therapy with stimuli responsive materials provide the opportunity to reduce systemic toxicities. Targeting cancer therapy employing ultrasound with phase-change nano-droplets (PCNDs) potentially has several distinct advantages over current cancer therapeutic modalities. PCND is a liquid perfluorocarbon nanoparticle that its liquid phase is transformable to gas phase upon ultrasound exposure. The vaporization phenomenon of PCND can selectively treat cancers mechanically by conjugating active targeting moiety to it. Here, we present that adenosine triphosphate (ATP) and high-mobility group box 1 (HMGB1), which are known as damage-associated molecular patterns (DAMPs), were released from a cultured cancer cell line by selective intracellular vaporization of PCNDs. The results indicate that not only destruction of cancer cells by intracellular vaporization but also anti-cancer immune response can be expected. The activated immune system can potentially promote systemic anti-cancer effect *in vivo*. The *in vitro* DAMPs release motivate further research toward the *in vivo* investigation.

Keywords: Ultrasound · Phase-change nano-droplet · Damage-associated molecular patterns

1 Introduction

During the last decade, ultrasound-responsive micro/nano-materials has attracted growing attention in the targeted-cancer therapy [1, 2]. Ultrasound provides the benefits of non-invasiveness, deep penetration (more than cm-order) and sub-mm to mm-order spatial controlling capability of ultrasound-beam-focusing that enables high spatial-temporal control of therapeutic activation. Previously, we described a new form of cancer therapy that physically treats cancer cells that employs a monoclonal antibody 9E5 conjugated to a phase-change nano-droplet (PCND) [3, 4]. PCND is a sub-micron sized

Y. Shiraishi et al. (Eds.): APCMBE 2020, IFMBE Proceedings 82, pp. 278–283, 2021.
https://doi.org/10.1007/978-3-030-66169-4_34

particle composed of a phospholipid shells and liquid state perfluorocarbon that vaporizes into microbubble upon ultrasound exposure [5]. 9E5 is an anti-epiregulin (EREG) antibody that targets the cell–membrane–expressed ligand of epidermal growth factor receptor [6]. The 9E5-conjugated PCNDs can be introduced inside the cancer cells via receptor-mediated endocytosis. In this condition, ultrasound exposure can induce intracellular vaporization that leads to rapid disruption of cell structure [3, 4].

Our therapeutic technology potentially increases anti-cancer immunity that its cytotoxic effect is selective to cancer cell. Moreover, chemical characteristics of antigen would be preserved without any alternation because cancer cells are destructed by mechanical vaporization. The activation of immunity would have several benefits including suppression of metastasis and recurrence that leads to increase of patients' quality of life. In order to trigger anti-cancer immunity, transduction of damage-associated molecular patterns (DAMPs), which act as a danger signal, to immune system is an important factor [7]. High-mobility group box 1 (HMGB1) which is a non-histone DNA binding protein that promotes inflammatory cytokine production such as type I interferon, adenosine triphosphate (ATP) which attract and activate antigen-presenting cells, and heat-shock protein 70 (HSP70) which is a molecular chaperon that promotes acquisition of antigens, are well-known DAMPs that promote anti-cancer immune response. Therefore, measurement of extracellular released DAMPs by intracellular vaporization is essential to evaluate the potential of our therapeutic technology as immunotherapeutic modality. Here we report that ATP and HMGB1, which are known as damage-associated molecular patterns, were released from a human colonic adenocarcinoma cell line by selective intracellular vaporization of phase-change nano-droplets.

2 Materials and Methods

2.1 Preparation of 9E5-Conjugated Phase-Change Nano-droplet

9E5 is an anti-EREG antibody that can selectively induce internalization of PCND. 9E5 hybridoma cells were intraperitoneally implanted in BALB/c nude mice, and ascites were obtained and purified on a Protein G column. 9E5 monoclonal antibody was prepared and biotinylated for conjugation to PCND as described previously [3]. All animals were maintained in accordance with the regulations set by the University of Tokyo, and all animal experiments were conducted following institutional guidelines. The animal study protocol was approved by the University of Tokyo (#RAC120101).

Biotinylated PCNDs that internal composition is 1:1 mixture of perfluoropentane and perfluorohexane were used to ensure thermal stability (boiling point: 40 °C) and sensitivity to activation by ultrasound. A laser diffraction particle analyzer (LS13320, Beckman Coulter, USA) was used for measurement of the particle size distribution. Biotinylated 9E5 was conjugated to PCNDs via SA-AF647 (Life Technologies, USA), and 9E5-conjugated PCNDs were prepared as described previously [3]. Briefly, biotinylated 9E5 and SA-AF647 were mixed in phosphate-buffered saline (PBS) at concentrations of 0.4 μM each at a total volume of 50 μL and incubated for 15 min at room temperature to form AF647-labeled 9E5-SA conjugates. The biotinylated PCND dispersion was mixed with a solution containing AF647-labeled 9E5-SA conjugate at a volume ratio of 1:1.

The mixture was incubated for 30 min, and unconjugated 9E5 and SA-AF647 were removed by centrifugation.

2.2 Cell Culture

The human colonic adenocarcinoma cell line DLD-1 was selected as the high EREG-expressing cancer cell line. The conjugation of 9E5 to PCNDs permits selective targeting of PCNDs to high EREG-expressing cells and induces internalization. A silicon ring (inner diameter = 21 mm; outer diameter = 26 mm; thickness = 15 mm) was placed on a 35-mm culture dish to avoid dish-edge culturing because of difficulty in coherent focusing of ultrasound at the dish edge. DLD-1 cells were cultured at an initial density of 10^5 cells/dish with RPMI culture medium containing 10% fetal bovine serum. After a 1-day incubation at 37 °C in a humidified atmosphere with 5% CO2, cells were treated with PCNDs following ultrasound exposure.

2.3 Treatment of PCNDs to Cultured Cell

After a 1-day incubation, all the culture medium in the dish was aspirated, the cells were washed with PBS and 1 mL of RPMI-diluted 9E5-conjugated PCNDs solution was added to the cells. After a 3-h incubation with 9E5-conjugated PCNDs, all the solutions were removed. To remove non-attached/non-internalized PCNDs, approximately 10 mL of RPMI was added to the dish to fill it with the media. A dish cap was placed on top, and the dish was inverted because the density of PCND (1.6 g/mL) is higher than that of the media. This washing procedure was repeated twice; lastly, 2 mL of fresh RPMI was added. As negative controls for the experiments, the cells were incubated and washed in the same protocol without PCNDs.

2.4 Ultrasound Exposure

The ultrasound exposure system shown in Fig. 1 was used to induce intracellular vaporization of PCNDs. A programmable ultrasound imaging system (V1, Verasonics, US) was used to control 128 transducer elements inside a diagnostic ultrasound imaging probe (EUP-L73S, Hitachi, Ltd, JP) for delivering vaporization pulses to PCNDs (Fig. 1a). A custom-made dish holder connected to a single-axis motorized stage (ALZ-115-E1P, Chuo Precision Industrial, JP) was used to position the dish at 36 mm from the transmission surface of the ultrasound imaging probe. The pulse length was set to 5 cycles at 5 MHz with a peak negative pressure of 4.6 MPa (measured inside the dish where cells are seeded) (Fig. 1b). The focus of transmitted ultrasound was set at 38 mm from the transmission surface of the ultrasound imaging probe. The focus was electronically scanned from −1.20 to 1.20 mm (the center of the probe was set to 0 mm), with a 0.20-mm pitch in the lateral direction of the probe; the pulse repetition frequency was set to 500 Hz. The probe was positioned by a two-axis motorized stage (ALD-604-E1P, Chuo Precision Industrial) with a 0.30-mm pitch in the elevational direction and a 1.20-mm pitch in the lateral direction, both at an interval of 100 ms. Trigger signals were transmitted to the programmable ultrasound imaging system at each position to apply six sets of the above-described vaporization pulses. The water tank was filled with degassed water, and the temperature was maintained at 37 °C.

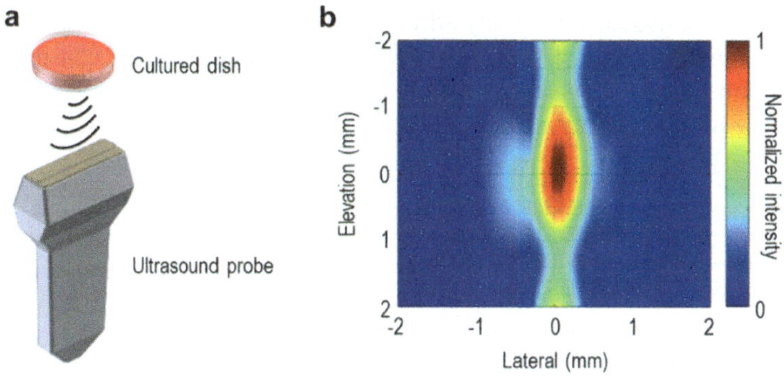

Fig. 1. Ultrasound exposure experimental setup to expose pulsed ultrasound to the cultured cells. (**a**) Schematic diagram of the system. Pulsed ultrasound was delivered inside the dish by the diagnostic ultrasound imaging probe connected to the programmable ultrasound imaging system (not shown in the diagram for simplicity). (**b**) Acoustic profile measured inside the dish by a calibrated hydrophone.

2.5 DAMPs Analysis

Extracellular released ATP, HMGB1 and HSP70, which are known as typical DAMPs, were quantitatively evaluated for ultrasound induced intracellular vaporized cells. Luciferin-luciferase luminescent reaction and sandwich enzyme-linked immunosorbent assays (ELISA) were used for evaluation of ATP and HMGB1 and HSP70, respectively. For ATP analysis, 100 μL of supernatant after ultrasound exposure was added to 96-well with the assay reagent ("Cellno" ATP ASSAY reagent, Toyo Ink Group, JP) and the luminescence was measured by a plate reader (ARVO X3, Perkin Elmer, DE). For HMGB1 and HSP70 analysis, HMGB1 ELISA (Fuso, JP) and HSP70 ELISA (Abcam plc, UK) were used, respectively. The 10 μL of diluted supernatant was added to an antibody-coated 96-well and the absorbance at 450 nm was measured by the plate reader.

3 Results and Discussion

The levels of extracellular released DAMPs are shown in Fig. 2. Results indicated that ATP (Fig. 2a), which can potentially attract and activate antigen-presenting cells, and HMGB1 (Fig. 2b), which can promote secretion of pro-inflammatory stimulators, were released outside the cells with ultrasound exposed PCNDs treated cells, while extracellular HSP70 (Fig. 2c) was not confirmed. However, HSP70 was not detected even with freeze-thaw cycled DLD-1 cells (extracellular HSP70 level = 0.36 ± 0.02, $N = 3$). Therefore, we assumed that expression level of HSP70 is originally low for DLD-1 cells. Note that significant bioeffects were not confirmed for the ultrasound exposed cells without treatment of PCNDs.

Although HSP70 was not released from intracellular vaporized DLD-1 cells, ATP and HMGB1 were released. These results indicated that the inflammatory response at the local cancer sites can be evoked by selective intracellular vaporization. The acute inflammatory response can potentially initiate activation of anti-cancer immune response with antigen presentation [8]. Moreover, activated immune system can promote systemic anti-cancer effect. This effect can be enhanced by combining our technology with conventional immunotherapy which targets immune checkpoint.

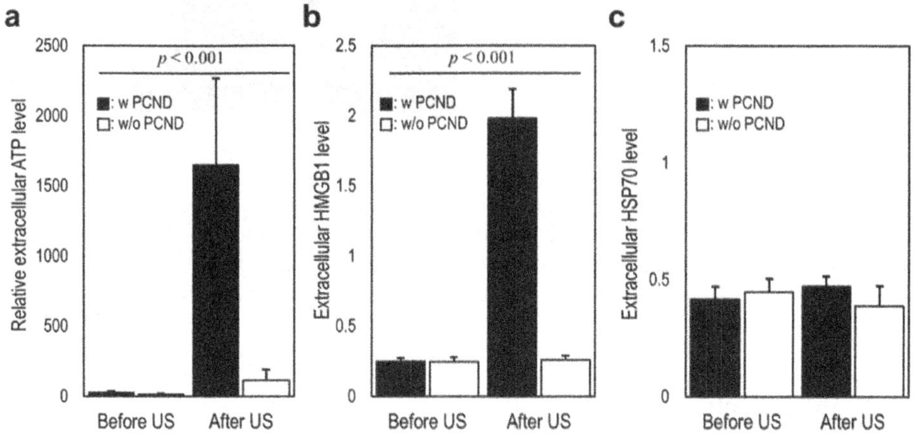

Fig. 2. Extracellular released DAMPs from intracellular vaporized DLD-1 cells. Black and white bars represent DLD-1 cells treated with and without 9E5-conjugated PCNDs, respectively. Significant levels of extracellular (**a**) ATP and (**b**) HMGB1 were detected for DLD-1 cells treated with 9E5-conjugated PCNDs after ultrasound exposure. Meanwhile, the level of (**c**) HSP70 was low. The bars represent *mean* ± *standard deviation* ($N = 4$). Analysis of variance (ANOVA) was used to establish the significance between the different experimental groups. The Tukey method was applied to evaluate the significance of differences between the groups.

4 Conclusions

Here, we present that ATP and HMGB1, which are known as DAMPs, were released from the cultured cancer cell line by selective intracellular vaporization of PCNDs. The results indicated that not only destruction of cancer cells by intracellular vaporization but also anti-cancer immune response can be expected. *In vitro* DAMPs release motivate further research toward the *in vivo* investigation of its therapeutic effect.

Acknowledgment. The authors thank Kenichi Kawabata and Rei Asami for valuable discussions and for providing PCNDs. This work was partially supported by a grant for the Translational Systems Biology and Medicine Initiative from the Ministry of Education, Culture, Sports and Technology of Japan, grant-in-aid from Olympus Corporation, and Japan Society for the Promotion of Science (JSPS) (18H05956).

References

1. Mura, S., Nicolas, J., Couvreur, P.: Stimuli-responsive nanocarriers for drug delivery. Nat. Mater. **12**, 991–1003 (2013)

2. Chen, W.T., Kang, S.T., Lin, J.L., Wang, C.H., Chen, R.C., Yeh, C.K.: Targeted tumor theranostics using folate conjugated and camptothecin-loaded acoustic nanodroplets in a mouse xenograft model. Biomaterials **53**, 699–708 (2015)
3. Ishijima, A., Minamihata, K., Yamaguchi, S., Yamahira, S., Ichikawa, R., Kobayashi, E., Iijima, M., Shibasaki, Y., Azuma, T., Nagamune, T., Sakuma, I.: Selective intracellular vaporisation of antibody-conjugated phase-change nano-droplets in vitro. Sci. Rep. **7**, 44077 (2017)
4. Ishijima, A., Yamaguchi, S., Azuma, T., Kobayashi, E., Shibasaki, Y., Nagamune, T., Sakuma, I.: Selective intracellular delivery of perfluorocarbon nanodroplets for cytotoxicity threshold reduction on ultrasound-induced vaporization. Cancer Rep. **2**, e1165 (2019)
5. Ishijima, A., Tanaka, J., Azuma, T., Minamihata, K., Yamaguchi, S., Kobayashi, E., Nagamune, T., Sakuma, I.: The lifetime evaluation of vapourised phase-change nano-droplets. Ultrasonics **69**, 97–105 (2016)
6. Lee, Y.H., Iijima, M., Kado, Y., Mizohata, E., Inoue, T., Sugiyama, A., Doi, H., Shibasaki, Y., Kodama, T.: Construction and characterization of functional anti-epiregulin humanized monoclonal antibodies. Biochem. Biophys. Res. Commun. **44**, 1011–1017 (2013)
7. Galluzzi, L., Buqué, A., Kepp, O., Zitvogel, L., Kroemer, G.: Immunogenic cell death in cancer and infectious disease. Nat. Rev. Immunol. **17**, 97–111 (2017)
8. Castano, A.P., Mroz, P., Hamblin, M.R.: Photodynamic therapy and anti-tumour immunity. Nat. Rev. Cancer **67**, 535–545 (2006)

Reconstruction of a Vascular Bed with Perfusable Blood Vessels Using a Decellularized Porcine Small Intestine for Clinical Application

Yusuke Tobe[1,2], Katsuhisa Sakaguchi[1,2(✉)], Jun Homma[2], Kazunori Sano[3], Eiji Kobayashi[4], Hidekazu Sekine[2], Kiyotaka Iwasaki[5], Tatsuya Shimizu[2], and Mitsuo Umezu[5]

[1] Graduate School of Advanced Science and Engineering, Waseda University, Tokyo, Japan
y.tobe.haru@moegi.waseda.jp, katsuhisa@aoni.waseda.jp
[2] Institute of Advanced Biomedical Engineering and Science, Tokyo Women's Medical University, Tokyo, Japan
[3] Tokaihit Co., Ltd., Fujinomiya-shi, Shizuoka, Japan
[4] Department of Organ Fabrication, Keio University School of Medicine, Tokyo, Japan
[5] Department of Modern Mechanical Engineering, Waseda University, Tokyo, Japan

Abstract. Engineering three-dimensional tissues *in vitro* for regenerative therapy is highly desired. Vascular beds with connectable arteries and veins play crucial roles in fabricating three-dimensional tissues *in vitro* and improving the survival rate of transplanted three-dimensional tissues *in vivo*. Here, we developed a method to reconstruct a new vascular bed that could be implanted into humans for clinical applications. Porcine small intestine with an arteriovenous loop was selected as a vascular bed skeleton and decellularized using transvascular perfusion with both detergent and enzyme. Subsequently, the decellularized intestinal graft was re-endothelialized with human cells to allow blood perfusion into the vascular bed. The small intestinal graft was successfully decellularized without severe damage to the burst pressure of the graft, and the decellularized intestinal graft could be transplanted without a severe inflammatory response in rats for two weeks. In addition, the decellularized intestinal graft was partially vascularized with human endothelial cells in seven days. Therefore, the reconstructed intestinal vascular bed may serve as a human transplantable vascular bed with potential in clinical investigations.

Keywords: Decellularization · Vascular bed · Vascularization

1 Introduction

Please Recent development of stem cell and tissue engineering technology has prompted regenerative medicine as a novel treatment method. Cell sheet-based tissue engineering utilizes cell sheets, which are two-dimensional monolayer cells collected from a temperature-responsive culture dish [1, 2]. A monolayer cell sheet and three-dimensional

Y. Shiraishi et al. (Eds.): APCMBE 2020, IFMBE Proceedings 82, pp. 284–292, 2021.
https://doi.org/10.1007/978-3-030-66169-4_35

(3-D) tissue fabricated by simply stacking cell sheets were applied to various organs [3]. To treat heart failure, previous studies have reported that cell sheets derived from myoblasts transplanted onto the damaged heart area improved cardiac function in both animal experiments and human clinical trials [4–6]. However, the cell sheets lack sufficient vascular structure for nutrient supply and waste removal, so the thickness of engineered 3-D tissue is limited [7].

A vascular bed with an arteriovenous loop was developed in rat and porcine models to counter problems such as hypoxia, nutritional deficiency, and waste product accumulation inside multilayered cell sheets. Sekine et al. demonstrated that blood vessel structures within the stacked cell sheets were formed by culturing the cell sheets on the vascular bed using the bioreactor system, and they successfully fabricated tissues with a thickness above 200 μm [8]. Furthermore, cell sheets on the vascular bed reconnected to the host blood vessel showed the highest *in vivo* survival rate compared to transplantation without anastomosis to host vessels or without a vascular bed [8]. This was considered because blood was supplied to the transplanted tissue via the blood vessels of the vascular bed immediately after transplantation. These data indicated that a vascular bed including the anastmosable artery and vein played a crucial role in solving issues such as nutrient deficiency by enhancing blood vessel formation in the 3-D tissues *in vitro* and allowing efficient transplantation of 3-D tissues fabricated *in vitro*. We also demonstrated that a large-scale vascular bed was successfully generated with porcine small intestine, which had an abundant vascular network and could easily be prepared into the graft with anastomosable blood vessels [9]. However, it is not practical to transplant 3-D tissues fabricated by the vascular beds in humans because the vascular beds are derived from rat and porcine tissue. Therefore, there is a clear need to construct a human transplantable vascular bed to realize a clinical application using this technology.

To that end, we developed a new method to reconstruct a human implantable vascular bed using decellularized tissue. Decellularized tissues are biomaterials derived from human, porcine, and bovine tissues from which cells have been removed, and they can be transplanted to humans without serious rejection even in heterologous animal tissues. Several decellularized materials such as dermis, small intestine, and pericardium have been put to practical use in Europe and the United States [10]. Here, we chose decellularized porcine small intestine as the skeleton to engineer the large-scale human transplantable vascular bed.

2 Methods

All animal experiments were approved by the Ethics Committee for Animal Experimentation of Tokyo Women's Medical University and performed according to the Guidelines of Tokyo Women's Medical University on Animal Use.

2.1 Intestinal Graft Harvesting from Pigs

Here, we demonstrate a human transplantable vascular bed fabrication method using decellularized porcine small intestine (Fig. 1). An intestinal graft with an artery and vein was obtained from a 2 to 3-month-old pig's small intestine to fabricate an implantable vascular bed. A 5–6 cm long segment of small intestine including the arteriovenous loop supplying blood to the area of the tissue was harvested.

2.2 Perfusion-Decellularization of Porcine Small Intestinal Graft

Decellularization was performed by sequentially perfusing sodium deoxycholate (DCA) and DNase-I through the artery and intestinal lumen of the small intestinal graft to eliminate the nuclei causing rejection in the human body. The harvested small intestinal graft was placed in the custom-made chamber (Tokai Hit, Shizuoka, Japan). The chamber was connected to the bioreactor system (Tokai Hit, Shizuoka, Japan) consisting of a roller pomp. The small intestinal graft was automatically decellularized by perfusing the artery and intestinal lumen with 4wt% DCA/MQ at 25 °C for 4 h, PBS at 25 °C for 30 min, and 2000 kU DNase-1/PBS at 37 °C for 4 h. The decellularization process was modified from the previous report [11]. Finally, the intestinal graft was flushed with PBS at 4 °C and stocked in a refrigerator.

2.3 Decellularized Intestinal Graft Evaluation

The decellularized intestinal grafts were fixed in 4% paraformaldehyde at room temperature before embedding in paraffin for morphological observation. Hematoxylin and eosin (HE) and Hoechst staining were performed to detect cell nuclei. Immunostaining with collagen IV and fibronectin antibodies was performed to detect extracellular matrix. The native porcine small intestine served as a control.

The amount of DNA in the native and decellularized intestinal graft was measured using fluorescent dye binding with a Quant -iT PicoGreen dsDNA assay kit.

The burst pressure of vascular network within the native and decellularized intestinal graft was measured. After the vein of the graft was ligated, India ink was perfused through the artery of the graft until the vascular structure burst. The pressure was measured using a pressure transducer.

2.4 Decellularized Intestinal Graft Implantation into the Rat

The decellularized intestinal graft was implanted into the dorsal subcutaneous layer of the rat to evaluate the immunogenicity of the engineered decellularized graft. The epidermis of the back of the 8-week-old rat was cut, and the muscular layer was exposed. The gamma sterilized decellularized small intestinal graft was transplanted and fixed in the muscle layer using sutures. As a control, the porcine small intestine sterilized using only γ-irradiation was used. The tissue was macroscopically observed and explanted two weeks after transplantation. The explanted samples were fixed in 4% paraformaldehyde and embedded in paraffin. Finally, HE staining and immunostaining with CD3 antibody, which is a T cell maker, and Iba 1 antibody, which is a macrophage marker, was performed to evaluate engineered graft immunogenicity.

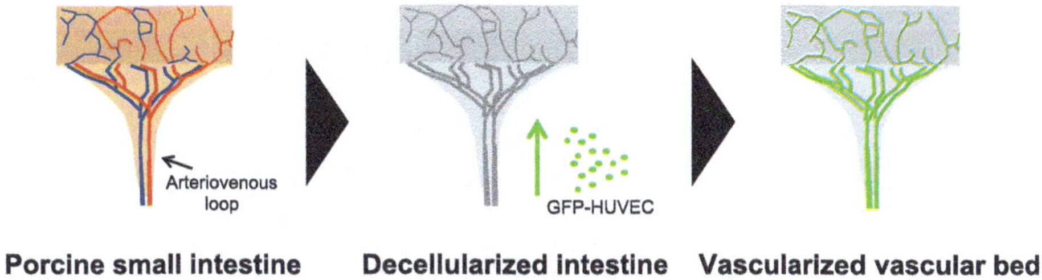

Fig. 1. Schematic illustration of intestinal vascular bed fabrication method. Porcine small intestine was decellularized to engineer the acellular scaffold of small intestine. Subsequently, the decellularized intestine was re-endothelialized with GFP-HUVECs to obtain a vascularized intestinal vascular bed.

2.5 Vascularized Intestinal Vascular Bed Fabrication

Human vascular endothelial cells were seeded into blood vessels of the decellularized intestinal graft to fabricate a vascularized vascular bed that could be perfused with blood (Fig. 1). For human vascular endothelial cells, GFP-positive human umbilical vein endothelial cells (GFP-HUVECs) were selected in this study. 1.0×10^7 GFP-HUVECs were suspended in 10 mL of culture medium and intravascularly injected into the artery of the graft. The graft was placed in a glove box maintained at 37 °C and cultured for one week by perfusing culture medium from the artery of the graft by the roller pomp to re-endothelialize the vasculature with GFP-HUVECs. After one week, the vascularized intestinal vascular bed was observed using a fluorescent microscope to detect GFP signals, and the tissue was fixed in 4% paraformaldehyde and embedded in paraffin. Finally, immunostaining with CD31 was performed to detect endothelial cells.

3 Results

3.1 Porcine Small Intestinal Graft Decellularization

Porcine small intestinal graft including the arteriovenous loop was harvested and placed in a chamber (Fig. 2a). The intestinal graft was decellularized by intravascularly and transluminally perfusing DCA and DNase-1 into the graft to obtain a skeleton of a human transplantable vascular bed. The intestinal graft was apparently transparent after decellularization (Fig. 2b). Morphological evaluation by HE and Hoechst staining of the native and decellularized graft demonstrated that cell nuclei were significantly removed and the original intestinal structure remained intact after the decellularization protocol (Fig. 2, c, d, e and f). Immunostaining confirmed that collagen IV and fibronectin were preserved in the decellularized intestinal graft (Fig. 2, g, h, i and j). In addition, the amount of DNA in both the native and decellularized intestinal graft was quantitated using the Quant -iT PicoGreen dsDNA assay kit. The amounts of DNA of the native and decellularized intestine were 138 ± 18.9 ng/mg and 37.3 ± 1.01 ng/mg, respectively (each $n = 3$) (Fig. 3).

The burst pressures of the native and decellularized small intestinal graft were also measured. The burst pressure of the native and decellularized intestine were 638 ± 86.3 mmHg and 609 ± 58.0 mmHg, respectively (each $n = 2$) (Fig. 4).

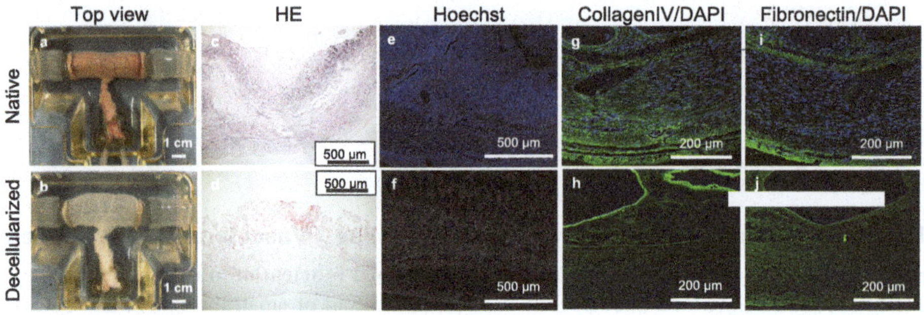

Fig. 2. Morphological observation of decellularized porcine small intestine. Native small intestinal graft placed in a chamber (a) was successfully decellularized using perfusion decellularization without severely damaging the tissue structure (b). HE and Hoechst staining shows that the decellularization process removes the cell nuclei throughout the tissue (c–f) and preserves the collagen IV and fibronectin of the small intestinal graft (g–j).

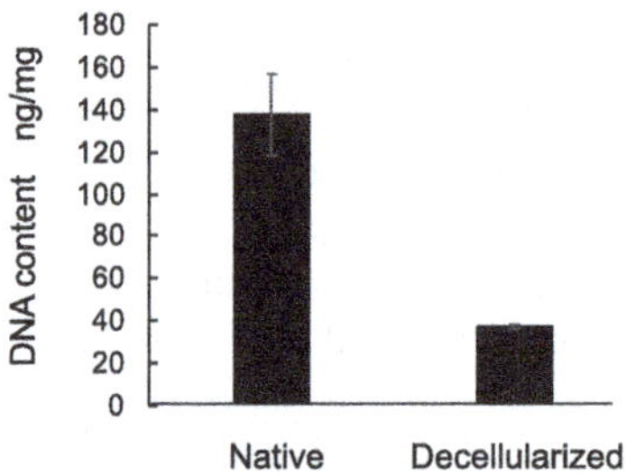

Fig. 3. DNA quantification of native and decellularized small intestine. DNA quantification confirms that the amount of DNA of the native small intestine decreased from 138 ± 18.9 ng/mg to 37.3 ± 1.01 ng/mg after decellularization (each n = 3).

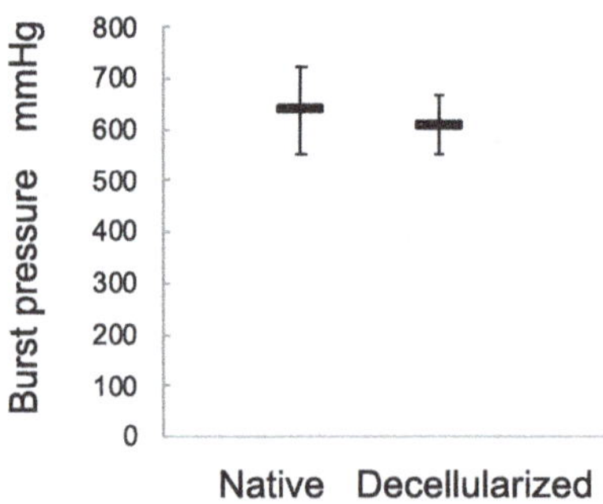

Fig. 4. Burst pressure of native and decellularized small intestine. Burst pressure measurement of both native and decellularized intestinal graft indicates that there is almost no difference in the burst pressure. The burst pressure of the native intestine is 638 ± 86.3 mmHg, and it is 609 ± 58.0 mmHg in the decellularized intestine (each n = 2).

3.2 Inflammatory Response to the Implanted Decellularized Graft in the Rat

Here, a decellularized porcine small intestine was selected as a scaffold to reconstruct a vascular bed that could be transplanted into humans. Therefore, it was essential that the engineered decellularized intestinal graft did not induce severe inflammation in other animal species. The decellularized intestinal grafts were implanted in rats for two weeks to confirm the inflammatory response to the decellularized intestinal graft *in vivo*. The 1-cm round tissues were transplanted onto the muscle layer (Fig. 5, a, b, c and d). Granulation-like tissue formed around the implanted native intestine after two weeks (Fig. 5e). On the other hand, nothing such was observed around the decellularized graft (Fig. 5f). HE staining showed that the thickness around the transplanted native intestine increased to 2 mm, but the decellularized graft did not increase (Fig. 5, g and h). Moreover, immunostaining of CD3 and Iba 1 showed that both T cells and macrophages infiltrated into the native intestine (Fig. 5, i and k), but neither T cells nor macrophages were more likely to be observed in the decellularized graft (Fig. 5, j and l).

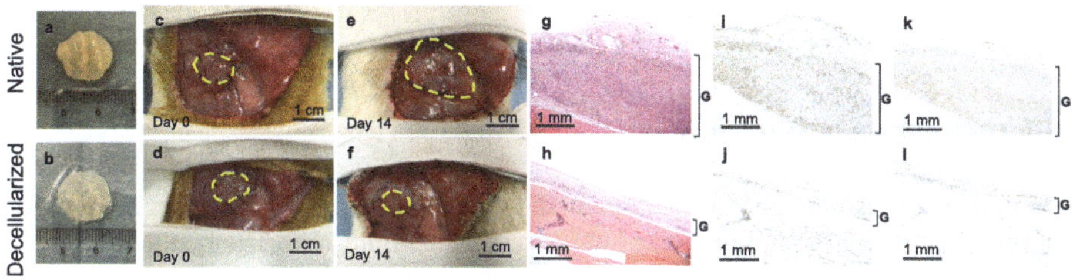

Fig. 5. Immunogenicity of the native and decellularized small intestine. One centimeter round shaped grafts (G) were prepared from native and decellularized intestine (a, b) and implanted into the dorsal subcutaneous layer of the 8-week-old rat (c, d). After 14 days of transplantation, granulation-like tissue was macroscopically observed around the implanted native intestine (e). HE staining also shows that granulation-like tissue with a thickness of 2 mm was formed (g). Immunostaining with CD3 (i) and Iba 1 (k) of the native intestine demonstrates that T cells and macrophages are remarkably infiltrated in the native intestine, indicating a severe inflammatory response. In contrast, the transplanted decellularized graft remains in its original form at the time of implantation as shown by macroscopic observation (f) and HE staining (h). Immunostaining with CD3 (j) and Iba 1 (l) of the decellularized intestine confirms that small T cells and macrophages exist, indicating that the decellularized graft is successfully transplanted without a severe inflammatory response.

3.3 Histological Observation of the Vascularized Intestinal Vascular Bed

GFP-HUVECs were seeded through the vasculature of the decellularized intestinal graft to re-endothelialize the graft and reconstruct the vascularized intestinal vascular bed so that it could be perfused with blood. The vascularized intestinal graft was prepared using perfusion culture for one week after seeding of 1.0×10^7 cells of GFP-HUVECs. Macroscopic observation and fluorescent imaging of the vascularized intestinal graft showed that GFP-HUVECs existed throughout the intestinal graft (Fig. 6, a and b). Immunostaining with CD31, which is a specific vascular endothelial cell marker, demonstrated that some blood vessels were covered with a single layer of GFP-HUVECs like in the living body (Fig. 6c).

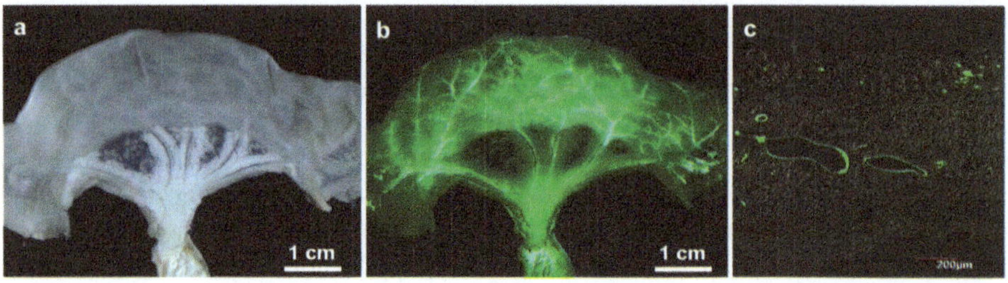

Fig. 6. Morphological evaluation of the vascularized intestinal vascular bed. GFP-positive HUVECs distributes throughout the tissue as confirmed by macroscopic and fluorescent observation of the vascularized intestinal vascular bed (a, b). Immunostaining with CD31 indicates that GFP-HUVECs attaches to the vascular wall and re-endothelializes the lumina (c).

4 Conclusion/Discussion

This study demonstrated the fabrication method of the vascular bed using decellularizion and subsequent re-endothelializion of the porcine small intestine. The histological evaluation and DNA quantification of the decellularized small intestinal graft indicated that the small intestinal graft was successfully decellularized using DCA and DNase-1 (Fig. 2 and 3). It was confirmed that the small intestinal graft decellularized by the method satisfied the minimal criteria compared to the minimal criteria of the decellularized tissue previously reported [10]. In addition, there was almost no difference between both burst pressures (Fig. 4). Simsa et al. demonstrated that decellularization by Triton-X, tributyl phosphate, and DNase-1 did not affect the mechanical properties of the porcine vena cava because elastin and collagen contributing to mechanical stability were preserved [12]. Therefore, our decellularization protocol preserved elastin and collagen content of the small intestinal graft because our present results of burst pressure did not change after decellularization. These data suggest that the decellularization protocol effectively removed graft cell nuclei without severely damaging the extracellular component and the acellular skeleton of the small intestinal graft was obtained.

Regarding immunogenicity, T cells and macrophages were less prevalent in the decellularized graft compared to the native graft in implantation test to the rat (Fig. 5). Mirmalek-Sani et al. demonstrated that decellularized liver biopsy transplanted into a rodent showed the presence of M2 macrophages around the decellularized specimen and the absence of activated T cells and M1 macrophages [13]. T cells and M1 macrophages are well known to be specifically present in the inflammatory reaction area, and M2 macrophages are known to be present in the tissue repair area. Here, the specimen showed that T cells were less accumulated in the decellularized graft. This indicated that the macrophage infiltration in the decellularized tissue was by M2 macrophages, and the immunogenicity of the porcine small intestine was decreased. Therefore, our results suggest that the decellularized small intestinal graft is a material that may have potential for clinical investigation.

Regarding reconstruction of the vascularized small intestinal graft, the decellularized intestinal graft was re-endothelialized using perfusion culture within seven days (Fig. 6). Endothelial cells seeded through the artery were distributed from the proximal to distal retained vascular structure of the decellularized intestinal graft, and the cells attached to both the artery and vein. These results indicated that the decellularized intestinal graft could serve as the skeleton of the vascularized intestinal vascular bed. However, it was also confirmed that there were some blood vessels not fully re-endothelialized (Fig. 6). Cebotari et al. showed that the tissue decellularized using detergents needed to be washed repeatedly to remove detergent toxicity and effectively attach the endothelial cells on the decellularized pulmonary valve scaffold [14]. Moreover, decellularized tissue using sodium dodecyl sulfate (SDS) and Triton-X showed high cell adhesiveness and was reportedly well recellularized in various organs [15–17]. Future optimization of rinsing time for the decellularized graft and decellularization method of the intestinal graft using SDS and Triton-X may improve cellular adhesiveness of the decellularized intestinal graft and allow stable re-endothelialization of the decellularized intestinal graft. The fully vascularized intestinal vascular bed may be perfused with blood for a long time *in vivo*, allowing transplantation of large-scale tissue fabricated *in vitro*.

In conclusion, we demonstrated a promising strategy to reconstruct a human implantable vascular bed derived from porcine small intestine using a decellularization and re-endothelialization technique. It had been indicated that the porcine small intestinal graft decellularized by the protocol might retain the native extracellular components, could be transplanted into other animal species without a severe inflammatory response, and served as a scaffold of the vascularized intestinal vascular bed. Hence, the vascularized intestinal graft may be indicated to serve as a transplantable material that allows large-scale tissue survival *in vivo* for regenerative therapy.

Acknowledgments. This research was supported by the "Development of innovative manufacturing technology for three-dimensional tissues and organs based on cell sheet engineering" from the Japan Agency for Medical Research and Development (AMED, https://www.amed.go.jp/en/index.html; grant no. JP17he0702249) and JSPS KAKENHI Grant Number JP 18J22398. We would like to thank Editage (www.editage.com) for English language editing.

Conflict of Interest Declaration. Kazunori Sano is an employee of Tokai Hit Co., Ltd (Japan). Tatsuya Shimizu is a member of the scientific advisory board and a shareholder of CellSeed Inc. (Japan). Tokyo Women's Medical University received research funds from CellSeedInc. Tokyo Women's Medical University received a research grant from Tokaihit Co., Ltd, Japan.

References

1. Okano, T.: A novel recovery system for cultured cells using plasma-treated polystyrene dishes grafted with poly(N- isopropylacrylamide). J. Biomed. Mater. Res. **27**(10), 1243–1251 (1993)
2. Shimizu, T.: Cell sheet engineering for myocardial tissue reconstruction. Biomaterials **24**(13), 2309–2316 (2003)

3. Shimizu, T.: Cell sheet-based tissue engineering for fabricating 3-dimensional heart tissues. Circ. J. **78**(11), 2594–2603 (2014)
4. Memon, I.A.: Repair of impaired myocardium by means of implantation of engineered autologous myoblast sheets. J. Thorac. Cardiovasc. Surg. **130**(5), 1333–1341 (2005)
5. Sawa, Y.: Tissue engineered myoblast sheets improved cardiac function sufficiently to discontinue LVAS in a patient with DCM- report of a case. Surg. Today **42**(2), 181–184 (2012)
6. Sawa, Y.: Safety and efficacy of autologous skeletal myoblast sheets (TCD-51073) for the treatment of severe chronic heart failure due to ischemic heart disease. Circ. J. **79**(5), 991–999 (2015)
7. Shimizu, T.: Fabrication of pulsatile cardiac tissue grafts using a novel 3- dimensional cell sheet manipulation technique and temperature-responsive cell culture surfaces. Circ. Res. **90**(3), e40–e48 (2002)
8. Sekine, H.: In vitro fabrication of functional three-dimensional tissues with perfusable blood vessels. Nat. Commun. **4**, 1399 (2013)
9. Inui, A.: Generation of a large-scale vascular bed for the in vitro creation of three-dimensional cardiac tissue. Regen. Ther. **11**, 316–323 (2019)
10. Crapo, P.M.: An overview of tissue and whole organ decellularization process. Biomaterials **32**(12), 3233–3243 (2011)
11. Totonelli, G.: A rat decellularized small bowel scaffold that preserves villus-crypt architecture for intestinal regeneration. Biomaterials **33**(12), 3401–3410 (2012)
12. Simsa, R.: Effect of fluid dynamics on decellularization efficacy and mechanical properties of blood vessels. PLoS ONE **14**(8), e0220743 (2019)
13. Mirmalek-Sani, S.H.: Immunogenicity of decellularized porcine liver for bioengineered hepatic tissue. Am. J. Pathol. **183**(2), 558–565 (2013)
14. Cebotari, S.: Detergent decellularization of heart valves for tissue engineering: toxicological effects of residual detergents on human endothelial cells. Artif. Organs **34**(3), 206–210 (2010)
15. Ott, H.C.: Perfusion-decellularized matrix: using nature's platform to engineer a bioartificial heart. Nat. Med. **14**(2), 213–221 (2008)
16. Uygun, B.E.: Organ reengineering through development of a transplantable recellularized liver graft using decellularized liver matrix. Nat. Med. **16**(7), 814–820 (2010)
17. Doi, R.: Transplantation of bioengineered rat lungs recellularized with endothelial and adipose-derived stromal cells. Sci. Rep. **7**(1), 8447 (2017)

Therapeutic Medical Engineering

Tactile Cue by Touching the Agonist Muscle Increases the Muscle Activity During Arm Curl Exercise

Kazushige Oshita[1,2](✉)

[1] Department of Human Information Engineering, Okayama Prefectural University, Okayama, Japan
oshita@ss.oka-pu.ac.jp
[2] Department of Sports Science, Kyushu Kyoritsu University, Fukuoka, Japan

Abstract. This study aimed to investigate the effect of tactile stimulation by touching the agonist muscle during the arm curl exercise. Nine healthy males (age, 20–35 years) performed five repetitions of arm curl under two different focus instructions; for the external focus condition, participants were instructed to lift the bar, whereas for the internal focus condition, the participants' biceps brachii was lightly touched by the investigator and they were instructed to concentrate on muscle contraction at the touched point. Muscular activity in the biceps brachii was evaluated during the exercise. The muscle activity in the internal focus condition was significantly greater than that of the external focus condition ($P = 0.01$, $d = 0.19$). This result suggests that internal focus caused by an individual's consciousness, as well as a focus attention through touching the agonist area increased the agonist muscle activity.

Keywords: Focus attention · Internal focus · Light touch

1 Introduction

Human movements are caused by muscle contraction, thereby generating muscle force. For instance, flexing movement of the elbow joint is caused by contraction in the biceps brachii, which is the generating force in the muscle. Agonist muscle contraction increases with heavier external resistance against movement. External resistance can be created using dumbbells, barbell, elastic tube, own body weight, water, etc. Depending on the resistance load, the agonist muscle fitness (i.e., strength, power, and/or mass) is expected to increase, and this is called "resistance training".

To gain muscle fitness by resistance training, exercises based on several principles should be performed. "Overload" is one of the basic principles [1], in which exercises must cause the muscles to work harder continually as the bodies adopt to existing workouts. Therefore, exercise and training that goes beyond normal levels of muscle contraction is required to gain muscle fitness [1]. Previous studies investigated attention during exercise reported that muscle contraction intensity differs depending on the individual's focus when lifting an external resistance, even with the same load of resistance

Y. Shiraishi et al. (Eds.): APCMBE 2020, IFMBE Proceedings 82, pp. 295–300, 2021.
https://doi.org/10.1007/978-3-030-66169-4_36

[2, 3]. Vance et al. [4] investigated the agonist muscle activity during arm curl (flexion of elbow joint) exercise under two different focus instructions while lifting a weight: concentration on contracting the biceps brachii (internal focus attention, INT) or concentration on lifting the weight (external focus attention, EXT). The INT produced higher electromyographic (EMG) activity in the agonist muscle, i.e. biceps brachii, than the EXT, even for the same weight [4]. Therefore, understanding the agonist muscle group during exercise is an important factor in enhancing the effect of resistance training.

However, individuals performing fewer exercises or those with lesser experience usually face difficulties in understanding the agonist muscle during resistance training. Previous studies on the knowledge of back squat exercises [5] or trunk muscle exercises [6] reported that almost all novices in resistance training or individuals without a previous experience of strength training under professional supervision did not properly understand which body part(s) was being trained by the exercise. Even in single-joint exercises [7], more than 90% of the participants with no experience of performing knee-extension exercise answered that the exercise not only trained the anterior thigh part, but also other parts, including the trunk, posterior thigh, and upper limb. Almost 80% of the participants who performed knee-extension exercise without prior experience of strength training under professional supervision felt that the exercise was appropriate to train the anterior thigh as well as other parts.

To solve this problem of misunderstanding of the agonist muscle during exercise, the instructor occasionally touches the agonist muscle area during exercise, and instructs the client with cues such as "concentrate on contracting of this (touching) part". This suggests that INT during exercise could be replaced by contact (tactile) stimuli. Some researchers have reported that light tactile stimuli to the body part can enhance the perception of body orientation, and this improves the human movement or sense of equilibrium [8–13]. If INT can be replaced by contact stimuli, it could lead to advances in healthcare technologies. For instance, human movements can be easily analysed using accelerometers or motion capture. To improve movement, tactile stimulation can be applied to the agonist muscle area at the appropriate time (e.g., by attaching a vibrator to that part, etc.) based on the motion analyze. If such a tactile stimulation can improve movements or enhance muscle activity, this technology could promote rehabilitation, training effects, prevention of sarcopenia, and athletic performance. Therefore, this study aimed to investigate the effect of tactile stimulation by touching the agonist muscle during the arm curl exercise.

2 Methods

2.1 Participants

Data were obtained from nine healthy males (age, 20–35 years) with no current or previous medical history of musculoskeletal system disorders. Each participant identified the preferred dominant arm with which he would write words. This was the right side for all participants. Before participating, all participants provided informed consent after being explained the study purpose. This study was approved by the Human Ethics Committee of the Graduate School of Human Development and Environment, Kobe University.

2.2 Materials and Procedure

Arm curl exercises were conducted using a preacher-curl machine on the non-preferred arm of the participants. Muscular activity in the biceps brachii was evaluated during the arm curl exercise. Surface EMG data were collected from the biceps brachii muscles to assess muscle activity. On the non-preferred arm, bipolar surface electrodes with an EMG amplifier (ID2PAD; Oisaka Electronic Equipment Ltd., Japan) were placed over the muscle belly at a 2-cm inter-electrode distance. To decrease skin impedance affecting the electrical signal, the skin was wiped by gentle abrasion with a skin preparation gel and isopropyl alcohol swabs. The signals were amplified 500 times and acquired at a sampling frequency of 1 kHz using a data logger with an analogue-to-digital converter (LP-MS1002; Logical Product Corporation, Japan).

2.3 Experimental Protocol

A maximum voluntary muscle isometric contraction (MVC) test was initially performed in the direction of elbow flexion at 90° joint angle. The participant sat on a seat of the preacher-curl machine while his non-preferred arm held a bar. The bar was fixed to the ground by a metal chain through a strain gauge (FG-5100, Lutron Electronic Enterprise Co., Ltd., Taiwan). Participants were instructed to attempt to flex their arm against the fixed bar. During the MVC tests, the visual feedback of the force signals was displayed on a digital indicator and verbal encouragement was provided by the investigators. Participants were asked to perform the MVC test twice for 5 s with a 30 s pause between tests. The participants were allowed to reject an effort that they deemed as not "maximal." The highest peak force was selected as the MVC force.

After the MVC tests, the participants performed an arm curl task weighted with a dumbbell. Mass of the dumbbell was equivalent to the estimated 50% MVC force. Participants performed five repetitions of arm curl with full range of motion under two different focus instructions, with a 5 min rest period between sets. For the EXT, participants were instructed to lift the bar, whereas for the INT, the participants' biceps brachii was lightly touched by the investigator and they were instructed to concentrate on muscle contraction at the touched point (Fig. 1). During these tasks, participants were asked to look straight ahead and concentrate on the bar or touched point. The movement time of the arm curl was prescribed by using a digital metronome (MA-1, KORG Inc., Japan) that was set to produce a click every 1.5 s. Participants were asked to perform their movements in synchrony with the metronome signal, i.e., the end of each upward and downward motion should coincide with one signal. The order of each focus condition was randomised across the participants. The weight mass and movement tempo of this study were based on a previous study that investigated the effects of focus attention on muscle activity during arm curl [4]. Before the MVC tests, a practice session to familiarise the participants with the arm curl task was conducted for approximately 5 min.

2.4 Data Analysis

The stored EMG data were processed using the waveform analysis software SPCANA (ver. 4.92). After band-pass filtering (1–500 Hz), the root mean square of the EMG signal

Fig. 1. A schematic diagram of the arm-curl experiment in the internal focus attention condition.

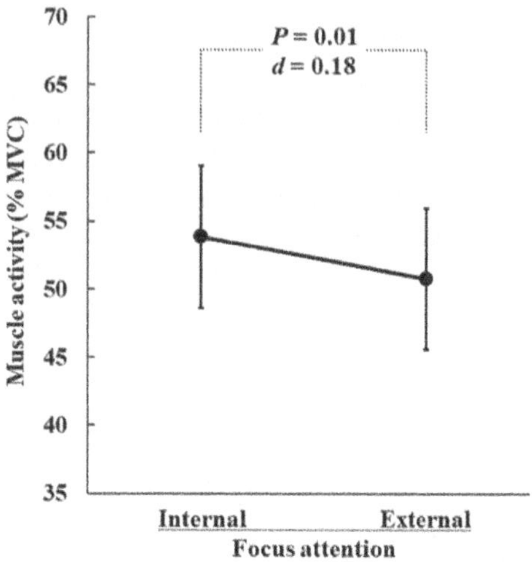

Fig. 2. Muscle activities in the biceps brachii during arm-curl exercise under internal or external focus attention

(RMS-EMG) in each task was calculated. For the MVC task, RMS-EMG was calculated every 0.2 s, and the maximum 0.2 s interval RMS-EMG value from the two MVCs was recruited as the MVC value. In each arm curl task, a 0.5 s period in the middle portion in each upward portion selected to calculate the RMS-EMG value. Only 2–4 repetitions in each focus condition were selected for analysis, and the average RMS-EMG value was calculated. The average RMS-EMGs in each arm curl task were normalised to the RMS-EMG in the MVC task (%MVC) for muscle activity evaluation.

Muscle activities (%MVC) in each arm curl task were evaluated using a paired *t*-test. Furthermore, the size of the effect (Cohen's *d*) of %MVC between the two focus conditions was calculated.

Value of $P < 0.05$ was considered statistically significant. These analyses were performed using JSTAT (ver. 20.0 J) software. Data were presented as mean values ± standard errors of the mean (SEM), unless otherwise stated.

3 Results and Discussion

The average %MVC (Fig. 2) in the INT focus condition was significantly greater than that of the EXT focus condition ($P = 0.01$). Regarding the focus effects, the difference in %MVC between the two focus conditions presented small-sized effects ($d = 0.19$). This result is consistent with the results of previous studies [2, 3] showing that increased muscular activity is observed in the INT as compared to the EXT. Although this study did not measure the triceps activity, i.e. antagonist muscle activity, a previous study reported that associated increase in the antagonist muscle activity was in line with the agonist muscle activity during INT, and that this suggests interference between the antagonist and agonist muscles during human movements [4]. Therefore, higher force production becomes difficult in an individual due to this interference when INT is practiced, and muscle activity in the agonist muscle might increase. This effect is observed not only in force or movement output, but also in endurance performance. Marchant et al. [14] reported that the INT reaches exhaustion faster than EXT during repetitive exercise. Such a phenomenon where an increased voluntary muscular activity is not accompanied with gross force or movement output is suggested as increase in "noise" in the motor control network [4, 14, 15]. Furthermore, this study revealed that INT caused by an individual's consciousness, as well as a focus attention through touching the agonist area increased the agonist muscle activity.

Conversely, increased agonist muscle activity in INT meant that it hampers movement economy as compared to that in EXT. A previous study also reported that angular velocity during arm curl exercise is executed faster in EXT than in INT conditions. Moreover, INT is observed to have detrimental effects on muscle power production [16] and variety of sports skills or performances [15, 17, 18]. Therefore, if individuals desire to improve movement skills or performance, i.e., accuracy, speed, and power production, there is an argument that they should adopt EXT. However, if muscle growth or strength gain is the desired goal, then the individual should employ INT [2, 3], even if increased muscle activity is slight (small effect) as shown in the present study. Additionally, this study suggests that tactile stimuli to the agonist area will also be effective in increasing the agonist muscle activity.

Our present results provide evidence that tactile stimulus to the agonist area from an instructor is effective in increasing agonist muscle activity during exercise. These results will be also helpful in the development of motor learning or assisting in improving movement. For example, decreasing gait ability in the elderly leads to falls. If tactile stimuli (by attaching a vibrator etc.) can provide stimulation of the agonist muscles at appropriate times through gait analysis, the individual will be able to understand which muscle(s) should be used at the time. Such a system might be improved gait performance and prevent falls. It could even become a useful tool for learning movement from the perspective of long-term effects and possibly apply to a variety of human motions. For such a technology to be realised, future studies identifying the exact mechanisms of focus attention, including tactile cues during human motion, will be necessary.

Acknowledgements. Research supported by KAKENHI (Grant-in-Aid for Scientific Research C, Grant Number JP18K06455).

References

1. Stone, M., Plisk, S., Collins, D.: Training principles: evaluation of modes and methods of resistance training. Sports Biomech. **1**, 79–103 (2002)
2. Neumann, D.L.: A Systematic review of attentional focus strategies in weightlifting. Front. Sports Act. Living **1**, 7 (2019)
3. Schoenfeld, B.J., Contreras, B.: Attentional focus for maximizing muscle development: The mind-muscle connection. Strength Cond. J. **38**, 27–29 (2016)
4. Vance, J., Wulf, G., Töllner, T., McNevin, N., Mercer, J.: EMG activity as a function of the performer's focus of attention. J. Mot. Behav. **36**, 450–459 (2004)

5. Oshita, K., Hagiwara, G., Monma, T., Tsuno, T., Koizumi, K., Oyama, Y., Yamaguchi, K., Tashiro, T., Funatsu, K.: Association between previous experience of strength training under appropriate supervision and knowledge of squats among university students who exercise regularly. Int. J. Sport. Health Sci. **14**, 61–71 (2006)
6. Oshita, K., Matsuzaki, T., Hikita, A., Miyazaki, R., Hagiwara, G., Tashiro, T., Tsuno, T., Higuchi, Y., Funatsu, K.: Consciousness of training body part(s) by showing a picture of various exercises in trunk muscles. Int. J. Sport. Health Sci **17**, 135–147 (2019)
7. Oshita, K., Matsuzaki, T., Hikita, A., Miyazaki, R., Hagiwara, G., Tashiro, T., Higuchi, Y.: A study of body-part(s) training consciousness using a picture of a single-joint exercise. Jpn. J. Phys. Fitness Sports Med. **67**, 393–401 (2018)
8. Stramel, D.M., Carrera, R.M., Rahok, S.A., Stein, J., Agrawal, S.K.: Effects of a person-following light-touch device during overground walking with visual perturbations in a virtual reality environment. IEEE RA-L **4**, 4139–4146 (2019)
9. Oshita, K., Yano, S.: Effects of haptic sensory input of a fluttering cloth on static and dynamic postural control. IFMBE Proc. **67**, 101–106 (2018)
10. Oshita, K., Yano, S.: Effect of haptic sensory input through a fluttering cloth on tandem gait performance. Hum. Mov. Sci. **55**, 94–99 (2017)
11. Oshita, K., Yano, S.: Effect and immediate after-effect of lightly gripping the cane on postural sway. J. Physiol. Anthropol. **35**(14), 1–8 (2016)
12. Oshita, K., Yano, S.: The effect of lightly gripping a cane on the dynamic balance control. Open Biomed. Eng. J. **9**, 146–150 (2015)
13. Boonsinsukh, R., Panichareon, L., Phansuwan-Pujito, P.: Light touch cue through a cane improves pelvic stability during walking in stroke. Arch. Phys. Med. Rehabil. **90**, 919–926 (2009)
14. Marchant, D.C., Greig, M., Bullough, J., Hitchen, D.: Instructions to adopt an external focus enhance muscular endurance. Res. Q. Exerc. Sport **82**, 466–473 (2011)
15. Zachry, T., Wulf, G., Mercer, J., Bezodis, N.: Increased movement activity accuracy and reduced EMG activity as the result of adopting an external focus of attention. Brain Res. Bull. **67**, 304–309 (2005)
16. Wulf, G., Zachry, T., Granados, C., Dufek, J.S.: Increases in jump-and-reach height through an external focus of attention. Int. J. Sports Sci. Coaching. **2**, 275–284 (2007)
17. Wulf, G., McConnel, N., Gartner, M., Schwarz, A.: Enhancing the learning of sport skills through external-focus feedback. J. Mot. Behav. **34**, 171–182 (2002)
18. Wulf, G., Lauterbach, B., Toole, T.: The learning advantages of an external focus of attention in golf. Res. Q. Exerc. Sport **70**, 120–126 (1999)

Gait Characteristics and Pain Assessment of Symptomatic Knee Osteoarthritis

Husnir Nasyuha Abdul Halim[1(✉)], Aizreena Azaman[1,2], and Azli Yahya[1]

[1] School of Biomedical Engineering and Health Sciences, Faculty of Engineering, Universiti Teknologi Malaysia, Johor Bahru, Malaysia
hnasyuha2@live.utm.my, aizreena@biomedical.utm.my, azliyahya@utm.my

[2] Sport Innovation & Technology Centre (SITC), Universiti Teknologi Malaysia, Johor Bahru, Malaysia

Abstract. The must-have clinical criterion or main symptom associated with knee osteoarthritis (OA) is knee pain. Gait alteration usually occur mainly to compensate the pain experienced at the knee joint. It consequently causes the deterioration of functional ability and affects activity of daily. The aim of the study is to extract kinematic gait features that characterize the knee OA gait and its association with pain assessment. Eight knee OA patients and five asymptomatic subjects were recruited and kinematic and spatiotemporal features were obtained using 3D Vicon motion analysis system during gait trial. There were 13 kinematic features extracted using Principal Component Analysis (PCA) and only two features which are hip flexion/extension phase shift and knee flexion/extension range of motion have significant correlation with pain assessment using ICOAP and KOOS questionnaire.

Keywords: Knee osteoarthritis · Gait characteristics · Pain assessment

1 Introduction

Knee osteoarthritis (OA) is the most common joint disease that affect approximately 240 million people globally and is ranked as the 11th leading cause of disability globally as it affects activity of daily living especially walking [1]. OA causes joint degeneration in which the decreasing of hyaluronic acid at synovial membrane that functions as the lubrication for joint space, changes in articular cartilage matrix components specifically collagen, and increasing activity of degradative enzymes [2] consequently causes structural changes in bone, including osteophytes formation and sclerosis informed consent were obtained from that lead to the progression of the disease [3]. The disease affects aging population and the risk factors includes obesity, gender; which women are more affected, history of injury, the effect of abnormal loading on the joint [4].

Clinically, knee OA is diagnosed based on its symptoms and by radiographic evidence of the anatomical changes of the joint. The radiographic image; x-ray also used for severity classification based on various grading systems, in which the most well-known

Y. Shiraishi et al. (Eds.): APCMBE 2020, IFMBE Proceedings 82, pp. 301–307, 2021.
https://doi.org/10.1007/978-3-030-66169-4_37

and established is Kellgren-Lawrenece grading system. However, knee OA diagnosis can be described based on two basis, either radiographic OA, the presence of changes in anatomical structure of the joint by radiographic observation or symptomatic OA the presence of pain and other symptoms of OA; and based on studies by Bedson et al. [5], there are discordance between those two. Because of that, there are needs to develop a method to characterize OA independently from radiographic classification. Moreover, severity classification based on symptoms and its effects could benefit knee OA intervention such as rehabilitation and physiotherapy specifically in terms of monitoring the patient progression and improvements after sessions of treatment.

The rapid advancement of technology has evolved to provide benefit in many sectors including healthcare, and one of the most noteworthy is application of machine learning on clinical data for diagnosis or characterization of diseases. Since the last decade, various researchers had investigating the application of various models and algorithms of machine learning to characterize knee OA. Most of the studies intention is to provide automatic diagnosis or classification of the disease based on x-ray, MRI or ultrasound images instead of manual observation and analysis by clinician. Recent studies had utilized gait data to accomplish the same purpose. Gait is locomotion or specifically refer to walking for mobility [6]. Gait analysis is the quantitative analysis of walking pattern and its usually acts as indicator for musculoskeletal disorders including OA [7]. However, gait produces complex, vast and multidimensional datasets so the feature extraction process is crucial in order to prepare the data for machine learning [8]. The features also need to be highly correlated to the symptom which is pain so that the classification will reflect symptomatic OA and its functional ability.

The objectives of this study are to apply principal component analysis (PCA) as feature extraction method for kinematic aspects of walking gait, to obtain spatiotemporal features and consequently to find the association with pain score from collected pain assessment. The features that have high correlation may have high reliability to be utilized as features in machine learning algorithm to indicate severity of this disease without depending on anatomical changes of the joint so that it is more feasible for symptomatic OA.

2 Methodology

2.1 Subject Recruitment

This study includes five asymptomatic subject and eight knee OA subjects. The protocol of this study was approved by the Medical Research and Ethics Committee (MREC), Ministry of Health (MOH) Malaysia with ID number NMRR-19-2586-46320(IIR) and informed consent were obtained from all subjects according to the guidelines from MREC. Inclusion criteria for both group of subjects were adults aged range 45–66 years old, able to walk without any mobility aid, and additional criteria for knee OA subjects were there must be presence of knee pain and other symptoms, attending physiotherapy sessions, and diagnosed by medical doctor with either unilateral or bilateral knee OA. Subjects were excluded if there were history of any surgery in either leg, were diagnosed with other musculoskeletal, neurological and cardiovascular problems, wearing any electronic medical devices such as heart pacemaker and any implants on body, and received corticosteroid injections, or had taken oral corticosteroids or antiinflammatories drug in the 24 h prior to testing.

2.2 Pain Assessment and Gait Test

Pain of knee OA subjects were assessed using subscale of pain from Knee injury and Osteoarthritis Outcome score (KOOS) questionnaire with nine questions and Measure of Intermittent and Constant Osteoarthritis Pain (ICOAP) questionnaire with 11 questions that were divided into two types of pain. Subjects were asked to walk for at least three trials on a 5 m walkway and the gait movement were recorded using Vicon Motion Analysis System (Oxford Metrics Inc., Oxford, U.K.). The attachment of 16 reflected markers on subject for Vicon recording were based on Plug-in Gait for lower limb model.

2.3 Data Pre-processing

Gait trial recording from Vicon were pre-processed using Vicon Nexus 1.8.2 and Polygon to obtain spatiotemporal data, and kinematic data of hip, knee and ankle on sagittal and frontal plane. The kinematic waveforms were time-normalized to 100% gait cycle. Data were taken from averaging data between right and left leg for normal subject and most affected leg for knee OA subjects.

2.4 Feature Extraction

For kinematic waveform, features were extracted using Principal Component Analysis (PCA). By utilizing PCA, the dataset is reduced to new uncorrelated variables called principal components (PCs) which is the eigenvectors of the covariance of a dataset [9]. The PCs describe the maximum variance within the dataset and ranked according to percentage of total variance so the first PCs indicate feature that have better distinctive criteria within the dataset [10]. Every PCs is describe using its coefficient, or also called loading vectors that are the magnitude of each sample for each component, and consequently the PC score which resulted from the orthogonal projection of the loading to each sample [10]. PC score is useful in understanding the influence of each PC loading to individual sample [11]. The first attempt of applying PCA on biomechanics gait data was by Deluzio et al. [12], where the authors successfully differentiate normal gait against osteoarthritic gait pattern features extracted from PCA model. Until recently, many studies had applied PCA onto biomechanics data as it does not neglect temporal information within the gait phases [10]. In this study, the kinematic features were represented by PC score of retained PCs in which that cumulatively explained 90% of the total variance. The PCs interpretation were done based on method by Deluzio et al. [13] that observe the shape of loading vector and waveform of representative extremes of the data.

2.5 Statistical Analysis

The extracted kinematic features from PCA and spatiotemporal features of subjects were tested its correlation using Spearman correlation analysis with two-tailed significant ($p < 0.05$).

3 Result and Discussion

Table 1 represent subject's demographic data which include ages, height, weight and BMI.

Table 1. Subject's demographic

	Mean (SD) n = 13
Age (years)	54.62 (8.16)
Height (m)	1.58 (0.08)
Weight (kg)	72.27 (11.07)
BMI (kg/m2)	29.01 (4.12)

The spatiotemporal features of normal and knee OA subjects were represented on Table 2. The mean and SD of cadence, step length, step width, stride length, and gait speed of knee OA group are lower than normalgroup while step time, stride time of knee OA are higher than normal group. Although there were no significant differences ($p > 0.05$) for all of the spatiotemporalparameters; most probably because of current smallsample size, the obtained results reach agreement with previous literatures [14–17].

Table 2. Spatiotemporal features of normal and knee OA subjects

	Normal (n = 8)	Knee OA (n = 5)
Cadence (steps/min)	92.42 (16.92)	86.00 (20.53)
Step length (m)	0.45 (0.04)	0.46 (0.14)
Step time (s)	0.69 (0.18)	0.72 (0.17)
Step width (m)	0.18 (0.06)	0.17 (0.04)
Stride length (m)	0.92 (0.07)	0.87 (0.26)
Stride time (s)	1.35 (0.27)	1.49 (0.45)
Gait speed (m/s)	0.71 (0.17)	0.65 (0.28)

Table 3 shows Principal Components (PCs) extracted and retained from kinematic data of all subjects. The retained PCs explained maximally the variation exist within the datasets of all subjects. From waveform of ankle, hip, and knee joint kinematic, there were 13 features extracted using PCA in which each PC are waveform characteristics were the features.

The interpretation of the retained PCs as gait kinematic features were done by observation on shape of loading vector waveform and its comparison with mean waveform of all subjects in which when the waveform entirely positive indicating magnitude or vertical shift feature, when the timing of changing positive to negative peaks in loading

Table 3. Numbers of retained PC for each kinematic, its cumulative explained variance and the interpretation of PC as kinematic features

	Cumulative variance (%)	Number of PCs retained	Features denoted as	Interpretation of PC/Kinematic features
Ankle dorsi/Plantar flexion	95.83	4	1PC1	Magnitude of flexion angle
			1PC2	Range of motion
			1PC3	Phase shift
			1PC4	Phase shift
Ankle inversion/Eversion angle	94.88	1	2PC1	Magnitude of inversion/eversion angle
Hip flexion/Extension	95.74	2	3PC1	Magnitude of flexion angle
			3PC2	Phase shift
Hip ab/Adduction	93.90	1	4PC1	Magnitude of abduction angle
Knee flexion/Extension	95.53	3	5PC1	Magnitude of flexion angle
			5PC2	Range of motion
			5PC3	Phase shift
Knee ab/Adduction	97.58	2	6PC1	Magnitude of abduction angle
			6PC2	Range of motion

vector waveform is in line with timing of peak appearance in mean waveform, signifying phase shift feature and when the loading vector waveform also alternating between negative and positive peaks, but the peaks are consecutively in line with peaks of mean waveform, indicating the PC as range of motion feature.

The features are also interpreted by observing characteristics difference between lowest and highest percentile of PC score of the data. The ankle dorsi/plantar flexion need the greatest number of PCs to cumulatively explained 90% of variation of the datasets, followed by knee flexion/extension angle with three PCs, hip flexion/extension and knee abduction/adduction angle with two PCs and the others that need only one PC two explained the variation.

Spearman correlation analysis revealed that all spatiotemporal features have no significant correlation with all pain score while 3PC2 has significant correlation ($p < 0.05$) with ICOAP constant pain score and 5PC2 has significant correlation with a ll three pain scores. In this study, two different pain assessment were used as the different types of pain assessment aspects could probably give more insights when tested its relationship to gait biomechanics. ICOAP questionnaire focused on quality of pain while KOOS pain focused on the intensity of pain during activity of daily living.

4 Conclusion

This study is preliminary study that investigates the feasibility of PCA as feature extraction for small number of samples for kinematic waveform of asymptomatic and different level of pain of knee OA subjects without depending on radiographic severity. For all kinematic waveform of the lower limbs joint, the first PCs which is the highest percentage of variance were all magnitude features. This indicate that the critical alteration of knee OA is the overall amplitude of joint kinematic although the two PCs which are hip flexion/extension phase shift and range of motion that have significant correlation with pain score.

The limitation of this study is the small number of subjects that causes the data to be not normally distributed. The study also did not utilize electromyography (EMG), the recording of muscle activity as the gait features that could provide more distinctive features of knee OA during gait. Future study could increase number of subject and record EMG and extract the features also using PCA.

Acknowledgement. This research was supported by UTM GUP Tier 2 (.18J11) grant.

References

1. Nelson, A.E.: Osteoarthritis year in review 2017: clinical. Osteoarthritis and Cartilage **26**(3), 319–325 (2018)
2. Bronner, F., Farach-Carson, M.C.: Bone and Osteoarthritis. Springer, London (2007)
3. Sofat, N., Ejindu, V., Kiely, P.: What makes osteoarthritis painful? The evidence for local and central pain processing. Rheumatology **50**(12), 2157–2165 (2011)
4. Palazzo, C., et al.: Risk factors and burden of osteoarthritis. Ann. Phys. Rehabilit. Med. **59**(3), 134–138 (2016)
5. Bedson, J., Croft, P.R.: The discordance between clinical and radiographic knee osteoarthritis: a systematic search and summary of the literature. BMC Musculoskeletal Disorders **9**(1), 116 (2008)
6. Introduction: Theory and practice in gait analysis A2 - Kirtley, Chris, in Clinical Gait Analysis. 2006, Churchill Livingstone: Edinburgh, pp. ix–xii (2006)
7. Whittle, M.W.: Methods of gait analysis. Gait Analysis (Fourth Edition), pp. 137–175. Butterworth-Heinemann, Edinburgh (2007)
8. Abid, M., Mezghani, N., Mitiche, A.: Knee joint biomechanical gait data classification for knee pathology assessment: a literature review. Appl. Bionics Biomech. **2019**, 14 (2019)
9. Phinyomark, A., et al.: Analysis of big data in gait biomechanics: current trends and future directions. J. Med. Biol. Eng. **38**(2), 244–260 (2018)
10. Astephen, J.L., Deluzio, K.J.: A multivariate gait data analysis technique: application to knee osteoarthritis. Proc. Inst. Mech. Eng. Part H: J. Eng. Med. **218**(4), 271–279 (2004)
11. Brandon, S.C.E., et al.: Interpreting principal components in biomechanics: representative extremes and single component reconstruction. J. Electromyogr. Kinesiol. **23**(6), 1304–1310 (2013)

12. Deluzio, K.J., et al.: Principal component models of knee kinematics and kinetics: Normal vs. pathological gait patterns. Human Movement Sci. **16**(2), 201–217 (1997)
13. Deluzio, K.J., Astephen, J.L.: Biomechanical features of gait waveform data associated with knee osteoarthritis: an application of principal component analysis. Gait and Posture **25**(1), 86–93 (2007)
14. Astephen, J.L., et al.: Biomechanical changes at the hip, knee, and ankle joints during gait are associated with knee osteoarthritis severity. J. Orthopaedic Res. **26**(3), 332–341 (2008)
15. Zeni, J.A., Higginson, J.S.: Differences in gait parameters between healthy subjects and persons with moderate and severe knee osteoarthritis: a result of altered walking speed? Clinical Biomech. **24**(4), 372–378 (2009)
16. Kaufman, K.R., et al.: Gait characteristics of patients with knee osteoarthritis. J. Biomech. **34**(7), 907–915 (2001)
17. Levinger, P., et al.: The application of support vector machines for detecting recovery from knee replacement surgery using spatio-temporal gait parameters. Gait and Posture **29**(1), 91–96 (2009)

Diagnostic Devices

PET-Laparoscope System with a Multi-layer Movable Detector: Feasibility Study with an Abdominal Phantom

Madhushanka R. Liyanaarachchi[1(✉)], Kenji Shimazoe[1], Hiroyuki Takahashi[1], Keiichi Nakagawa[1], Etsuko Kobayashi[2], and Ichiro Sakuma[1]

[1] The University of Tokyo, Tokyo 113-8656, Japan
rukshani@bmpe.t.u-tokyo.ac.jp
[2] Tokyo Women's Medical University, Tokyo 162-8666, Japan

Abstract. The PET-laparoscope system has been proposed to diagnose lymph node metastases intraoperatively in gastric cancer surgeries. The system uses two detector units, a fixed detector array and a movable detector probe to acquire PET images. The detector probe can be inserted into a patient's stomach through laparoscopic ports. A surgeon can move the detector inside the patient's body but the movement is limited. In this study the system was evaluated with an abdomen phantom so that the probe movement was realistic. Moreover, performance of an MLEM algorithm was investigated for the PET-laparoscope system. Experiments with the abdomen phantom confirmed that the system performs well when the probe movement is similar to that achievable during a laparoscopic surgery. MLEM algorithm provided better images. Image artifacts were reduced and the spatial resolution in lateral directions were better than 3 mm and the spatial resolution in depth direction was around 10 mm. Intraoperative PET image was overlaid on the laparoscopic video feed.

Keywords: Image reconstruction · Image overlay · Laparoscope · Lymph node metastases · Positron emission tomography

1 Introduction

Positron emission tomography (PET) identifies cancers based on the accumulation of a radiotracer injected into a patient. The most common tracer used in cancer screening with PET is fluorodeoxyglucose (FDG) which is a glucose analog. Cancer cells are highly metabolically active and absorb higher amounts of FDG thus radiotracer concentration is an indication of cancer. Conventional bulky PET systems are limited to preoperative use, but intraoperative diagnosis with PET is also required to identify margins of cancers and lymph node metastases in minimal invasive surgeries.

PET-laparoscope system has been proposed to diagnose lymph node metastases during laparoscopy-assisted stomach cancer surgeries [1]. The system uses two detector units as shown in Fig. 1 to acquire PET images. One is a fixed detector array connected to a surgical bed of a patient and the other one is a movable detector probe to be inserted

Y. Shiraishi et al. (Eds.): APCMBE 2020, IFMBE Proceedings 82, pp. 311–318, 2021.
https://doi.org/10.1007/978-3-030-66169-4_38

into a patient's stomach. After conducting simulation experiments which reconstructed images with better than 7 mm spatial resolution, a prototype detector was developed [8]. The detector probe was tracked by an optical tracking system (OTS) (Polaris Vicra OTS, Northern Digital Inc., Ontario, Canada). The spatial resolution of the PET-laparoscope system was further improved by using a pixelated multi-layer movable detector probe [2]. The fixed detector was a 7×7 array of $10 \times 10 \times 20\ \text{mm}^3$ Cerium-doped gadolinium aluminum gallium garnet crystals coupled with silicon photomultipliers (SiPMs). The Detector probe consisted of 4 layers of cross-shaped 12-pixel arrays with $2 \times 2 \times 3\ \text{mm}^3$ size gadolinium fine aluminum gallate crystals coupled with SiPMs. Detector prototype reconstructed images with 3 mm spatial resolution in lateral directions.

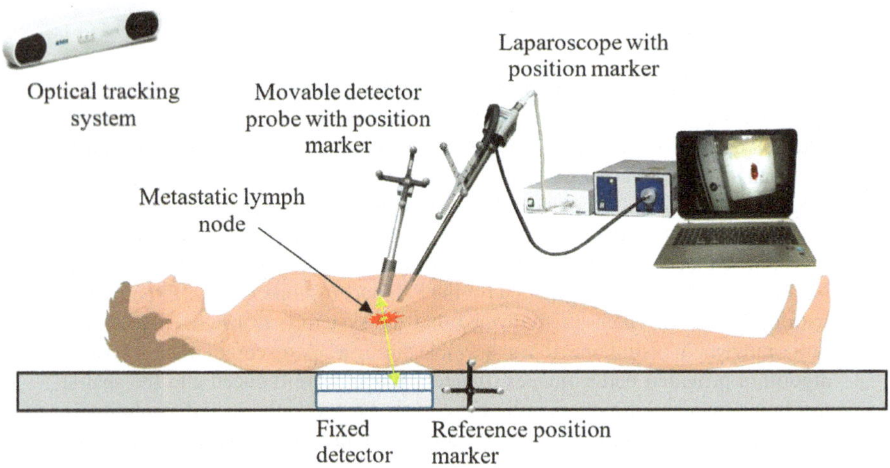

Fig. 1. Overview of surgical setup with the intraoperative PET-laparoscope system.

Since the detector probe is inserted through an incision on the abdomen, its movement is limited. This study presents an evaluation of the system using an abdomen phantom to investigate the performance of the system with realistic probe movements, and a demonstration of how the system will aid the surgeon. Laparoscopic video feed provides an inside view of the abdomen to the surgeon. When the intraoperative PET image is presented separately, it can be difficult for the surgeon to process information coming from different sources. Hence the intraoperative PET image will be overlaid on the laparoscopic video feed so that the surgeon can see the tissues to be removed and decide the positions of cuts easily.

In minimally invasive laparoscopic interventions, surgeons receive visual feedback through the laparoscopic camera, and additional intraoperative information is of great assistance to aid surgical interventions. Information from other imaging methods overlaid on laparoscopic video feed has become the state-of-the-art in computer-assisted surgeries. Vessel structures obtained intraoperatively from an X-ray C-arm have been augmented on the view of a laparoscope to aid laparoscopic liver resection [3]. Preoperative SPECT/CT or intraoperative freehand SPECT scans have been overlaid in the laparoscopic video feed of a robot-integrated fluorescence laparoscope for the surgical management of prostate cancer [4]. This study proposes overlaying intraoperative PET scans on laparoscopic video feed suggesting a new direction in computer-assisted surgical interventions.

In addition to the evaluation of the system using an abdomen phantom, the reconstruction algorithm of the system was improved. Previously images were reconstructed by a filtered back projection (FBP) algorithm. Since iterative reconstruction methods are considered to be more suitable when full angular coverage is not available, a maximum likelihood expectation maximization (MLEM) algorithm was introduced.

2 Methods

2.1 PET Image Reconstruction

Images from FBP algorithm and an MLEM algorithm were compared. An FBP algorithm which provides an analytical solution to the reconstruction problem was discussed for the PET-laparoscope system previously [1]. It is based on the central slice theorem and precise if all angles are available. FBP algorithm is fast which is essential for intra-operative imaging. Iterative reconstruction methods are considered to be more suitable when full angular coverage is not available because it does not have any assumption of full angular rotation. It takes an initial guess for the image and does progressive improvement. Performance of an MLEM algorithm was investigated for the PET-laparoscope system.

When the MLEM algorithm is applied to PET image reconstruction, it leads to the following iterative equation [5]. In the equation, f_j and p_i represent the activity in voxel j and the coincidence count in i^{th} projection respectively. The next estimate of voxel j; $\hat{f}_j^{(n+1)}$ is estimated based on the current estimate $\hat{f}_j^{(n)}$. The elements H_{ij} of the system model H, characterize the imaging system and represent the probability that an emission from voxel j is detected in i^{th} projection.

$$\hat{f}_j^{(n+1)} = \frac{\hat{f}_j^{(n)}}{\sum_{i'} H_{i'j}} \sum_i H_{ij} \frac{p_i}{\sum_k H_{ik} \hat{f}_k^{(n)}} \tag{1}$$

Since the detector probe is randomly moving its positions are not known beforehand. If the elements of the system model are calculated beforehand, interpolations have to be used since the actual positions of the probe during the scan would be different to the positions where calculations were done. Hence for the PET-laparoscope system the elements of the system matrix were calculated on the fly analytically and only detector geometry was considered in the calculation. Ray-driven forward projection and voxel-driven backprojection were used.

2.2 Image Overlay

As shown in Fig. 1, laparoscope and the movable PET detector probe are tracked by an OTS. A reference position marker is attached to the patient's bed. To overlay intraoperative PET images in laparoscopic video feed, both PET image and the laparoscopic image are acquired in one common coordinate system with respect to the reference position marker, so that the PET image can be directly superimposed on the laparoscopic image in real time.

An intraoperative PET image was acquired from coincidence data collected by the fixed detector array and the movable detector probe. The PET image was transformed to the reference coordinates. Metastases with high radioisotope concentration were segmented from the intraoperative PET image by thresholding. An Epiphan frame grabber (DVI2USB 3.0, Epiphan Video, Ontario, Canada) captured the video of laparoscopic camera in real time. 2D image plane of the laparoscopic camera was transformed into the reference coordinate system by the transformation given in Eq. 2, where scope marker is the position tool attached to the laparoscope.

$$T^{Reference}_{Image} = T^{Reference}_{ScopeMarker} T^{ScopeMarker}_{Camera} T^{Camera}_{Image} \tag{2}$$

Image to camera transformation was obtained by the pinhole camera model. To place the image plane with respect to the camera, intrinsic camera parameters were obtained by camera calibration. Camera to scope marker transformation which was the rigid transformation between the laparoscopic camera and the position tool attached to the laparoscope was obtained from hand-eye calibration. Scope marker to reference transformation was provided by the OTS.

Image overlaying was performed using SlicerIGT extension [6] of 3DSlicer open source software which is intended for development of image-guided therapy applications. SlicerIGT works with the Plus toolkit [7]. Plus toolkit communicates with OTS and the Epiphan frame grabber, synchronizes tracking data of the laparoscope and video images and relays those to 3DSlicer in real time. Segmented PET nodes were loaded into the 3DSlicer. PET image was overlaid on the laparoscopic view by transforming both images into the same coordinate frame of the reference marker which was attached to the surgical bed.

2.3 Evaluation Experiments

An abdomen phantom was used to evaluate the PET-laparoscope system by moving the detector probe in a similar manner to that achievable during a laparoscopic surgery. A ^{22}Na radiation source was placed inside an abdomen phantom and a laparoscope and the movable detector probe were inserted through incisions on the phantom as shown in Fig. 2 (a). The ^{22}Na source had less than 0.71 mm diameter and was inside a 3 mm diameter aluminum enclosure. Figure 2 (b) shows the dimensions of the experimental setup.

The ^{22}Na radiation source was scanned for 2 min. The detector probe was moved by hand during the scan. Figure 3 shows a movement of the detector probe recorded during a scan. In the figure the ^{22}Na source position is shown by the black sphere and probe positions and orientations are shown by 100 mm long arrows. Images were first reconstructed with the FBP algorithm. MLEM image was also obtained from the same data. After the intraoperative PET image was reconstructed, the source was segmented from the FBP image and overlaid on laparoscopic video feed.

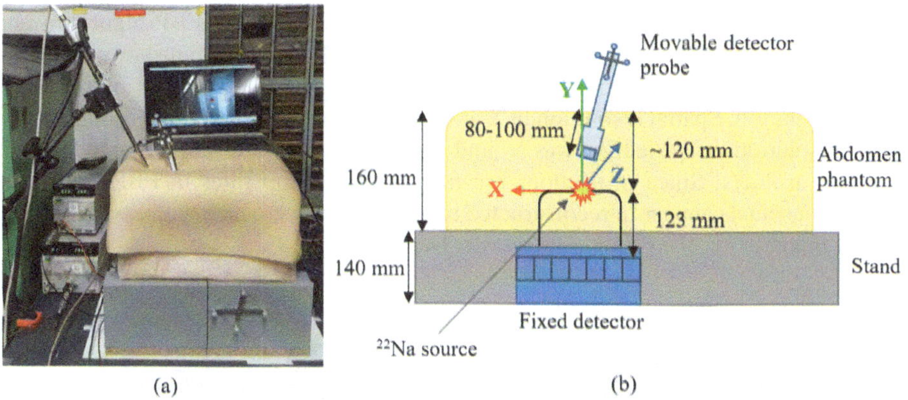

Fig. 2. (a) Laparoscope and the detector probe inserted into the abdomen phantom (b) Dimensions of the experimental setup

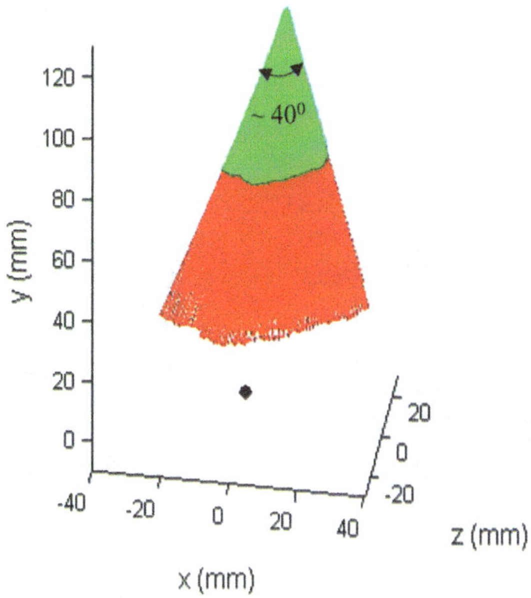

Fig. 3. A movement of the detector probe recorded during a scan

3 Results and Discussion

Full width at half maximum (FWHM) spatial resolution of the reconstructed images in the X, Y, and Z directions were 2.9 ± 0.2, 15.7 ± 1.6 and 3.5 ± 0.3 mm, respectively with the FBP algorithm. When the probe was moved in a manner similar to that achievable during a laparoscopic surgery, the system could reconstruct images well with about 3 mm spatial resolution in X and Z directions. Spatial resolution in Y direction was limited due to low angular coverage.

Figure 4 shows the reconstructed images acquired from the FBP algorithm and MLEM 5th iteration for same coincidence data obtained from the scan shown in Fig. 3. Table 1 lists the spatial resolution of the two images. MLEM algorithm improved the spatial resolution. Spatial resolution in X and Z directions were better than 3 mm and the spatial resolution in Y direction was around 10 mm. MLEM algorithm provided images with less artifacts. Since projections are filtered by a ramp filter in FBP algorithm, negativity artifacts appear in reconstructed images. There were no negativity artifacts in MLEM images.

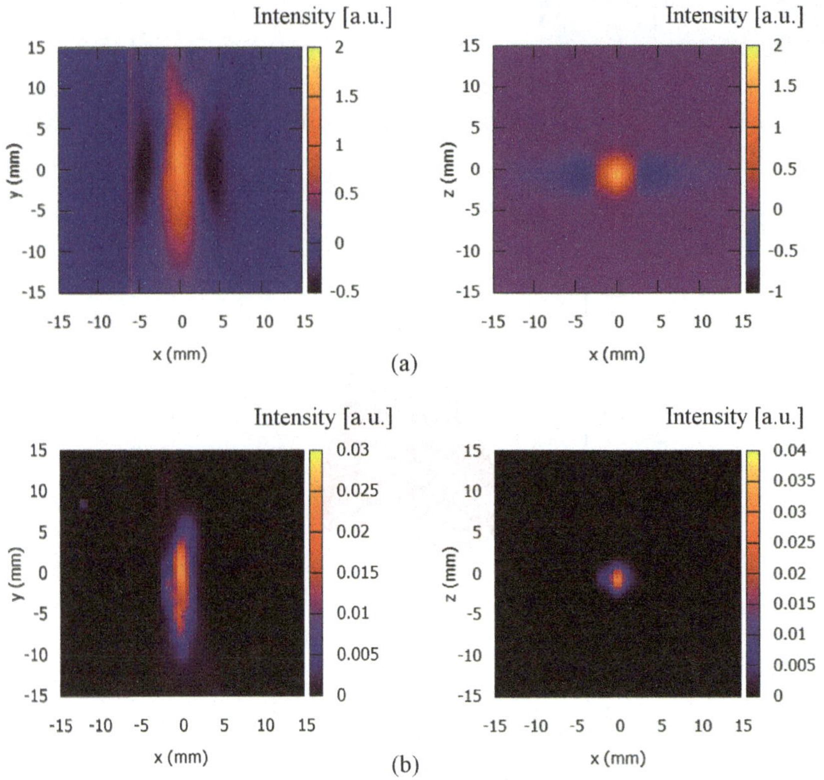

Fig. 4. (a) Image from FBP algorithm (b) Image from MLEM algorithm 5th iteration. For (a) and (b) left images show image slices through XY plane and the right images show image slices through XZ plane.

Two screenshots showing the PET image from the FBP algorithm overlaid on laparoscopic video for two orientations of the laparoscope are given in Fig. 5. Camera distortions were ignored in this work and might be a reason for the small overlaying error.

Table 1. Comparison of spatial resolution of the reconstructed images from FBP and MLEM algorithms.

Reconstruction method	FWHM spatial resolution (mm)		
	X direction	Y direction	Z direction
FBP	2.7	16.0	3.5
MLEM	1.7	10.0	2.9

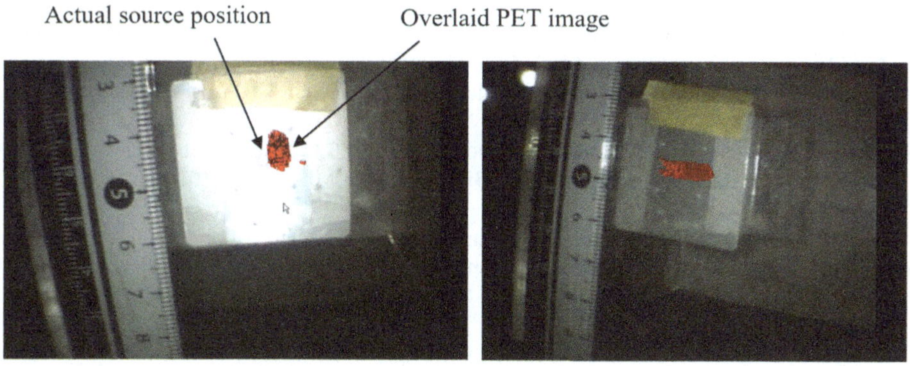

Fig. 5. Two screenshots showing the PET image overlaid on laparoscopic video for two orientations of the laparoscope

4 Conclusion

The system performed well when the probe movement was similar to that achievable during a laparoscopic surgery. Intraoperative PET images were overlaid on the laparoscopic video feed. MLEM algorithm provided better images. Image artifacts were reduced and the spatial resolution in lateral directions were better than 3 mm and the spatial resolution in depth direction was around 10 mm. This 3 mm spatial resolution is better than the typical spatial resolution of clinical PET systems which is about 4–6 mm. Better lateral resolution could be obtained by placing a detector with small crystals closer to the sources.

Acknowledgment. This work was partially supported by a grant for the Creation of Innovation Centers for Advanced Interdisciplinary Research Areas Program: Translational Systems Biology and Medicine Initiative funded by the Ministry of Education, Culture, Sports, Science and Technology, Japan, and Grant in Aid from Delta Electronics, Japan.

Conflict of Interest. Part of the study was supported by a research grant from Delta Electronics, Japan.

References

1. Liyanaarachchi, M.R., Shimazoe, K., Takahashi, H., Kobayashi, E., Sakuma, I.: Simulation study on system configuration refinement for the PET–laparoscope system—an intraoperative tool for surgical navigation in gastric cancer surgery. Biomed. Phys. Eng. Express **5**, 015002 (2018). https://doi.org/10.1088/2057-1976/aae902
2. Liyanaarachchi, M.R., Shimazoe, K., Takahashi, H., Kobayashi, E., Nakagawa, K., Sakuma, I.: Prototype detector for intraoperative PET-laparoscope system with a multi-layer movable detector. Nucl. Instr. Methods Phys. Res. Sect. A Accel. Spectr. Detect. Assoc. Equip. 162788 (2019). https://doi.org/10.1016/j.nima.2019.162788
3. Feuerstein, M., Mussack, T., Heining, S.M., Navab, N.: Registration-free laparoscope augmentation for intra-operative liver resection planning. In: Proceedings of SPIE 6509, Med Imaging 2007 Vis Image-Guided Proced 650915 (2007). https://doi.org/10.1117/12.706766
4. van Oosterom, M.N., Engelen, M.A., van den Berg, N.S., KleinJan, G.H., van der Poel, H.G., Wendler, T., van de Velde, C.J.H., Navab, N., van Leeuwen, F.W.B.: Navigation of a robot-integrated fluorescence laparoscope in preoperative SPECT/CT and intraoperative freehand SPECT imaging data: a phantom study. J. Biomed. Opt. **21**, 086008 (2016). https://doi.org/10.1117/1.jbo.21.8.086008
5. Alessio, A., Kinahan, P.: PET Image Reconstruction. Nucl. Med. **1**, 1–22 (2006)
6. Ungi, T., Lasso, A., Fichtinger, G.: Open-source platforms for navigated image-guided interventions. Med. Image Anal. **33**, 181–186 (2016). https://doi.org/10.1016/j.media.2016.06.011
7. Lasso, A., Heffter, T., Rankin, A., Pinter, C., Ungi, T., Fichtinger, G.: PLUS: open-source toolkit for ultrasound-guided intervention systems. IEEE Trans. Biomed. Eng. **61**, 2527–2537 (2014). https://doi.org/10.1109/TBME.2014.2322864
8. Liyanaarachchi, M.R., Shimazoe, K., Takahashi, H., Nakagawa, K., Kobayashi, E., Sakuma, I.: Development and evaluation of a prototype detector for an intraoperative laparoscopic coincidence imaging system with PET tracers. Int. J. Comput. Assist. Radiol. Surg. **16**(1), 29–39 (2021). https://doi.org/10.1007/s11548-020-02282-0

Improvement of Force Sensor Accuracy Using Redundant Outputs for Biomedical Applications

Masaru Higa(✉), Ryutaro Hayami, Masahiro Ono, Shizuki Nakamura, and Shiori Matsuda

University of Hyogo, Himeji, Hyogo 671-2280, Japan
higa@eng.u-hyogo.ac.jp

Abstract. While strain gauges are commonly used in a force sensor, we used multiple digital Hall-Effect magnetic sensors for developing a three-dimensional force sensor. We are aiming to develop a round shape three-dimensional force sensor. In order to get three-dimensional unknown force components, three independent outputs from the sensor are at least necessary. Four or more outputs however, would not be useless but redundant data sets, which could be useful to improve sensor accuracy. We calculated the unknown three-dimensional force values using nine outputs or less and estimated how many redundant outputs we should use to get good accuracy. As a result, when three outputs were used, the sensor accuracy was ranging from 8.2 to 15.7% depending on which outputs were used. When six outputs were used, the sensor accuracy was ranging from 5.5% to 5.6%. When nine outputs were used, the sensor accuracy was 4.1%. Although the best accuracy was acquired when we used all nine outputs, improvements were minimal when we used six or more outputs.

Keywords: Force sensor · Magnetic sensor · Design

1 Introduction

In biomechanical applications, most systems require force and displacement, especially three-dimensional measurements. For example, orthopedic surgeons would take benefits by use of an intra-articular load sensor for total joint arthroplasty [1]. For robotic applications, tactile sensors are used for gentle handling of an object [2]. A flat force plate is commonly used to take ground reaction forces and positions in gait analyses or in the field of sports biomechanics. Most of those force platforms use strain gauges for determining forces, moments and positions exerted on the platform along x, y and z axes. Although a strain gauge itself could be small and thin, it is an analog element and one or two dimensional in plane measurements. If we would develop a three-dimensional force sensor using strain gauges, exquisite deforming body structure that need some rooms would be required. While we can develop thin and small force sensor using capacitive sensors [3]. The capacitive force sensor however, is too sensitive to external forces, that means noise-signal ratios are not good enough to apply biomechanical field. The idea of using optical sensors have also been introduced in the biomechanical field. There

Y. Shiraishi et al. (Eds.): APCMBE 2020, IFMBE Proceedings 82, pp. 319–324, 2021.
https://doi.org/10.1007/978-3-030-66169-4_39

are commercially available products as well [4]. The sensing platform of them could be small, but light sources and surrounding electric circuits are required. Hence, it would be difficult to develop a small sensor structure using optical sensors. Magnetic sensors are recently used in force plate [5] or tactile sensors [2]. We use commercially available one of the small digital three-dimensional Hall-Effect sensors (MLX90393, Melexis, Belgium) to develop a three-dimensional round shaped force sensor. In order to get better accuracy, we used redundant number of the magnetic sensors inside the whole sensor structure. And we used all output to get better accuracy. To the best of the author's knowledge, it has not been investigated that how to utilize redundant sensor outputs from Hall-Effect sensors for the accuracy. Hence, the aim of this study is to investigate relationships between the number of redundant outputs and sensor accuracy.

2 Methods

2.1 Structure

We have developed a sphere-shaped three-dimensional force sensor (Fig. 1). In this structure, four Hall-Effect magnetic sensors (MLX90393) are implemented including one reference for geomagnetic cancellation. The three sensors are paired with each ferrite magnet (5 mm of diameter and 1.5 mm of thickness with 86 mT of surface magnetic flux density). The magnetic sensor outputs three-dimensional strength of magnetic flux density on its surface. There are a lot of assets of the Hall-Effect sensor. The first one is it is small enough to develop a s small force sensor (3 mm square and 1 mm of thickness). We were able to implement four of the Hall-Effect sensors with an electronic board, which has surrounding electric circuits, and enough power of batteries in the sphere structure (s$\phi = 32$ mm). The second asset is this sensor includes 16-bit AD converter. So, we do not need another external AD converter or amplifier as well. Overall electronic circuit was able to be simplified using I2C digital signals cables only. There are no analog signal cables, which could easily take noise (Fig. 2). The sphere-shaped structure was divided into two pieces and all Hall-Effect sensors are put on one side with random directions and all magnets were put on the other side of the structure separated by a deforming body. The Hall-Effect sensor has strong enough in resolution to measure geomagnetism. Since one reference sensor was put to detect the geomagnetism and subtract the values on it from the other sensor readings to avoid geomagnetism errors.

2.2 Sensor Calibration

Before using the developed structure as a force sensor, we needed calibrations to calculate a constant matrix, which is a kind of a spring constant of the structure. Based on a simple Hooke's law with linear correlation assumption, an unknown three-dimensional force vector exerted on the structure can be calculated as follows [6]:

$$\boldsymbol{F} = \boldsymbol{T}\boldsymbol{B} \tag{1}$$

where $\boldsymbol{F}$ is a (3×1) force vector exerted on the devise, $\boldsymbol{B}$ is a (9×1) outputs vector from the magnetic sensors and $\boldsymbol{T}$ is a calibration constant matrix (3×9). The purpose of this calibration is to get the constant matrix $\boldsymbol{T}$, which is unique value at each structure, and estimate error values of the structure. In this calibration, we applied known force vectors ($\boldsymbol{F_{app}}$) by varying both directions and values. The cone angle (θ) were varied from 0 to 45° in 15 degrees of interval, polar angle (ϕ) were varied from 0 to 315° in 45 degrees of interval, force values were varied from 0 to 100 N. The applied force values were decided based on ASTM E-4 standard, which describes calibration methods of a load sensor [7]. Angle definitions and a picture of this calibration are shown in Fig. 3. Since ith force vector ($\boldsymbol{F_{app,\,i}}$) has a corresponding output vector ($\boldsymbol{B}_i$), we made the applied force and outputs matrices: $\boldsymbol{F_{app}}$ (3 × n) and $\boldsymbol{B}$(9 × n). Where n is the number of applied force vectors. Using both $\boldsymbol{F_{app}}$ and $\boldsymbol{B}$ matrices, $\boldsymbol{T}$ was calculated by multiplying an inverse matrix of $\boldsymbol{B}$ ($\boldsymbol{B^{-1}}$) from the right side in the Eq. (1). Since $\boldsymbol{B}$ was not a square matrix, a pseudoinverse matrix of $\boldsymbol{B}$ was used instead. The number of equations (outputs) is nine and the number of unknowns is three, that means this is an overdetermine problem. The most likelihood values were calculated using least-square methods. Same to the situations when four or more outputs were used. For example, when six outputs were used, $\boldsymbol{T}$ was (3 × 6). When we used only three outputs, $\boldsymbol{T}$ was (3 × 3) matrix, and this is so-called determined problem. Once we got the matrix $\boldsymbol{T}$ at each condition, we calculated the applied force vectors ($\boldsymbol{F_{cal}}$) again using $\boldsymbol{B}$ and $\boldsymbol{T}$ matrices in the same Eq. (1). Then we estimated error values by differences between $\boldsymbol{F_{app}}$ and $\boldsymbol{F_{cal}}$ as follows [6]:

$$error(\%) = 100 \times \sum_i abs\left[\left(F_{i-calc} - F_{i-app}\right)/F_{i-max}\right]/n \qquad (2)$$

where F_{max} is the maximum value of the applied force at each direction. The above error values in percentiles were calculated by varying the number of outputs used.

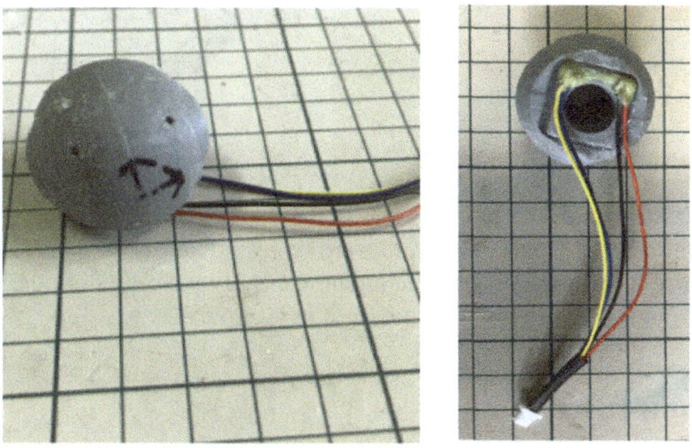

Fig. 1. These figures show outlook of developed three-dimensional force sensor. Although every component including batteries could be packed inside the structure, the cables are connected for testing and use of rechargeable batteries. The backdrop has 1 cm square of grid.

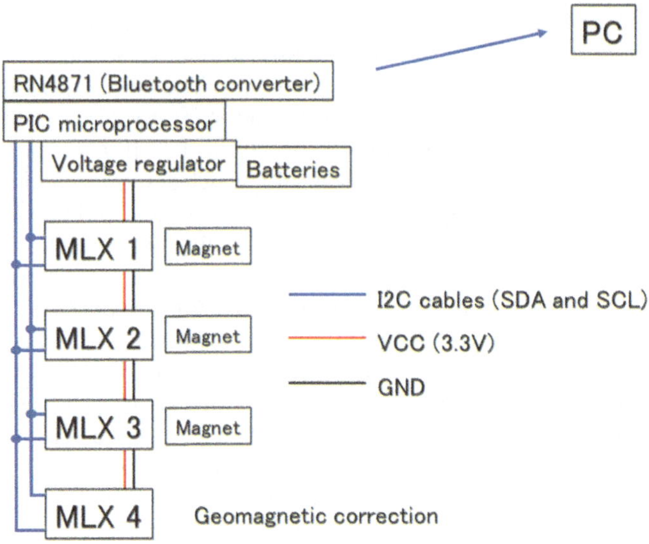

Fig. 2. A schematic illustration of electronic circuit is shown. Only important elements are expressed, while surrounding small elements such as resistances or capacitors are omitted. MLXs are representing each magnetic sensor paired with magnets. The 4th MLX is for geomagnetism correction.

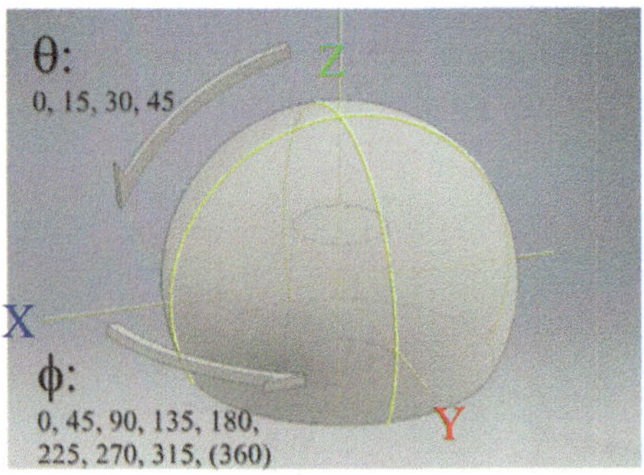

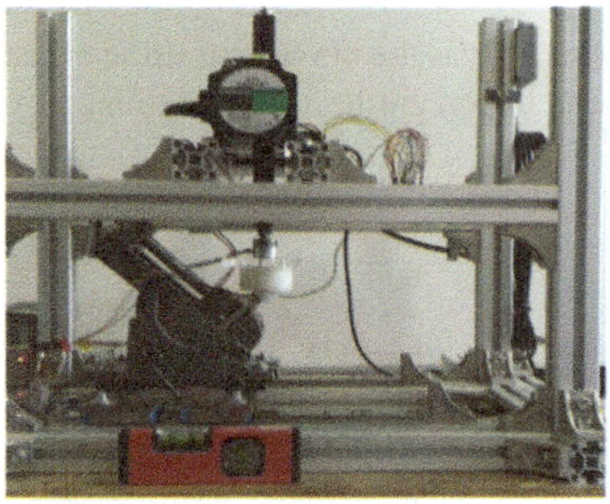

Fig. 3. The angle definitions and applied forces are expressed (left). The angles were varied using an angle vice and applied force values were monitored by a load cell. Since the angle vice is on two-dimensional linear sliders, they were not fixed in plane (right).

3 Results and Discussion

After the calibration procedures have been completed, the constant matrices $\boldsymbol{T}$ and error values were successfully calculated regardless of the number of the outputs used. When three outputs were used to calculate the exerted forces, the error values were 10.2%, 8.2% and 15.7%, depending on which outputs were used. When six outputs were used, the error values were 5.5%, 5.6% and 5.5% depending on which outputs were used. When all nine outputs were used, the error was 4.1%. These error values and relations with output signals were summarized in Table 1. Interestingly, when six outputs were used, three types of combinations showed almost same error values even though each three group had quite different error values. We did not treat all signals separately but consecutive three were bundled, because they were from one element. Although the best accuracy was acquired when we used all nine outputs, improvements from six to nine were minimal.

The number of outputs should be decided by taking into account of not only the accuracy but also manufacturing costs and measuring times. The sensor structure developed in this study was aimed to measure three-dimensional forces around 100 N. Although improving error from approximately 5% to 4% is good, simplifying the whole structure and data logging program by removing one Hall-Effect sensor (cutting three signals) would be better in our case. Our error values when six outputs were used might be accurate enough compared to the other studies such as a tactile sensor integrated by Hall-Effect sensors [2] or a joint force sensor integrated by strain gauges [8].

Table 1. Relationships between calculated error values and outputs used. The consecutive three outputs were bundled and used together all the time.

Output signals	Error values (%)				
	Used three	Used six			Used nine
1, 2, 3	10.2	5.6		5.5	
4, 5, 6	8.2		5.6		4.1
7, 8, 9	15.7				

References

1. Higa, M., Chang, C., Roche, C., Struk, A.M., Farmer, K.W., Wright, T.W., Banks, S.A.: Instrumented trial prosthesis for intraoperative measurements of joint reaction forces during reverse total shoulder arthroplasty. Sensors Mater. **30**(2), 1989–1996 (2018)
2. Tomo, T.P., Somlor, S., Schmitz, A., Jamone, L., Huang, W., Kristanto, H., Sugano, S.: Design and characterization of a three-axis hall effect-based soft skin sensor. Sensors **16**(4) (2016)
3. Paydar, O.H., Wottawa, C.R., Fan, R.E., Dutson, E.P., Grundfest, W.S., Culjat, M.O., Candler, R.N.: Fabrication of a thin-film capacitive force sensor array for tactile feedback in robotic surgery. In: Conference Proceedings: Annual International Conference of the IEEE Engineering in Medicine and Biology Society 2012, pp. 2355–2358. IEEE Engineering in Medicine and Biology Society (2012)
4. Liu, C., Correia, R., Ballaji, H.K., Korposh, S., Hayes-Gill, B.R., Morgan, S.P.: Optical fibre-based pulse oximetry sensor with contact force detection. Sensors **18**(11) (2018)
5. Cook, N.H., Carignan, F.J.: Multi-axis force platform (1998)

6. Bergmann, G., Graichen, F., Rohlmann, A., Westerhoff, P., Heinlein, B., Bender, A., Ehrig, R.: Design and calibration of load sensing orthopaedic implants. J. Biomech. Eng. **130** (2), 021009 (2008)
7. Standard Practices for Force Verification of Testing Machines, ASTM International, E4, pp. 13–22 (2008)
8. Heinlein, B., Graichen, F., Bender, A., Rohlmann, A., Bergmann, G.: Design, calibration and pre-clinical testing of an instrumented tibial tray. J. Biomech. **40**(Suppl 1), S4-10 (2007)

Combining Surface-Based Electromyography and Motion Sensing for Detection of Human Falling

Lawrence Ibarrientos[1(✉)], Harold Jay Bolingot[2], Lea Cristina Macaraig[1], Nathaniel Libatique[1], and Gregory Tangonan[1]

[1] Ateneo Innovation Center, Ateneo de Manila University, 1108 Quezon City, Philippines
lawibarrientos7@gmail.com

[2] Kyushu Institute of Technology, Wakamatsu Ward, Kitakyushu City, Fukuoka 808-0196, Japan

Abstract. The study of falling, especially for the elderly, is a very critical problem in geriatric care and medicine. This study investigated the potential role of a surface-based electromyography (sEMG) and inertial measurement unit (IMU) sensor package for analyzing and detecting humans falling. Previous studies have only relied on motion sensors only, such as accelerometers and gyroscopes, detecting falling, which completely ignores the muscular force reaction of extremities during the precise moment before a person actually falls. This experiments conducted in this study show promising results of combining the sEMG and accelerometer signals to describe the muscular activity generated as the reflex reaction of the subjects in their arms as they grab onto something to prevent the fall. Further simulations of falling were conducted at a foam pit, specifically deliberate falling and surprised falling. Similar findings were obtained in this scenario where the subjects did not have anything to grab onto upon falling, which cancan be attributed to the fact that the subjects feel the same sensation of reflex and panic whether deliberately falling or surprised falling.

Keywords: Myo Armband · Motion-based · Geriatric care · Fall detection

1 Introduction

For many people, falling is a fatal incident in their daily lives, such as when slipping inside the bathroom, or when climbing the stairs. This happens most especially to elderly people who have mobility problems, especially on their lower extremities. Usually, these accidents cause hip injuries, and a study by Stevens et al. [1], shows that hip injuries and fractures are the most common type of injury among elder adults in the US. Due to the frequency of hip injuries caused by slipping and falling, hip-replacement and hip-treatment are about $10–20B a year industry.

A study by Morita, Yasumoto, Taki, Fujimoto, Suwa and Arakawa [2], entitled Beacon-Based Time-Spatial Recognition toward Automatic Daily Care Reporting for Nursing Homes, uses a more digital method of tracking geriatric patients around the

Y. Shiraishi et al. (Eds.): APCMBE 2020, IFMBE Proceedings 82, pp. 325–331, 2021.
https://doi.org/10.1007/978-3-030-66169-4_40

nursing home. However, this study only uses beacons as a tracking device for the nursing patients.

Wearable sensors, especially smartwatches, are prevalent today and companies implement motion sensors such as accelerometers and gyroscopes to detect physical activities of the user. These sensors are also used for fall detection; however, accelerometers and gyroscopes are susceptible to false positives [3] especially when the user's motion is similar to that of falling and the device interprets this as an accident but in reality, no falling happened to the user. A study by Ngeo, Tamei, and Shibata [4] explores the presence of electromechanical delay in sEMG signatures. Electromechanical delay occurs when sEMG signatures increases its peaks before actual movement. In this study, electromechanical delay is measured and explored in simulated falling experiments. Specifically, the objective of this study is to use the Myo armband to measure and analyze the EMG and acceleration of participants when simulating falling.

2 Methodology

This study used the Myo Armband to collect the EMG and accelerometer signals in the forearm of subjects. Developed by Thalmic Labs, the Myo Armband has eight sensors that detect the sEMG, acceleration, and the position of the arm of the person wearing it. The Myo Armband's EMG signal is an ADC Muscle Activation Unit with a range of −128 to 127 [5]. In Bolingot's study [6], they used the Myo Armband (Fig. 1) to determine the effect of fatigue on laparoscopic training of surgeons. Their results show that the amplitude of the sEMG Muscle Activation Unit have decreased when applied with fatigue, compared to non-fatigued participants where the amplitudes of the graphs was larger.

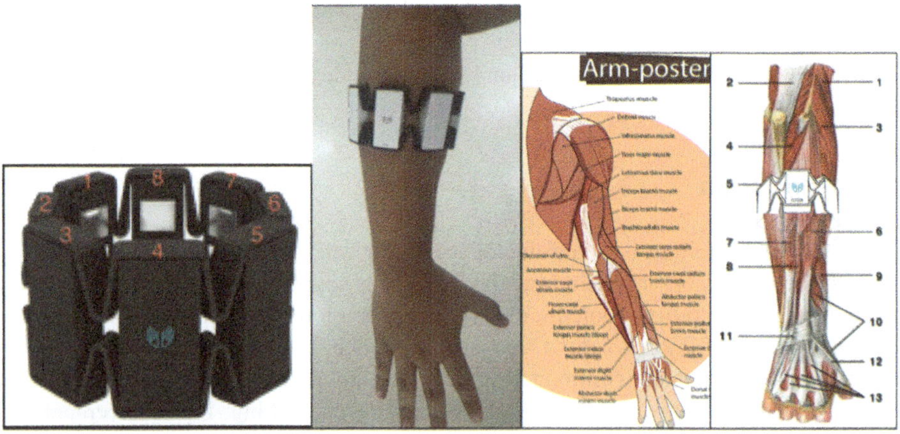

Fig. 1. (Left) The Myo Armband, developed by Thalmic Labs, (Middle) The Myo Armband worn on the forearm, (Right) The muscle groups the Myo Armband is detecting and the corresponding placement of the Myo Armband on the forearm

In accordance with the research objective, this study employed the Myo Armband to record EMG and accelerometer signals during different scenarios of falling. The proponets asked for participants' consent to do two methods of falling simulation. Note that the Myo Armband was always worn by the participants all throughout the data collection. The first method was using an office chair where unbeknownst to the participant, the lever of the chair was pulled such that the chair subsequently and immediately lowers down and surprises them (Fig. 2). This simulates a reflex reaction of the arm (specifically on the forearm, where the Myo Armband is worn). For the control, the participants were asked to do a similar motion of raising their arm (Fig. 3). This was done to compare the sEMG signatures for a surprise reaction compared to an intentional motion.

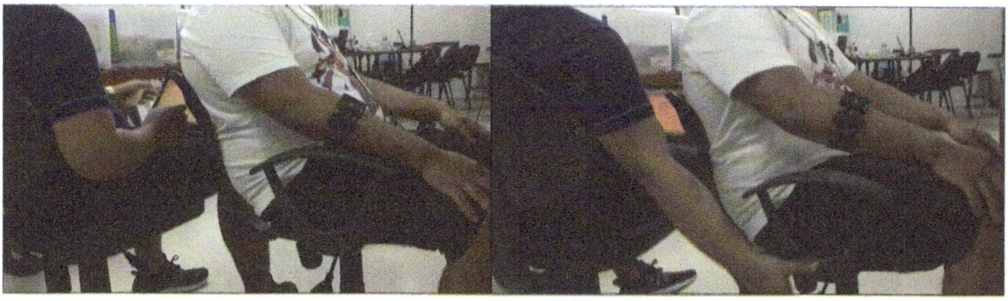

Fig. 2. (Left) Participant unaware of the time of the pulling of the lever (Right) Surprised participant showing an arm reflex motion after the lever was pulled

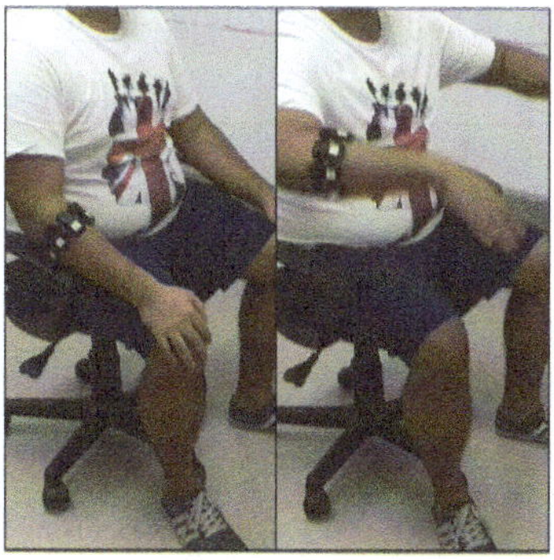

Fig. 3. Participant intentionally swinging their arm

The other method was done in a safe foam pit, where an emergency response team was present during the duration of the data collection (Fig. 4). The control data for this method is for the participants to fall on their own into the foam pit.

For the last method, the participants were pushed into the foam pit and were unaware of the timing of the push. These two methods are used to compare the difference between a surprise fall and an intentional fall.

This study collected all eight (8) sEMG Muscle Activation Unit from the eight electrodes of the Myo Armband; the x-, y- and z-acceleration signals were also collected.

Fig. 4. (From left to right) Participant intentionally falling into the foam pit; Participant pushed into the foam pit

This study used the C++ Program developed by Thalmic Labs to collect the signals detected by the Myo Armband. MatLab was then used for the illustration of the graphs and the calculation of the RMS.

3 Results

Using MatLab, the EMG signals for the participant were offset to compare the amplitude of intentional movement versus surprised falling simulation (Fig. 5 and 6). MatLab was also used to calculate the RMS values of each graphs. The RMS of the EMG signals from the intentional trials and the surprised falling simulation trials are shown in Fig. 7.

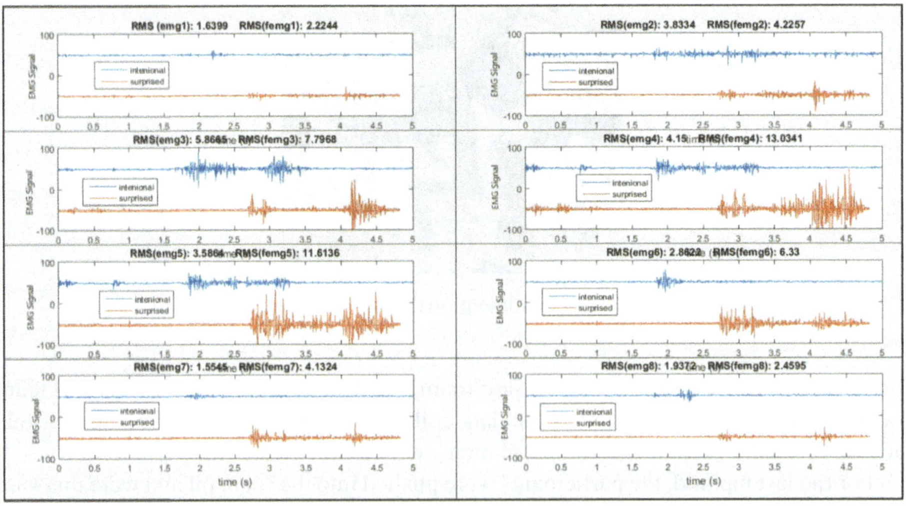

Fig. 5. Offset EMG graphs of one of the participants' intentional arm swings (blue) and surprised chair falling simulation (orange) with corresponding RMS values for both.

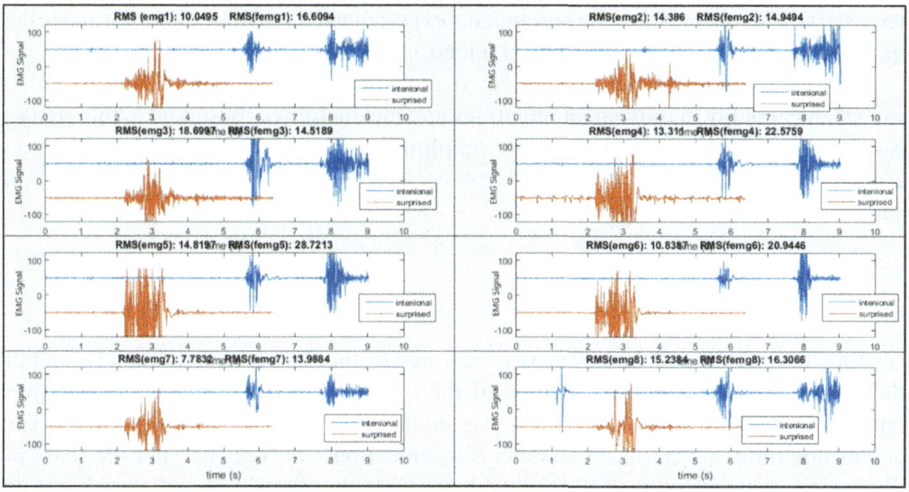

Fig. 6. Offset EMG graphs of one of the participants' intentional foam pit fall (blue) and surprised foam pit fall (orange) with corresponding RMS values for both.

It can be observed that the values of the RMS of the EMG signals in the surprised falling are larger than that of the RMS of the EMG signals in the intentional falling. This presents a differentiating factor between the two motions, as one is an intentional movement while the other is a reflex reaction from the forearm. However, the difference of RMS values in the foam pit falling trials were smaller than that of the office chair trials. Note that for the foam pit falling trials, both intentional falling and surprised push falling involved actual falling of the participants. Thus, even the intentional fall yielded

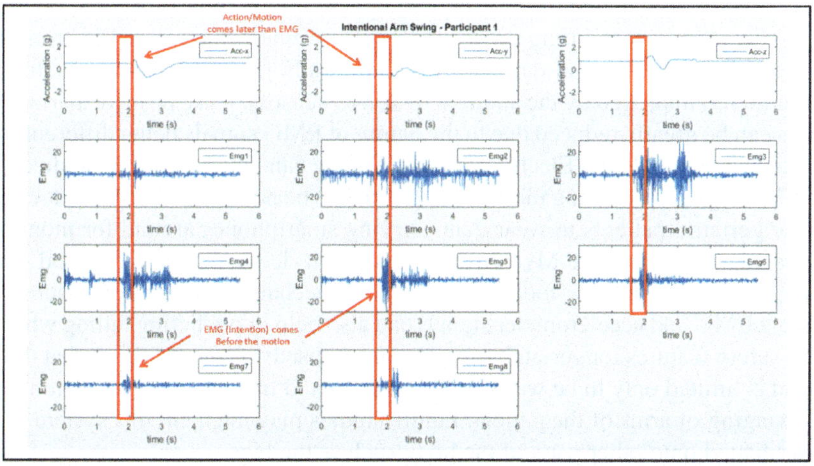

Fig. 7. Electromechanical delay is detected as the initial muscular activity right before a movement has occurred.

increased sEMG peaks, since the participants experienced the falling sensation and reflex reaction were also triggered despite the intention.

As mentioned earlier, the study of Ngeo, et al. [4] highlighted the presence of electromechanical delay in movement intentions and this factor was observed in this study.). As seen in Fig. 7, increased peaks in the amplitude on the EMG signals already started before any movement was done, which can be seen from the x-, y-, and z-acceleration signals.

4 Discussion

Using the Myo Armband, participants were asked to simulate falling through two methods. The results from the first method of the office chair falling simulations showed that the amplitude of the EMG signals was smaller than the surprise falling trials. This denotes that reflex reactions are distinct from intentional movements. For the foam pit falling trials, the amplitude of EMG signals for both intentional falls and surprised falls showed increased magnitude. This can be attributed to the fact that for both trials, the participants experienced actual falling. This thus confirms that EMG signal acquisition can be used to characterize a reflex reaction of a person falling, and from there EMG signals can be used for wearable devices and nursing homes. As seen in Fig. 7, electromechanical delay was confirmed when the graphs for the SEMG were aligned to the graphs of the acceleration, which signifies that the body already had the intention to move before actual movement.

5 Conclusion

This study was able to collect data from the Myo Armband and showed that there is a difference between intentional movement and muscle reflex reaction movement. This study was also able to show that during a reflex reaction movement, there is an Electromechanical Delay Electromechanical delay [4]. Luque's study [3] and Yasumoto's study [2], can be furthered improved by the addition of sEMG sensors. False positive alarms of fall detectors can be greatly reduced due to the nature of EMG signals being different from a normal movement and then Electromechanical delay can also be used to predict falling due to EMG signals preceding movement. It can also be noted that wearable devices are easier for geriatric patients to wear than carrying smartphones around for monitoring. The advantages of using the Myo Armband is that it decreases the likelihood of false positive alarms, it is an easy tool to wear. Electromechanical delay as measured from combining EMG and accelerometer signals can also help in predicting falling which can improve safety features for geriatric care. However, disadvantages of this is that the Myo Armband is limited only to be worn on the forearm and from the reflex reaction related to the swinging of arms of the patient. Future studies may augment this system to have sensors situated onto other parts of the body such as the legs.

Conflict of Interest. The authors declare that they have no conflict of interests.

References

1. Stevens, J., Corso, P., Finkelstein, E., Miller, T.: The costs of fatal and non-fatal falls among older adults. Inj. Prev. **12**, 290–295 (2006)
2. Morita, T., Taki, K., Fujimoto, M., Suwa, H., Arakawa, Y., Yasumoto, K.: Beacon-based time-spatial recognition toward automatic daily care reporting for nursing homes. J. Sens. **2018**, 1–5 (2018)
3. Luque, R., Casilari, E., Morón, M., Redondo, G.L.: Comparison and characterization of android-based fall detection systems. Sensors **14**, 18543–18574 (2014)
4. Ngeo, J., Tamei, T., Shibata, T.: Continuous and simultaneous estimation of finger kinematics using inputs from an EMG-to-muscle activation model. J. Neuroeng. Rehabil. **11**, 122 (2014)
5. Visconti, P., Gaetani, F., Zappatore, G., Primiceri, P.: Technical features and functionalities of Myo Armband: an overview on related literature and advanced applications of myoelectric armbands mainly focused on arm prostheses. Int. J. Smart Sens. Intell. Syst. **11**, 1–25 (2018)
6. Bolingot, H., Tangonan, G., Abrajano, G., Libatique, N., Salvana, M., Reyes, D., Ngeo, J., Shibata, T.: Simulated laparoscopic training and measurement systems based on a low-cost sEMG and IMU armband. In: 39th IEEE Engineering in Medicine and Biology Conference, Jeju, South Korea (2017)

Wireless ECG Electrode to Improve Tele-cardiac Monitoring Usability

Norma Hermawan[1,2](✉), Achmad Arifin[1], Fauzan Arrofiqi[1,2], Mohammad Nuh[1], and Yoshifumi Saijo[2]

[1] Institut Teknologi Sepuluh Nopember, Keputih, Sukolilo, Surabaya 60111, Indonesia
norma.hermawan@bme.its.ac.id

[2] Tohoku University, 6-6-12, Aramaki Aza Aoba Aoba-ku, Sendai, Miyagi 980-8579, Japan

Abstract. This paper describes wireless biotelemetry system implementation in order to improve the flexibility of existing tele-cardiac monitoring system. The proposed method is an integration of single-lead ECG front end circuit and TCP/IP based wireless communication System on Chip (SoC) running 3 parallel tasks including modified Amplitude Zone Time-Epoch Coding (AZTEC) data reduction algorithm. The electrode wireless node is designed to minimize the size and power consumption meanwhile maintaining acceptable performance of the instrument. On the other hand, an ECG server is implemented as a TCP/IP data receiving unit featuring data recovery algorithm with store and forward capability. This feature has enabled integration with existing tele-cardiac monitoring system, in addition to raw ECG data logging for manual analysis. The experiment has demonstrated the implementation feasibility of ECG acquisition system with commercial ECG front-end IC and an on-chip ADC. A moderate performance and acceptable error have been observed. The proposed system provides a potential improvement in telemedicine service.

Keywords: ECG · Wi-Fi · Aztec · Biotelemetry · Cardiac analysis

1 Introduction

Heart disease has been one of world major cause of death in the world. Methods that have been developed to accurately predict and diagnose heart disease mostly rely on electrocardiogram (ECG) reading and clinical data simultaneously [1]. This prediction technique requires continuous monitoring of the patient's ECG that lead to developments of various wearable monitoring devices [2]. A tele-cardiac monitoring system has been developed to address geographic problem in treating heart disease in Indonesia [3]. In this system, a remote instrument acquires ECG signal from a subject, extracts the cardiac parameter and sends it to the receiver over HF or VHF radio communication. The user interface in the receiver displays cardiac parameters for cardiologist analysis. The flexibility of tele-cardiac monitoring has been enhanced by adding Short Message Service (SMS) reporting feature that sends monitoring summaries to the cellular phone by periodic or request base SMS [4].

Y. Shiraishi et al. (Eds.): APCMBE 2020, IFMBE Proceedings 82, pp. 332–339, 2021.
https://doi.org/10.1007/978-3-030-66169-4_41

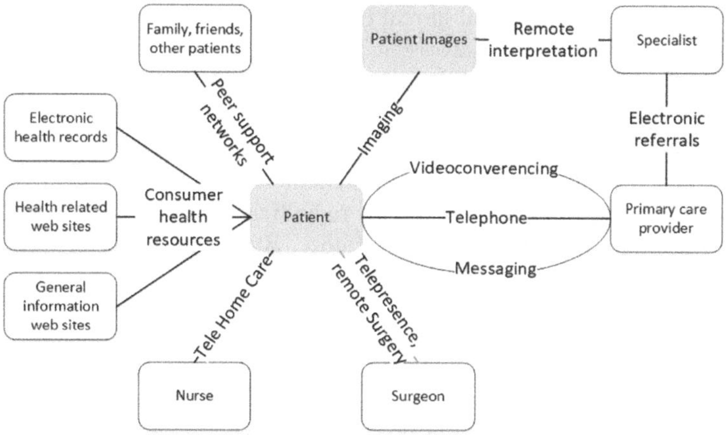

Fig. 1. In the data interconnection among telemedicine entities, tele-cardiac monitoring system gives a contribution in tele-home-care service.

1.1 Tele-cardiac Monitoring in Telemedicine

Figure 1 illustrates the usage of electronic communications to bridge patients to various health resources in telemedicine scheme [7]. The tele-cardiac monitoring and reporting system exists in the telemedicine as one of tele-homecare system implementation. It provides a method to connect patient and health providers over amateur radio channel to extend the coverage of established communication network as well as adding redundancy and thus improve reliability.

In tele-cardiac monitoring systems, ECG signals are processed on the remote instrument before being sent to the receiving device for display (Fig. 2). ECG may be transmitted in the form of raw signal or the temporal parameter extraction results, depending on the quality of communication channel and the desired data fidelity. A data extraction process on the remote instrument plays an important role in limiting the bandwidth requirements. The AX.25 amateur packet radio protocol is used. To improve flexibility, the monitoring summary is forwarded to cellular phone through SMS, since it is a basic feature which is supported by all cellular operator networks. However, this tele-cardiac monitoring and reporting system architecture requires cabling to link the ECG electrode and instrumentation amplifier circuit. Even though such system is capable to operate

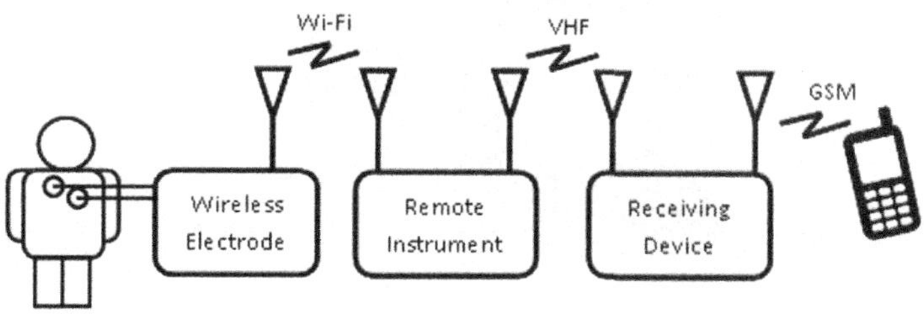

Fig. 2. Tele-cardiac monitoring and reporting scheme

remotely, its usage is not practical in real cases since instrumentation cabling inhibits patients from normal daily activities.

1.2 Wi-Fi Based Wireless ECG Electrode

In order to make tele-cardiac monitoring system more practical for daily use, the ECG acquisition instrument is separated from the radio transmitter module. Standard short-range data transmission methods such as Zigbee, Bluetooth, Wi-Fi or other wireless communication protocols are common. Furthermore, commercial IoT device such as Wi-Fi module and the ECG instrumentation amplifier front-end microchip are available to support rapid implementation. Different data reduction techniques can be employed namely Huffman algorithm, Amplitude Zone Time Epoch Coding (AZTEC) [5], the Turning Point (TP) algorithm, Coordinate Reduction Time Encoding System (CORTES) algorithm [6] and FAN algorithm. In this work, an implementation of wireless ECG electrode using commercial embedded Wi-Fi module and ECG front-end circuit with modified AZTEC compression technique is described.

2 Methods

2.1 Low Cost Wi-Fi Microchip and Single Lead ECG Front End Integration

The proposed wireless ECG electrode hardware is based on integration of a single lead ECG front-end Analog Devices AD8232 and low-cost Wi-Fi microchip Espressif® ESP8266 as shown on Fig. 3. Both test devices were chosen for cost effectiveness and availability reason. The Analog Devices AD8232 circuit is supplied with 3 V from a CR2032 battery. It is configured as an instrumentation amplifier cascaded with 0.5 Hz two-pole high-pass filter and a 40 Hz low-pass filter on the subsequent stage. The instrumentation amplifier provides a gain of 100, while filter stages give 11, resulting in total gain of 1100. An opensource evaluation circuit of ESP8266 is used for signal acquisition as well as signal processing and transmitting device.

Fig. 3. Wi-Fi ECG biotelemetry hardware

On the other side, a commercial PC with built-in Wi-Fi feature is used as a data receiver. A GUI based software is developed to receive and decode the compressed TCP packets. This receiver is designed to store the received data to local drive and forward packet to the subsequent tele-cardiac monitoring channel.

2.2 Software Implementation

A specific purpose firmware is developed for the ESP8266 Microcontroller Unit (MCU). The lack of sufficient documentation on underlying hardware architecture and memory map has restricted the development to rely on provided Non-OS type Software Development Kit (SDK). The compiled binary file was uploaded by Non-OTA method with a firmware download tool. RTOS SDK was avoided since it takes noticeable amount of computation resource for a relatively lightweight program.

The firmware implementation is split into three different tasks: sampling task, main task and transmit task. The sampling task is responsible for signal acquisition at 5 ms sampling period reading the analog input of ESP8266. At the same time, main task ticks at 1000 ms running AZTEC data reduction algorithm and save the result to send buffer. Sender task pulls data from send buffer and put it to the TCP stack every 2000 ms. Eventually, the data in TCP stack are transmitted to the receiving computer over Wi-Fi by network socket.

2.3 Amplitude Zone Time Epoch Coding (AZTEC) Scheme

AZTEC algorithm works by continuously examine the difference between a maximum value (v_{max}) and a minimum value (v_{min}) of ECG samples $\{v_i\}_0^m$. v_{max} is updated when a higher value occurs in the signal, as well as v_{min} which is updated on lower value occurrence. When the difference between v_{max} and v_{min} is greater than a threshold K, the set is considered as an AZTEC *line* L which is specified by its values,

$$v(L) = \frac{v_{max} + v_{min}}{2} \tag{1}$$

and duration

$$\tau(L) = n \tag{2}$$

The succeeding samples are evaluated to obtain the AZTEC line series $\{L_i\}$. The ith line L_i is stored as a *plateau* if its duration is greater than a threshold T,

$$\tau(L_i) > T \tag{3}$$

On the other hand, an *extremum* is characterized by gradient inversion of subsequent line values.

$$\{v(L_{i+1}) - v(L_i)\}\{v(L_i) - v(L_{i-1})\} < 0 \tag{4}$$

Any group of lines $\{L_i\}_1^{m-1}$ that does not contain any plateau of extremum is called as *slope* (S) with a duration of

$$\tau = \sum_{i=1}^{m-1} \tau(L_i) \tag{5}$$

This sequence of plateau, extremum and slope is a representation of a complete AZTEC transformed ECG [5].

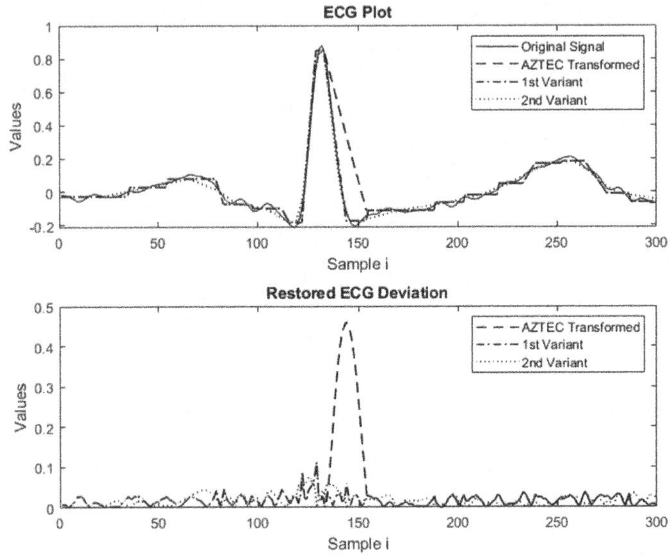

Fig. 4. Graphic plot of compressed ECG and its absolute deviation against original signal

2.4 Algorithm Modification

The decompression process of AZTEC transformed ECG involves plateau and slope restoration into flat or sloping graph. Experiments with original AZTEC algorithm have shown a large deviation of restored ECG near QRS events (Fig. 4). This deviation is mainly caused by algorithm failure to determine correct timing to produce plateau or extremum. Algorithm modifications are carried out to improve this shortcoming.

The first modification variant is changing extremum detector into a turning detector, so that Eq. (4) can be reformulated to

$$\{v(L_{i+1}) - v(L_i)\}\{v(L_i) - v(L_{i-1})\} \leq 0 \tag{6}$$

The second variant of modification is removing the slope detection so that the output is only represented by a sequence of value and duration which are specified by (1) and (2). Both variants have improved error and simplified compression algorithm with lower reduction ratio as a trade-off.

3 Results

3.1 Hardware Performance

The measurement result of the hardware matched the data specification. The input voltage range of 10-bit precision SAR ADC is 0 to 1.0 V. A voltage divider in the evaluation circuit raises the upper limit of ADC range to 3.2 V. Based on this result, interfacing ESP8266 and AD8232 is accomplished without further requirement to adapting components. The output swing of the AD8232 is 0.1 V to +(Vs − 0.1) V, where Vs is power supply voltage. ESP8266 has an operating voltage of 2.5 V to 3.6 V, while AD8232 supply range is 2.0 V to 3.5 V. In a measurement with custom simulator program which receives and displays the signal into the monitor, the output of AD8232 swings between 110 to 200 for 8 bits MSB of ESP8266 ADC.

3.2 Wireless Transmission Performance

The AZTEC compressed ECG were received, unpacked and plotted into graphic in real time. A software was developed to process accurate data receiving and estimating data loss during transmission. Some missed data were observed in 2000 ms interval which is equal to the transmission task tick time. The missing signal length were varied with an average of 100 ms. This result has indicated that sampling task has been interrupted by the data transmission task, even though it is implemented in separated task. It can be deduced that the TCP stack processor is not an independent module to the main processor and consequently interfere the ADC performance. Therefore, the multitasking does not run optimally since TCP stack process has blocked the other tasks.

3.3 Testing with Subjects

The prototype was tested to measure 11 voluntary normal subjects in relax condition. The participants were ranging from both gender and at average age of 20. The ECGs

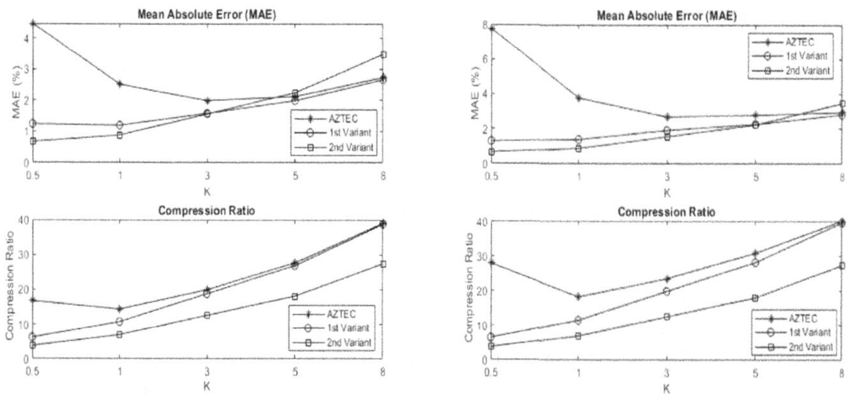

T = 10 T=20

Fig. 5. Modified AZTEC reduction ratio test with real ECG data

were recorded for 10 s using ESP8266 ADC with 200 Hz sampling rate, preprocessed by 1000 Hz oversampling. AZTEC algorithm and its modification variants were employed to compress the acquired data. Figure 5 compares Mean Absolute Error (MAE) and compression ratio of two different plateau thresholds T. The graphics were plotted by varying the level threshold K. It can be seen that the value of MAE and compression ratio are correlated with K and T. Generally, the 2nd AZTEC modification variant gives good MAE profile at lower K and higher T values.

4 Conclusion

A Wi-Fi based Wireless ECG Electrode has been implemented. Experiment results have shown a moderate performance of the ECG front-end circuit. The basic multitasking works in the evaluated Wi-Fi SoC device with drawbacks on relatively low rate transmission and data missing due to limited task separation functionality and computation power of the microcontroller. The 1st variant of AZTEC modification gives up to 3.42% error improvement when keeping compression ratio degradation below 13, while the 2nd variant gives 3.78% improvement at an equal degradation level. AZTEC modification provides options to negotiate the compression ratio with error profile.

Acknowledgment. This research was funded by Institute of Research and Community Service (LPPM) Institut Teknologi Sepuluh Nopember (ITS), Surabaya, Indonesia, under the scheme of *Penelitian dan Pengabdian Masyarakat Dana Departemen* year 2018. We want to thank the students of Biomedical Engineering Department, ITS, class of 2015, who have contributed to this research.

References

1. Ravish, D.K., Shanthi, K.J., Shenoy, N.R., Nisargh, S.: Heart function monitoring, prediction and prevention of heart attacks: using artificial neural networks. In: 2014 International Conference on Contemporary Computing and Informatics (IC3I), Mysore, pp. 1–6. IEEE Press (2014). https://doi.org/10.1109/IC3I.2014.7019580
2. Rocha, V., Borza, P., Correia, J., Gonçalves, G., Puscas, A.: Wearable computing for patients with coronary diseases: gathering efforts by comparing methods. In: IEEE International Conference on Automation Quality and Testing Robotics (AQTR), Cluj-Napoca, pp. 339–347. IEEE Press (2010). https://doi.org/10.1109/AQTR.2010.5520712
3. Hermawan, N., Retistianto, M.F., Arifin, A.: Sistem Tele-cardiac Monitoring: Ekstraksi dan Transmisi Parameter Temporal Sinyal Jantung melalui Kanal Radio. In: Seminar Nasional Inovasi Dan Aplikasi Teknologi Di Industri (SENIATI), Malang, pp. B47.1–6. ITN (2017)
4. Hermawan, N., Arifin, A., Retistianto, M.F.: Sistem Pelaporan Tele-cardiac Monitoring Melalui Layanan Pesan Singkat (SMS). J. Nasional Teknik Elektro **7**(3), 183–190 (2018). https://doi.org/10.25077/jnte.v7n3.576.2018
5. Cox, J.R., Nolle, F.M., Fozzard, H.A. Oliver, G.C.: AZTEC, a preprocessing program for real-time ECG rhythm analysis. IEEE Trans. Biomed. Eng. **BME-15**(2), 128–129 (1968). https://doi.org/10.1109/TBME.1968.4502549

6. Abenstein, J.P., Tompkins, W.J.: A new data-reduction algorithm for real-time ECG analysis. IEEE Trans. Biomed. Eng. **BME-29**(1), 43–48 (1982). https://doi.org/10.1109/TBME.1982.324962
7. Brennan, P.F., Starren, J.B.: Consumer health informatics and telehealth. In: Shortliffe, E.H., Cimino, J.J. (eds.) Biomedical Informatics: Computer Applications in Health Care and Biomedicine, p. 513. Springer, New York (2006). https://doi.org/10.1007/0-387-36278-9_14

Author Index

A
Ab Patar, Mohd Nor Azmi, 141
Abdu Salam, Muhammad Naim, 141
Agarie, Yukio, 93, 102
Ahmed, Faiza, 257
Ahsan, Monjurul, 159
Aki, Fumitaka, 272
Arifin, Achmad, 332
Arrofiqi, Fauzan, 332
Ashrafuzzaman, Md., 159, 257
Azaman, Aizreena, 111, 135, 301
Aziz, Muhammad Fariz Shafiq Abd., 135
Azuma, Takashi, 278

B
Barma, Shovan, 231
Bilal, Sagida M. A., 135
bin Abdul Hamid, Muhamad Fikhri, 126
bin Mohd Ramli, Mohd Hanif, 126
Bocchi, Leonardo, 83
Bolingot, Harold Jay, 25, 31, 325

C
Caballas, Kerwin, 25
Chang, Wei, 215
Che Zakaria, Noor Ayuni, 126

D
Dai, Lulin, 181
Dao, Quy-Thinh, 202

F
Fabbri, Rachele, 83
Fan, Bowen, 60
Fan, Yingwei, 3, 215

G
Gherardelli, Monica, 83

H
Halim, Husnir Nasyuha Abdul, 135, 301
Hanafusa, Akihiko, 93, 102, 118, 141
Hatano, Asuka, 47, 52, 60, 188
Hayami, Ryutaro, 319
Hazim Jamaluddin, Muhammad Amir, 141
Hazmi, Helmy, 118
Hermawan, Norma, 332
Hida, Mitsumasa, 153
Higa, Masaru, 319
Homma, Jun, 284
Hoshijima, Masahiko, 52
Hossain, Md. Sakib Abrar, 159
Hu, Chengquan, 3, 215

I
Iadanza, Ernesto, 83
Ibarrientos, Lawrence, 25, 325
Ikebata, Hayato, 93
Inoue, Yusuke, 77
Ishak, Nurhanna Zulaikha, 118
Ishijima, Ayumu, 278
Islam, Nabil, 159
Ito, Satoshi, 249
Iwasaki, Kiyotaka, 284
Izumi, Satoshi, 47, 52, 60, 188

Y. Shiraishi et al. (Eds.): APCMBE 2020, IFMBE Proceedings 82, pp. 341–343, 2021.
https://doi.org/10.1007/978-3-030-66169-4

J
Jamali, Annisa, 118
Jamaludin, Mohd Syahmi, 93
Ji, Honglei, 240

K
Kaewlee, Thitikorn, 38
Kakimi, Kazuhiro, 278
Kaneko, Ryotaro, 188
Kang, Hongxiang, 3, 215
Kawamura, Taiki, 93
Kawanaka, Haruki, 67
Kinoshita, Toshibumi, 10
Kitagawa, Kodai, 153
Kobayashi, Eiji, 284
Kobayashi, Etsuko, 278, 311
Koshiji, Kohji, 223

L
Lee, Hokyoo, 141
Li, Jia Wen, 231
Li, Yangxi, 215
Liao, Hongen, 3, 215
Libatique, Nathaniel Joseph, 25
Libatique, Nathaniel, 31, 325
Lin, Chin-Yuan, 175
Liu, Xin, 231
Liyanaarachchi, Madhushanka R., 311
Lu, Bing-Yuh, 175
Lv, Yi, 240

M
Macaraig, Lea Cristina, 25, 31, 325
Maeda, Eriko, 47, 60
Mak, Peng Un, 231
Matsuda, Shiori, 319
McCulloch, Andrew D., 52
Melillo, Paolo, 83
Mimi, Afsana, 257
Minakata, Shin, 10
Mitobe, Kazutaka, 272
Miyazaki, Sakurako, 60
Mizuno, Toshihide, 197
Mohamaddan, Shahrol, 118
Mohamed, Zulkifli, 126
Morishita, Yusuke, 60

N
Nagamune, Teruyuki, 278
Nagasaki, Takayuki, 153
Nakagawa, Keiichi, 311
Nakamura, Kazuhiro, 10
Nakamura, Shizuki, 319
Nakano, Sota, 153
Nawata, Kan, 47, 60
Nishinaka, Tomohiro, 197
Niwa, Yusuke, 67
Nuh, Mohammad, 332

O
Oguri, Koji, 67
Ohnishi, Kengo, 93, 102
Okamatsu, Shogo, 153
Okamura, Jun-ya, 181
Okinaga, Tomoki, 223
Ono, Masahiro, 319
Ono, Minoru, 47, 60
Oshita, Kazushige, 295
Otsuka, Hiroshi, 93, 102
Ouchi, Shohei, 249

P
Pun, Sio Hang, 231

R
Ritthipravat, Panrasee, 38

S
Saijo, Yoshifumi, 332
Sakaguchi, Katsuhisa, 284
Sakuma, Ichiro, 47, 60, 278, 311
Salleh, Noor Aimie, 111
Salud, Ingrid, 25, 31
Sano, Kazunori, 284
Sekine, Hidekazu, 284
Shetu, Sabrina, 257
Shibasaki, Yoshikazu, 278
Shimazoe, Kenji, 311
Shimizu, Tatsuya, 284
Shiraishi, Yasuyuki, 77
Simonelli, Francesca, 83
Someya, Makoto, 52
Sulaiman, Maisarah, 111
Suzuki, Ryogo, 93

T
Takagi, Shu, 47, 60
Takahashi, Hiroyuki, 311
Tanaka, Hiroaki, 52
Tangonan, Gregory, 25, 31, 325
Tatsumi, Eisuke, 197
Tobe, Yusuke, 284
Tomii, Naoki, 47, 60
Tonthat, Loi, 272
Toyoshima, Hideto, 10
Tran, Van-Thuc, 202
Tsukihara, Hiroshi, 47
Tsukihara, Hiroyuki, 60
Tsukiya, Tomonori, 197
Tuan, Le Van, 93

U
Umezu, Mitsuo, 284
Uto, Yoshihiro, 181

V
Vai, Mang I., 231
Van Thuc, Tran, 102
Vu, Thuy-Nguyen, 202

W
Wada, Chikamune, 153
Wang, Gang, 181
Wang, Jiu Jiang, 231
Wang, Junchen, 240
Wang, Pan Ke, 231
Watanabe, Eiichi, 67
Watanabe, Takuya, 47

Y
Yabukami, Shin, 272
Yahya, Azli, 111, 301
Yamada, Akihiro, 77
Yamada, Yousuke, 181
Yamaguchi, Haruo, 60
Yamaguchi, Satoshi, 278
Yamamoto, Shinichiro, 118
Yamamoto, Shin-ichiro, 93
Yamamoto, Shin-ichiroh, 102, 202
Yamamoto, Takahiko, 223
Yamamoto, Yoshiyuki, 272
Yamamoto, Yusuke, 181
Yamauchi, Haruo, 47
Yambe, Tomoyuki, 77
Yang, Yu, 19
Yee, Low Cheng, 118
Yoshizawa, Makoto, 77
Yu, Yuan Yu, 231

Z
Zarin, Farahdiba, 257
Zhang, Shuang, 231
Zhang, Xinran, 215
Zhao, Zijian, 19

The manufacturer's authorised representative in the EU is Springer Nature Customer Service Centre GmbH, Europaplatz 3, 69115 Heidelberg, Germany. If you have any concerns regarding our products, please contact ProductSafety@springernature.com

Printed and bound by CPI Group (UK) Ltd, Croydon, CR0 4YY
30/09/2026
02225292-0004